The Performance of Social Systems

Perspectives and Problems

The Performance of Social Systems

Perspectives and Problems

Edited by

Francisco Parra-Luna

University of Madrid
Madrid, Spain

Kluwer Academic / Plenum Publishers
New York, Boston, Dordrecht, London, Moscow

Library of Congress Cataloging-in-Publication Data

The performance of social systems: perspectives and problems/edited by Francisco Parra-Luna.
p. cm.
Includes bibliographical references and index.
ISBN 0-306-46309-1
1. Social systems—Evaluation. 2. System theory. 3. Social values. I. Parra Luna, Francisco, 1937–

HM701 .P47 2000
301—dc21

99-054947

ISBN 0-306-46309-1

233 Spring Street, New York, New York 10013

http://www.wkap.nl/

10 9 8 7 6 5 4 3 2 1

A C.I.P. record for this book is available from the Library of Congress

Printed in the United States of America

Contributors

Walter Buckley is professor of Sociology at the University of New Hampshire. He is one of the world's leading sociologists on Social Systems Theory. Very well known are two of his books: "Sociology and Modern Systems Theory," Prentice Hall, New Jersey, 1967; and "Modern Systems Research for the Behavioural Sciences," Aldine Publ. Co., Chicago, 1968, which have been recommended references for systemic sociologists during the last decades. Prof. Buckley is Honorary President of the Research Committee on Sociocybernetics and Social Systems Theory of the Int. Soc. Associates.

Iris Bálsamo is a fellow of the National Academy of Sciences of Buenos Aires, Argentina. and author of several works on the explanation in social sciences, among them: "La Evolucion de la causalidad." Edit. E. Molina, A. Carreras y J. Puertas, Univ. de Zaragoza, 1995.

Bruce Buchanan is a retired physician who has had a career in medical practice, public health, psychiatry, and university teaching mainly in Toronto (Canada). He is the author of several works such as "A System of Social Significance," American Documentation, IX, 2; and "Assessing Human Values," Kybernetes, Sept. 1997.

Michael P. Byron, of the Adjunct Faculty at the Palomar Community College in California, has authored several articles on Systemic Learning published in the Journal of Social and Evolutionary Systems, Kybernetes and Systemica.

Vladimir Degtiar is Chief Research Fellow of the Management and Consulting Service, of the Group Decision Support Systems for management in Unity Science Co., and a Senior Research Fellow at the Moscow State University. He has published five books, among them "Evolutionary Stable Mechanism for Reduction of Conflicts...", Plenum Pub .Co. 1990, as well as over 50 papers in several professional journals.

J.L. Elohim is an electronic engineer, retired professor of the University of Mexico, and currently Research Fellow at the University of Vienna, Austria. He is author of numerous publications dealing with the applications of the systemic and cybernetic phenomena in the social sciences.

Moisés Garcia Jara is an advanced graduate student at the University Complutense of Madrid. He is developing methods for the optimisation of the Internet as a useful information system for social sciences.

Martin L.W. Hall is a Consultant at the Values Technology Center in Santa Cruz, California, and a Visiting Research Fellow at the University of Lincolnshire and Humberside. He is author of several books including "General Systems Theory and Cybernetics in the Real World," San Jose State University, California, 1992; and "Systems Thinking and Human Values," University of Lincolnshire. U-K., 1997.

Dario Menanteau-Horta is a professor of Rural Sociology at the University of Minnesota and currently Director of the Center of Rural Sociology and Community Analysis at the same university. He has also been professor at the University Austral of Chile where he was awarded the title of "Extraordinary Professor" in 1985. He is the author of several books and numerous papers on different aspects of the efficiency of social systems.

P. Nicolopoulos is Assistant Professor at the University of Piraeus, Greece. He belongs to the Board of the Research Committee on Sociocybernetics and Social Systems Theory of the Int. Soc. Associates. He is the author of several papers on the performance of political systems.

F. Parra-Luna is professor at the University of Madrid and has been the Director of the Instituto Universitario de Recursos Humanos in Madrid. He is the author of eleven books on social systems theory and its applications, among them "Elementos para una Teoria del Sistema Social," Univ. Complutense Madrid, 1983, and "El Balance Integrado de la Gestion Estrategica," Ed. Deusto, Bilbao, 1993. He was the founder and first President of the Working Group on System Theory of the I.S.A.

A. Sánchez Súcar is an Aeronautic Engineer in Iberia, S.A., Spain, where he is the Head of the Department of Ecology and Environment. He has been carrying out several studies about the recurrent environmental problems in big air companies.

Markus Schwaninger is a professor of Management at the University of St. Gallen, Switzerland. He is the author of more than 100 publications in six languages, including such books as "Integraler Unternehmungsplanung," Campus, 1994, and "Organizational Transformation and Learning: A Cybernetic Approach to Management," Wiley 1996 (co-authors: R. Espejo, W. Schuhmann, and R. Bilello).

Karl-Heinz Simon is a professor of the University of Kassel (Germany) and deputy director of the Center for Environmental Systems Research. He has coordinated several projects in energy planning, resources analysis, man-environmental interactions, and methodology of systems research and has published several works on these subjects.

Felix Tretter is a doctor at the District Hospital Haar/Munich in Germany. He has been engaged on studies and projects dealing with the definition and measurement of the efficiency of hospitals, both public and private.

Raf Vanderstraeten is professor at the Faculty of Social Sciences of the University of Utrecht (The Netherlands) and has been Visiting Fellow at the University of Bielefeld in Germany. He is the author of several works on the efficiency of the educational system in Belgium.

J.P. van Gigch is Professor Emeritus at the California State University, Sacramento, California. He is one of the most well known authors on the application of the systemic approaches to the management of sciences and other social systems. Among his numerous books we find the celebrated "Applied General Systems Theory," Harper & Row, N.Y. 1978.

Gerard de Zeeuw is Full Professor at the Faculty of Mathematics and Computer Science of the University of Amsterdam, where he teaches "mathematical modelling and innovation of complex social systems." He is also one of the most well known authors on social systems theory. He is Fellow of the Netherlands Institute for Advanced Studies, Past-President of the Dutch Systems Group, and Honorary member of several scientific societies. He has published numerous works on Systems Theory and Cybernetics.

Foreword

WALTER BUCKLEY
Professor of Sociology, University of New Hampshire, Durham, New Hampshire

All liberal democracies–political democracies built on a capitalist base of business oligarchy (or plutocracy in some cases)–require, in principle, a periodic accounting or assessment of how well the social system as a whole is doing in meeting the general values, needs, and desires of its citizens. To the extent that those in positions of authority or power desire to regulate the workings of the societal system, a system based on feedback of policy decision and implementation results must be put in place. Thus there must be an assessment of the system's Performance measured in terms of the ratio of Outputs to Inputs, to use Parra Luna's terms.

Most businesses attempt to do this for their own performance, and are usually required by investors to do so. For this purpose a substantial staff is usually devoted to gathering data to measure profits and costs in relation to investment inputs–a job that is relatively easy in terms of the very narrow definitions given to the costs and income (for example, difficult to assess human and environmental costs are not normally figured in.) On the social system level, however, global performance measures have hardly been attempted–partly because they have not been welcomed by many political and business interests (businesses in particular have strongly resisted any accounting made by outsiders), and partly because they are seen as very difficult if not impossible to develop and implement. And this in spite of the fact that each year, in many countries, a "State of the Union" message is given by the head of State. In addition, the study of such global societal measures is made difficult by the fracturing of academic disciplines studying aspects of society from several, often incompatible, perspectives. This unfortunate isolation–firmly sanctioned by academic tradition and the business control of universities via the encapsulated trustee system–makes it almost impossible for social research to reflect the real systemic nature of complex societies.

However, some performance measures have been attempted, even required, in restricted areas of society. Programs have been set up to provide an assessment of the potential or actual consequences of certain government projects and programs, for example in the United States via social or economic "impact statements," and to provide feedback accounting for completed programs via certain "assessment" procedures. Also relevant is the development of various "indicators" aimed at keeping tabs on the state of society in certain respects. In addition are the many "cost-benefit" procedures, and the once popular "operations research" movement which attempted to carry design and performance analysis, primarily military, to its most precise technical state. But these have all been very limited in scope and often only half-heartedly pursued. And in particular they were not aimed primarily at the needs and values of the common man.

It is in this global performance context that the present volume is placed. The charge to the authors of the included papers is oriented "in the necessary direction of addressing social system PERFORMANCE and describing the Value Systems produced as a whole," in the words of the editor. Difficult indeed! The definitions of the main concepts involved are notoriously difficult to pin down in any operational way without losing much of the core meanings commonly attached to them. And especially problematic is the editor's attempt at a "standardised VALUE SYSTEM taken in its entirety." In complex multi-cultured societies one may ask if such a standardised value system is possible to define for the varying and conflicting values, wants and needs of the several ethnic, cultural and class subgroups. It will clearly be a difficult job to define adequately the key terms and justify the types of measures and mathematical techniques necessary to get at the global performance of social systems in a meaningful way.

But such an important–even crucial–attempt at developing such measures must be undertaken in today's increasingly interrelated, yet more and more disorganized social world. Such a courageous attempt presented in this book is, in Parra Luna's word, a "step, albeit small, exploratory and contingent." Let us hope it will be carried forward by many others.

Acknowledgments

I wish to express my deepest appreciation to all the above authors for their valuable contributions to this volume. My thanks as well to Ms M. Clark for her technical help in preparing the manuscript.

Contents

Page

PART II: SOME MODELS FOR DEFINING SOCIAL PERFORMANCE

Introduction

FRANCISCO PARRA-LUNA
Facultad de Ciencias Políticas y Sociologia. Universidad Complutense de Madrid

From cradle to grave, human beings are generally obliged to live in mutual interdependence, a Fact of history that may be related to the so-called natural systems characteristic of the animal kingdom. Gradually, individuals' realisation that they are better able to meet their needs if they co-operate with one another has given rise to "artificial" social systems, "artificial" insofar as they are human "inventions" or institutions. In both cases the *leit motiv* of such co-operation is purely and simply the enhancement of human endeavour or the overall efficiency of tasks: in a word, higher performance.

Although cumulative social experience provides ample evidence of the central nature of the notion of Performance, it has been incomprehensibly semi-ignored by social science (with the exception of Management Sciences, more or less obliged by the existence of a competitive marketplace to deal with the issue) and in particular by Sociology and Political Science, the two disciplines that should be most interested in this concept. Nonetheless, the notion is to be found at the hub of any societal activity. Indeed, if two or more individuals co-operate in a task (from making love to setting up a business) solely and exclusively to better satisfy their needs, appetites or interests, the concept that defines, denotes or measures to what extent such satisfaction is reached, i.e., the ideal of "global efficiency" (as the quotient resulting from RESULTS / EFFORTS, BENEFITS / COSTS or OUTPUTS / INPUTS), is a compulsory reference for any decision and, naturally, for any scientific theory on society viewed as a suite of global interests. The notion of performance in this regard is victim (or fruit) of a curious paradox: the idea must be felt to be so obviously central and important that it can be ignored. When someone decides to buy a soft drink to quench their thirst they have no need to be aware of any prior OUTPUT (perspective of pleasure or comfort) / INPUT (cost and inconvenience) computation to calibrate their PERFORMANCE; rather, they react semi-automatically, completely oblivious of the concept. It is in social science that such oblivion

The Performance of Social Systems
Edited by Francisco Parra-Luna, Kluwer Academic/Plenum Publishers, 2000

leads to perplexity: if efforts to describe the many social phenomena, processes and adaptations taking place internally in the organised societies that we call "social systems" ignore the concept, the result is the commission of a rather substantial epistemological-methodological error, consisting of EXPLAINING something without knowing what it is, without having defined, much less measured it beforehand. If what any social system pursues is to obtain the best or largest possible OUTPUTS from the smallest possible INPUTS (a notion essential to good performance), what we social scientists undertake, by contrast, is to furnish arduous and complex explanations (based on complicated internal transformation processes) of (?).... of what? We simply don't know.

What we do have is a reasonably valid knowledge of (objective and subjective) social indicators that can operationalise OUTPUTS, which we owe, among others, to the movement appearing in the sixties (from the early Gross, Bauer and Biderman reports to the valuable contributions made by the journal *Social Indicators Research*, not to mention the socio-economic indicators of major national and international agencies and bodies such as the UN, UNESCO, WTO, WHO, EU, OECD, etc.); such systems do not, however, propose standardised and comparable indicators for relative levels of development (the UNRISD -- United Nations Institute for Social Development – studies are an exception), which would entail explaining the VALUE SYSTEMS pursued or reached by the different countries or units compared. Before broaching the notion of PERFORMANCE, then, we should be obliged to first describe such suites of indicators insofar as they constitute Systems of Values; but we do not do so, to begin with because we are unable to recognise them, we have not defined them systematically nor recorded them as such, and hence we are unable to compute the degree of Performance. We find, then, that we are using a series of explanatory variables without knowing which variable to explain and commission social science with the paradoxical task of explaining something unknown to it. We do not know why we are doing what we do, nor do we as scientists understand which SYSTEM OF VALUES governs our own actions. We are, for that reason, even less able to distinguish between the various Value Systems pursued by different groups of people, by those in power and those who are not; nor can we identify the inevitably axiological origin of the conflicts arising among the most significant socio-political groups in any society (economic, religious, trade union, military, and so forth) in their respective attempts to impose their conscious or sub-conscious VALUES. This is like driving a car without knowing where we're going (description of goal or aims), and yet being constantly concerned about the state of repair of the vehicle as a means (explanation) of getting there.

I do not mean to say with all this, that social science does not generally address individuals' needs and the values such needs satisfy. Indeed, any number of specialised sociologies (Health, Labour, Economic, Political,

Sports, etc.) ultimately (although implicitly) deal with nothing but specific Needs/Values. What can be asserted is that none of the known branches of social science (of which this author is aware) addresses the VALUE SYSTEM, whether pursued or reached, as an inter-related whole, a balance of the best Output possible among a series of (at times contradictory or negatively correlated) values, as an "axiological profile" which should tend to be optimised within the limitations to which the entire social system is subject. Such a standardised VALUE SYSTEM, taken in its entirety, is what is wanting as an object of analysis and major component of the notion of performance.

In political science, for instance, rulers are interested in knowing not only whether the people accept their policies (political effectiveness), but also the relative degree of welfare reached (health, economic development, distributive justice, freedoms, conservation of nature, etc.), in terms that are comparable to other political units (political efficacy). Political scientists, however, seem to have disregarded such ultimate aims of any political activity. And instead of stressing this sociological rapprochement (the man in the street's opinion), they tend to interpret the "outputs" of the political system to be the "formulation of rules", "enforcement of rules" and "judiciary function" (Almond), or in even more abstract and scantly operative terms, they consider the outputs to be "Decisions" and "Actions" (Easton). Political science (Dahl, Eckestein, for instance) seems to have given in long ago to the impossibility of defining "performance". Dahl reduced the difficulty of measuring political performance down into three problems: how to deal with "values" (need for theoretical models); how to find reliable data (validity problem); and how to elaborate a methodological model suitable for quantitative data. Eckestein, in turn, identified five major problems: 1) non-existence of adequate data; 2) operationalisation or validity problem; 2) data reliability; 4) need to incorporate the time factor and 5) weighting system. Today, most of these problems have been solved in most developed countries. However, a glimpse at the *International Political Science Abstracts* for recent years shows that the notion of "Political Performance" barely appears and that there is hardly a reference to notions such as "Efficacy", "Effectiveness" or "Efficiency" as applied to social systems.

Naturally, I am aware that a series of INTERSUBJECTIVE AGREEMENTS among experts is requisite to the computation of the performance of any political, business or organisational system. There must, for instance, be general agreement on: 1) the definition of the final aim or utility sought in the computation to be made; 2) the definition of the temporal and spatial scope; 3) the valid input and output indicators for the aim pursued; 4) the relative weight of such indicators; 5) the data sources to be used; and 6) the different operations or mathematical treatment to be employed. Operational definitions in social science are not perfect, but neither are bricks and yet the houses made from them are useful. Where such prior agreements cannot be reached (although I believe they can be, on the basis of the substantial amount of data and methodologies – such as Delphy - that has accumulated as grounds for such agreements), no computation of the notion is possible. But the changes occurring in the last decade (proliferation of historical data series, growing standardisation and computer programs) make the venture seem to appear feasible. Given, furthermore, that the "Inputs" do not usually pose insoluble operational problems in most systems, the focus would have to be on the "OUTPUTS" as a configured axiological system or as a unified suite of standardised and interconnected indicators that enable us to perceive the Value System pursued/reached, as well as the differences between the two.

Such axiological unity, geared to reaching a balance, might, perhaps, contribute to avoid or at least mitigate the enormous contrasts between those who have too much and those who have barely anything at all; between unbridled economic growth in some countries and their enormous social problems, between sophisticated technology and senseless warfare, between the over-exploitation of natural resources and the impoverished planet we may leave to future generations. Such imbalance could be systematically and routinely addressed at the individual level of any social system, from each family of the UN.

The attempt to take a step, albeit small, exploratory and contingent, in the difficult direction of addressing social system PERFORMANCE and explaining the Value Systems produced as a whole, is the underlying motivation for the authors of the papers in this book who, from a variety of approaches, concentrate on the idea of performance. Some of them approach the question from a global view of the whole complex system; others focus on specific areas in the social system. A brief description of each of their contributions follows:

Martin Hall defines the central concern of organisations to be Values, insofar as they should satisfy the needs of the individuals comprising such organisations. Pursued or core Values are what nourish and motivate organisational operation, they are what stand up to recurrent chaotic situations and, ultimately, define the effectiveness or performance of social systems. Martin Hall's approach is systemic and therefore places values in the framework of an inter-related suite of factors that form patterns in terms of the values most commonly sought by individuals. Information systems are thus formed that contribute to reducing operational complexity and channel systemic activities and functions. Finally, Martin Hall views the measurement of values as the key to understanding how and to what extent values are produced in organisations. This is what would enable an analysis of rational decision-making processes, consistent as far as possible with the ultimate aims of the organisation as well as with the individual aims of its members. Martin Hall is aware of the difficulties of measuring such intangibles, in terms of both the weakness of the tools available and the diversity of concepts involved, but attributes two major objectives to such measurement: a) information on the organisation's internal discrepancies; and b) the move up from mere intuition to a more accurate understanding of the situation, on which grounds certain levels of social consensus may be achieved.

Bruce Buchanan's article likewise lays society's foundations on its Values in so far as they are tools for meeting human needs, recognising their strategic value in the analysis of social systems. Buchanan, quite logically, levels some criticism against considering Value Systems as a definition of the world as it is. Rather, he sustains that they represent things as we see them through the various methodologies and approaches used, in particular where measurement is concerned. In this regard he lists five aspects that any measurement of the performance of social systems must cover: 1. to deal with the effectiveness of the system (whether individual or social); 2 goals that validly reflect real human needs; 3. goals that take account of systemic limitations; 4. methods that take into account political factors; and 5. aims that do not foreclose on novel and creative possibilities. Buchanan draws necessary distinctions between Needs, Values and Ideologies as well as between "Effectiveness" (achieving the effects or outputs that are required to provide for its continuing existence), "Efficiency" (basically the input/output ratio) and "Performance" (which must also include a comparison of what is achieved to what is needed and intended). Finally, Buchanan discusses and particularises on some of the most relevant aspects of Parra-Luna's "Referential Pattern of Values", a paper summarised below.

As in previous works, Parra-Luna sustains that what he calls Universal Values (limited in number), to be distinguished from Cultural Values (the number of which is at least as large as the number of known cultures), may play a highly relevant role in sociological theory and in particular in towards an Axiological Systems Theory able to quantify concepts such as Change, Deviation, Socialisation and others. The author suggests an operational system for differentiating inputs, outputs, the transformation process and the respective feedbacks and regulation processes, all through new concepts with an axiological content. What social systems NEED or EXPECT, what they TRANSFORM, what they objectively and subjectively ACCOMPLISH and what, finally, they REGULATE or control are but, could only be (since they are in response to human needs), VALUE SYSTEMS. Social system analysis would comprise a process consisting of the four following phases: 1. sociologisation (system seen from the standpoint of the man on the street); 2. axiologisation (its main prime materials are Values); 3. cybernetisation: the Value Systems produced are controlled (not always anti- or pseudo-democratically); 4. reflexivisation (both aims and modifications of aims are constantly self-referenced). The Social system is defined by Parra-Luna to be a Transforming entity (T) whose inputs (X) and outputs (Y) are permanently controlled by a more or less effective Regulating process (R). The most basic notion of Performance would be defined as: T = Y/X where “Y” is always a complex System of Values and “X” represents the effort or cost needed to achieve it.

Gerard de Zeeuw deals with the problem of observation in social science from the standpoint of so-called Second-order cybernetics; his point of departure is the myriad situations in everyday life that are readily understood and generally accepted to validly represent a kind of “reality”, even though such reality is a “construct”. For Zeeuw the problem arises when the issue to be defined is more complex, such as for instance, everything concerned with organised communities. In this sense, such observations, including certain measurements taken to be accurate reflections of scientific concepts, may be so well established that they need not change, but they may also be subject to change in view of new formulations or developments. Any observation, in this regard, would be contingent and according to Zeeuw no possible change would make the thing in question less “real”. According to this author, “most of the time such changes will be developed to deal with immediate problems. They may also be the result of disciplined enquiry – as in the case of accountancy, and also of course of economists, sociologists, social geographers and statisticians. It is in the latter case that major questions concerning performance measures arise. A number of difficult problems of this kind are then considered. These include: firstly how to recognise what makes observations of performance belong together. Secondly how to increase the authority of what belongs together. Thirdly how one may construe measures of performance that have value as observations to guide action. And fourthly how one may increase the quality of such measures.”

J. Van Gigch, always concerned about the operationalisation and quantification of systemic values (he writes that “measurement is one of the central elements of the Evaluation phase of the Systemic Paradigm”) and also aware of the necessity of using subjective and objective social indicators, presents some epistemological and methodological rules for testing the performance of social systems. Given the complexity of the subject, Van Gigch begins by defining what is meant by “measurement” and then proceeds to an analysis of the difference between Numerals and Numbers, the Conditions for Order, the law of Addition, or the difference between Quantities and Qualities. On these grounds, Van Gigch further analyses policy validation, the problem of identification of outputs, the objectives and measures of Effectiveness or the relationship between states and flows. The Van Gigch paper is, then, a cardinal contribution in terms of the purposes of this book, insofar as it stresses the “measurement of and problems” surrounding such intangibles as Value Systems.

On the basis of the problems characteristic of the observation of social systems outlined by G. van der Zoueen, Parra-Luna proposes a general model of society susceptible of being formalised and quantified. The model maintains the four-dimensional structure envisaged by the author in the paper summarised above, i.e.: INPUTS-TRANSFORMATION-OUTPUTS-REGULATION, against a backdrop critical of the excessive emphasis that social system leaders (the elite) tend to place on – primarily subjective – Regulation processes. The description of the model is completed with a suggestion for a system of empirical indicators capable of validly recording both people's Needs (N) and Objective (Y(O)) and Subjective (Y(S)) Outputs. This system of indicators seeks to represent the accomplishments of a typical modern Nation-State, but can and should be suitably adapted to the characteristic needs of each system analysed. Such a suite of indicators corresponds to a General Pattern comprising the following nine Values: Health, Material Wealth, Security, Knowledge, Freedom, Justice, Conservation of Nature, Quality of Activities and Prestige. In this regard a list of indicators (that the author has been applying to the Social Accounting System for business concerns) is attached in an annex to illustrate the variegated operational content that such overall Values can hold. It will be noted that here the model covers ten Values instead of the usual nine, in view of the characteristic features of private enterprise, in which the Value POWER acquires prominence in response to the need to compete on a free market.

Nicolopoulos' article presents a model which coincides in essence with the above, but which highlights the processes that introduce instability in social systems, with an emphasis on systems of a political nature and kinds of leadership. According to Nicolopoulos, political leadership based on the sociocybernetic model must formulate and implement the following policies, capabilities and work: 1) compiling data and information as accurate as possible on the environment (natural, social and political); 2) improving on the "problematique" relating to the "absolutism" of perspectives; 3) processing, analysing and interpreting the data and information collected; 4) organising the existing "matter" and "energy" in the political system and its environment; 5) developing a flexible and effective adaptation and response capability with regard to the respective environment, based on the "input-output-feedback" model; 6) striking a balance between adaptation to environment and attainment of goals; and 7) perceiving the trend of change in society which may generate new types of structures and institutions. The model formalises the Input/Output relationship in terms of "W" (value system produced), with inputs represented by the sum of "t" (time) and "c" (cost).

For Karl-Heinz Simon the concept of performance of social systems is discussed mainly from a macro-sociological point of view. First, he deals with the concept of society and the idea of evaluating the performance of social systems. Parsons' AGIL scheme as well as social goal functions are discussed. Then, the so-called "orientor's approach" is taken as a methodological basis and several functions of society are analysed with respect to criteria which could serve as an evaluation perspective. Some ideas are also presented about how to quantify criteria loadings and what kind of results are achievable following the approach. For Simon the quantification of indicators would the critical next step which is often not taken. Up to now, in most cases only an intuitive association of values is undertaken, e.-g., on the basis of marks. Investigations of self-organisation processes in virtual environments (Krebs/Bossel) produced first results which could be useful for further improvement of the approach.

Markus Schwaninger's article turns on the self-control sub-system in social systems. The author sustains, and rightly, that the sociocybernetic approach must necessarily deal with system performance if the point of departure is the definitions used in cybernetic science, regarded as the science for the control of system effectiveness; in this connection he conceives the global efficiency of such systems in terms of the suitability of the social control exerted on them, control which should not mean a lack of transparency or internal democracy. This explains the author's distinction between "intrinsic" and "extrinsic" control and his insistence that the former should prevail over the latter, in keeping with the idea of autopoyetic systems. Drawing on several theoretical attempts to broach the notion of Performance, the author suggests a control model that combines the theory of systemic self-control with the English systemic school's Feasible Model approach.

Vladimir Degtiar's paper presents a comprehensive review of the importance of Diagnosis for the development of social systems in crisis, in which the condition and attitudinal arrangements of both individuals and organisations must be examined. Ailing social organisations can only be treated with a view to improving their performance from this dual individual/organisational perspective. Degtiar deals with the process of development towards Performance in a formalised manner and in accordance with a transitional logic that addresses different conditions. He sustains that prior to any diagnosis a mechanism must be created to pool individual interests to form social interests, which in turn entails three fundamental steps: "a) the creation of non-manipulated mechanisms to aggregate individual into collective decisions; b) the creation of mechanisms to diagnose the level of individual competence and the formation of personnel structures for the social control mechanism; c) development of the architecture of splitting an object into sub-objects at all levels of an enclosure and architecture of the social control mechanism, determining the interaction of separate subjects at the decision-making level". The explanation of both autopoyetic (due to internal structures) and allopoyetic (due to the interaction between the internal structure and the environment) processes is of crucial importance. Degtiar's work also presents a table of common social aims for organisations in a manner very similar to the models outlined by Hall, Buchanan and Parra-Luna mentioned above. He finishes by regretting the non-application of these principles of diagnosis, social control and definition of organisational performance in the case of Russia today.

Iris Balsamo describes the usefulness of the systemic approach (which she feels to be the most appropriate scientific method for analysing complex social systems) for reviewing the causes informing the performance of social systems, highlighting four structural fields that she terms: 1. change of "status" or structural change; 2. destructive change; 3. productive interactions; and 4. destructive interaction. These, in turn, give rise to four different meanings of the scientific law which the author, following M. Bunge, defines as follows: objective, nomological, monopragmatic and meta-nomological. Balsamo also considers causal law to explain social system performance from the Aristotelian perspective, with its four kinds of causal laws: a) formal; b) efficient; c) material; and d) final, and models possible combinations of the four. From the perspective of second order cybernetics and in this explanatory context of causes, Balsamo stresses the growing importance of outputs in complex (self-determined and self-referenced) systems as compared to the lesser relevance of inputs, further supporting the action of systemic autopoyesis, in line with Maturana and Varela's well-known concept.

In the third group of papers included hereunder (dedicated to specific applications of the concept of performance in social systems), A. Sánchez Sucar, head of Iberia Air Lines' Environmental Development department discusses his company's efforts to prevent the environmental damage typically caused by the air carriage business. The steps taken can be summarised as follows: Firstly the "physical environment" variable is positioned in relation to the company's other socio-economic variables to consider them as a systemic whole. Secondly, he suggests an accounting system that would provide accurate information based on the experience and data compiled by other industry companies and studies conducted by world-wide organisations such as the United Nations Environmental Programme, the Environmental Protection Agency or the Securities and Exchange Commission. And thirdly, in order to accommodate the inevitable system of values represented by company outputs, he integrates such values through formalised vectors that denote the company's overall performance from the specific perspective of SUSTAINABLE DEVELOPMENT. Sánchez Sucar's theoretical discussion of the issue, including a series of models broken down into departments and/or operations undertaken, is supported by empirical data which compare Iberia's performance to that of other similar companies in terms of Sustainability.

Raf Vanderstraeten applies the notion of Performance to Belgium's education system. His aim is to measure and explain the overall performance of the system during the period following the second world war. In the first part of his paper, the author presents figures on education system outputs and the increasing demand for education. The body of the article is devoted to an interpretation of these figures, although this interpretation does not merely focus on the particular characteristics of educational planning and policy. A brief review of society in general shows that similar patterns were emerging in other areas, such as with respect to medical care, legislation or even political decision-making. In fact, most social systems were confronted with expanding demands. This phenomenon indicates that higher expectations and increasing participation were perhaps not so closely related to education and educational policy as such, but were primarily characteristics of modern society, and hence open to general sociological analysis. In this context, Vanderstraeten uses the systemic theoretical framework developed by Niklas Luhmann together with empirical data for some relevant variables to analyse the Belgian education system from 1960 to 1990.

Felix Tretter, a clinical doctor working in a German hospital, deals with this kind of social system as an organisation which must seek higher efficiency or performance. But for the author there is a huge gap between practical work in social systems and the theoretical foundations of the design for action. Furthermore, he feels that social research neglects problems which arise in practical work in social systems, such as, for instance, the management of hospitals as organisations. For Tretter, then, theoretical social science also faces the problem of how to adapt social research to the needs of some organisations. Moreover, empirical data on organisations goes largely unheeded in theoretical social science. In view of all this, there is an obvious need for better communication and relationship between data and theory to control and improve the efficiency of specific social systems (the sociocybernetic approach). On these grounds F. Tretter models and applies his sociocybernetic approach to the operation of a hospital as an organisation.

Dario Menanteau's paper applies the notion of performance to the United States of America. From a triple perspective – individualist, structural and the theory of limited resources – the author stresses the importance that variables such as health, education, life expectancy, suicide or income distribution, which while not strictly economic are closely related to the economy, have for the new concept of performance, insofar as they are Values that meet people's complex needs. But it is with regard to income distribution that Menanteau furnishes empirical data indicative of a certain imbalance among various segments of the U.S. population, and more specifically in the state of Minnesota. For the author, income distribution (in a market-driven economy) represents the most visible mechanism by which a social system shares its resources, benefits and rewards among its members. Income distribution is, therefore, a powerful indicator of socio-economic performance for the satisfaction of social needs. Menanteau's study illustrates the extent to which performance of the social system can be evaluated by measuring and analysing numerous social and individual needs different from economic growth in any society. In this case, estimates of income inequality for localities (counties) and measurement of social well-being in the state of Minnesota are proposed as performance indicators. His data confirm that development in the United States, usually defined in terms of economic expansion, has resulted in uneven growth, increased poverty and greater inequality.

Mike Byron deals with the international political system, which is for the author a complex adaptive "learning" system. Pursuant to this enquiry the issue of whether this learning system is best characterised by between-group conflict only, or by a temporally varying combination of between-group conflict and between-group co-operation, is investigated. To accomplish this aim, potential causal relationship(s) between inter-state trade and inter-state warfare, in the context of historically increasing levels of systemic power, are also investigated by the author. The modern world system is hypothesised to evince crisis-driven learning behaviours with respect to time. To test these propositions with empirical data, their underlying theoretical bases are operationalised into a 10-nation BASIC language computer simulation of the world system. Two classes of model systems, one premised upon realist power-denominated systemic assumptions, along with another premised upon the hypothesis that international trade plays a key role in determining the trajectory of the world system, are evaluated in this context. Key model variables are adjusted to allow for methodical testing of hypothesised economic interaction (interstate trade) effects upon warfare frequencies of a statistical significance. Empirical data for real word warfare frequencies are utilised to allow for a potential falsification of the model derived findings.

For J.L. Elohim, Humanity seems to be confronted at present with the most serious crisis that it has ever had to face, a crisis largely begotten by short-sighted and narrow-minded civilised intentions, which have, rather paradoxically, ensued from scientific advances and technological innovations. Such developments, intelligently conceived and supported by highly developed educational activities, have nonetheless ignored, *de facto*, that the aim of such action should be to provide a reasonable degree of satisfaction for the basic needs of billions of human beings. Such questions have not been well handled to date. At present, most systems in nearly every society oblige most people to continue to idealise how their individual and collective needs may be gratified, in keeping with assumed Values that in practice are only partially or even only scarcely taken into account. Elohim assumes that we do not need empirical data to show that the performance of the planet Earth, as a pseudo-system, is declining. However, global society as a prospect obliges us to assume that all human endeavour may yet be re-steered to allow each and every human being to systematically learn how everyone might contribute to making the evolution of the human venture sustainable, on the basis of ecological, ethological and ethical values and bearing in mind that our vantage is from the Earth itself.

Finally, M. García Jara undertakes to apply the notion of performance, in terms, always, of the Input/Output ratio, to the general operation of the INTERNET. In this regard, he analyses the benefits (the new System of Values that is inevitably forming) and drawbacks (including the cost and time entailed) that have been forthcoming to date, to conclude that overall Internet performance is highly positive and the expectations it is generating are in principle exciting but disquieting as well. Nonetheless, for García Jara the net of nets still needs to be more widely extended, systematised and structured if it is to afford larger portions of the population knowledge in less time and at a lower cost.

By way of conclusion, I would merely like to thank all the colleagues cited above for their invaluable collaboration, and in particular my friend Professor Walter Buckley, a teacher for many of us, for his foreword, in bringing to light a series of ideas around his concept of "global efficiency" or "performance", which is (or should be) ordained to play an increasingly central role in certain social sciences, such as Sociology, Political Science, Economics and especially business and public administration Management Science. The underlying assumption in all our papers is that social organisations are essentially composed of specific people (as opposed to certain formalist excesses), with constantly renewed specific NEEDS and in permanent pursuit of VALUES to satisfy those needs; moreover, in keeping with the core idea of Cybernetics, the "science of the effectiveness of action" (Couffignal), the ultimate aim of such systems is to achieve all this at the lowest possible cost, energetically speaking. Hence the need for the notion of SOCIAL PERFORMANCE.

F. Parra-Luna

Madrid, Spring 1999.

Part I

A World of Values

Chapter 1

Systems thinking and human values

Towards understanding performance in organizations

MARTIN L. W. HALL
University of of Lincolnshire and Humberside and Values Technology Center, Santa Cruz, California

1. INTRODUCTION

The idea of chaos in organizations is nothing new. Many people have expressed the idea of describing human and organisational interaction as chaotic and complex (Goldstein, 1994; Goerner, 1994). But a key feature to understanding this chaos and complexity is to see the patterns that emerge from these interactions. Humans tend to interact with each other through the choices and decisions that are made. If one can understand the pattern of decision making of the individuals and the organisation, then the true goals of the organisation can be made apparent. This is a very powerful concept for allowing organisations to understand themselves and become more effective. Values are the filter by which we make these decisions (Ellis and M.L.W. Hall, 1994). Understanding values therefore is a key to unleashing the complexity and power of an organisation. Systems thinking and human values are the context for the tools and methodologies used to unleash and hopefully harness the power in these organisations.

For example, if we have a group of people and we all see a tree, we may all agree that it is a tree (information). But what the use of the tree is or how it may be applied has to do with the values of the individuals. Let's assume we all have *very* different values as we approach the tree. One may see *history* or *beauty*. Another may see *shade*. Another *warmth*. Another *food*, etc. It is through these values that we have a relationship to the information

The Performance of Social Systems
Edited by Francisco Parra-Luna, Kluwer Academic/Plenum Publishers, 2000

and this makes it knowledge and hopefully more meaningful to us. How we change through life may also affect how we view this information (the tree). It is important to note that no one way is correct. It has to do with where we are at a particular point in time or more specifically in our lives.

Human values measurement and systems thinking principles can become a part of the system for understanding performance in organisations. Tools and methodologies can be used to help make sense out of the complexity in the modern organisation. Core issues include such elements as decision-making, organisational culture and the usefulness of effective values measurement techniques and their applicability in a systems framework.

Links will be made between values and organisational structure and process, which will be centered around how to specifically and practically use both individual and organisational values to transform organisations toward a more effective and efficient reality.

2. VALUES AND COMPLEXITY

The tree example talks about values as if only one value is acting at a time. In reality, a person may have several values that are priorities to him or her. These are all present as we make decisions. The complexity of the process means that many times the decision making process is different for each person. But at the same time, there is a higher chance that one might have values in common with someone else that is making decisions around the same issue. While we are all diverse beings, we can also have similar or congruent priorities with others around us. Complexity with congruence. Complexity with congruence can help create simplicity.

The congruence around common values allows us to look for the patterns. These patterns are about making relationships.

As we enter into an age where there is too much information, we need to use filters. Internet or intranet search engines are an example of this. Search engines help us to filter using keywords or phrases. It is this implicit filter that we use to reduce the complexity of the information. Rather than take in all the information (impossible), we use these simple filters. The keywords are the way in which we connections or relationships to the information. The degree to which the information becomes meaningful to us, it turns from information to knowledge.

A next step filter could be one that helps us view information through a values-based filter (we already do this but without the benefit of technology -- selecting keywords for an Internet search are values decisions. In that the words are some how meaningful to the information we want to find.). If a level of the filter uses the values that are important to a person's or an organisation's success, it is going to greatly help the knowledge extraction process. By connecting to it or filtering the information through values we are increasing the probability of the information becoming meaningful. Meaningful information is knowledge. It will focus the knowledge that is most useful first. We may do the search in the normal internet search fashion, but information that is most important (most relevant or meaningful) will be viewed first. What we would have is a knowledge engine.

How we interact with the world, and how we create our perceptions of the world are very important for how we understand and define our existence. How we define ourselves and how others see us will greatly affect our ability to interact in teams, organisations and other social situations. It is our values and its link to consciousness that allows us to build and hold relationships.

We tend to interact with people that have values that are congruent with our own. Systems thinking to a large extent is about managing these relationships:

How human beings act together greatly affects how systems will respond, a better understanding of this response will aid methodologies that may be used intervene

A person's ability to look at an item systemically is greatly enhanced by their ability to understand the environment around them (their level of consciousness)

One idea is that values and the valuing process are a way in which human behaviour in organisations can be made more explicit, especially from the way in which we as individuals interact with the world. Particularly from the point of view that what we value affects our behaviour (we enact behaviour to satisfy our values (needs and desires)). A key manner in which this is experienced is during the decision making process. In many ways, the decision making process is a valuing process. One is making an item a priority over another. When making choices, one is valuing an item as more important (in the moment) than something else. This will cause a person to enact values or a valuing process that is much more related to those emotions. An example of this might be how feeling fear might produce a very security-oriented or survival-based response.

By understanding values, we have a way of understanding human behaviour. The literature is very clear in this area (Maslow, Beck and Cowan, B. Hall, Graves, et al). Values can be a way that people use to describe their current state of consciousness (if one values global justice rather than putting food on the table -- these statements would suggest a different conscious focus for an individual – it potentially gives more meaning to the starving artist metaphor).

In any event, what does this mean and how can it help systems thinking or organisations, in general? Most systems methodologies (possibly all) are problem solving or decision making methodologies. Therefore, if values are affecting how we make decisions and improve performance they will affect how systems methods get executed. All too often the systems methodologies presume a type of mechanical decision making process that is not necessarily taking a valuing or value priority process into account. This will lessen the effectiveness of the methodologies, because it is ignoring one part of the system, the human beings. People will make decisions using values whether the values are explicit or not.

In the way in which people look for patterns in chaotic situations to help understand them, values measurement helps people understand the dynamics of organisational behaviour by identifying patterns of value-based decision making and relationships.

Historically, the difficulty in introducing the human-relations dimension is that there have been as many approaches to human relations as there are understandings of the human being (Jackson, 1991). This made human relations measurement difficult. If one does not have a consistent method for understanding behaviour it can make the human relations aspect of organisational interventions a guessing game. This ambiguity has fuelled organisational consulting for many years. But what is more important is that it has people viewing organisational performance as mysterious. This point of view has created a desire to "control" the organisation rather than "harness" the power of it.

Chaos and complexity theory is about recognizing patterns in the seemingly unexplainable (Gleick, 1987), and using these patterns to gain greater understanding. Values measurement is about understanding and recognizing patterns of human and organizational behavior. In other words, values (and values measurement) can help one to understand some of the seemingly chaotic behaviour that goes on in organisations. Values measurement in organisations is about measuring culture. Measuring culture brings organisational analysis of performance to a new level.

3. SYSTEMS AND VALUES

The link between a systems view of the organisation with the values and decision making of the individuals within that organisation make for a powerful synergy that in many ways is like opposites working together similar to the Ying and Yang of Chinese culture. Each element is useful, but the real power is when they are brought together and are used in an integrated fashion.

They (systems and values) operate together in a complementary manner. Systems methodologies have had weaknesses in connecting with the individual. This is where values are strong. Values are weak on connecting with the organisation (system). This is an obvious strength of systems methodologies. This appears to be due to the fact that values theory tends to focus on the individual and how they fit into the system, and systems theory tends to focus on the organisation and how individuals might fit into it. The challenge is making sure that the integration of these two perspectives is done smoothly and effectively. Decision making as one of the primary connecting elements between values and systems methodologies should be understood.

It is this connection through decision making that allows values and systems elements to inform each other. It is also a key issue because decision making is how both individuals and organisations interact with their environment. After decision making, the concept of worldviews also becomes extremely important. It is the worldview that creates the context in which decision making is made. A person's approach and understanding of the world will affect their decision making ability. Values help to form worldviews; and systems methodologies use worldviews as a context for creating a mechanism for use within the particular methodology.

The area of values and systems (specifically organisational interaction) that I will group under the term, "Value Systems Theory" or (VST). The main idea of VST being that values information and systems concepts work in concert in an organisation. The genesis of this idea has been explored somewhat (B.P. Hall, Graves, Beck and Cowan, M.L.W. Hall) but the methodological and systemic aspects have been particularly weak. These theories primarily seeing the individual as affecting the organisation through the interaction of groups of individuals (individual-centered interaction).

This investigation will be used to support the development of an argument that a very important unifying factor in organisations is "human values". By exploring recent research in the areas of values and systems science, and by utilising the concept of a "universal set" of values (B.P. Hall, 1986, 1993), a linkage between system purpose and human motivation will be proposed. This linkage will be made primarily through the concept of decision making and the concept of worldviews will be critically examined in an effort to show how intervention in organisational problem situations may be more effectively guided.

As Hofstede (1991) noted, the concept or ability to measure values has in the past been weak and not helpful as an organisation tool. Maslow (1954) and Rokeach (1979) began to promote measurement as a necessity since values, motivation and decision making were linked to human and organisational priorities. Hall and Tonna (Hall, 1986) since the early 1970s were able to move this concept (of values measurement) forward by developing standard definitions of values, and from this validated empirical measurement tools were created. From this work computerized instrumentation was developed. The computerized instrument is now being exercised in a number of global institutions (Hall, 1993). It continues to be very useful in the area of organisational development.

A proposition of this chapter is that a "new unified science of values" be introduced. This will provide a consistent and measurable way of dealing with the human relations component of an organisation. B. Hall and Tonna propose a theory whereby all human values are selected from a universal set of 125. This theory is based on over 25 years of human and organisational research into why people make the judgments or decisions they do. If one uses a definitive set of values, dealing with the human relations aspect of organisations can become easier by allowing certain elements to be viewed in a scientific, logical manner. This allows the organisational behaviour patterns to be much more obvious. This unified method for understanding values when used with other organisational methodologies creates a holistic approach to solving problems that might not always be self-evident.

Decision-making is based on a combination of scientific (or otherwise "factual") data and values (Robertshaw, Mecca and Rerick, 1978). Factual data are the core elements needed to make the decision. Values can be looked as a filter by which we make decisions (or judgments) based on the factual data. The study of values are also cross-disciplinary, and similar to the study of systems science, pulling resources from many disciplines (Rokeach, 1979). This perspective is necessary to identify the values patterns that would not otherwise be apparent in an organisation.

Values measurement is a key to understanding how values are understood within an organisational context. It allows one to see how the pockets of decision making in an organisation can be congruent as well as diverse. And how different parts of the organization can have different values to another part but overall the decision-making can be congruent. And when the decision making is not congruent a values clarification process can be very helpful. Core values are the center to understanding these patterns of congruent and non-congruent decision making as an overall expression of social systems performance.

4. VALUES MEASUREMENT: THE TOOLS

One very important asset that has been gained from the scientific and industrial periods of history is the importance of measurement. Modern accounting is a clear example of this. But first we must be clear that any measurement system is limited within its purpose. It is a way for us to increase our understanding of a given circumstance. For example, accounting can tell us a lot about our financial affairs, how we are coping financially up until yesterday. But it is very poor at telling us what how we might do in the future. But it is a string indicator of the direction we are moving: forward or backwards.

In the same, way values measurement is a measure of the meaning making dynamic of an organization's culture. It's purpose is two fold:

To inform us about gaps and discrepancies in the organization: states of non-alignment.

To make tacit information explicit. By doing this with values measurement it moves values from what we feel we know intuitively to what can be confirmed and known explicitly by a group as a shared consensus experience. Once the values are explicit then we are more in control of where we are going. It is a critical methodology for empowerment of any management system and central to values deployment.

Taming the chaos and understanding the performance of the organization is what this is all about. Identifying organizational values can lead to understanding the issues that most critically face the organization.

Organizational management techniques when used in conjunction with values measurement can help focus the organization and more importantly "tame" or "harness" the power of the organization. One cannot easily control this power or sanitise the uniqueness. But it does give one insight and also suggests how one might focus resources and energy.

Values measurement of both the individual and the organisation, allows for views of the organisation to be made from different perspectives. Looking at organizational values on a human (individual) scale allows one to see patterns emerge depending on position of the observer. For example, the values can be looked at in terms of developmental, cyclical and dialectical orientations. These are simply different perspectives or levels of resolution for looking at organizational behavior. What once seemed eminently chaotic may actually have a sense of elegance and order.

Human beings when studied closely (by some) can seem to be some type of irrational decision-making machine. If not irrational, at least very complex. When you put a group of these people together it would seem that they would be unable to make coherent group decisions, let alone get decisions made inside large organizations. Yet people can and do make good and coherent decisions daily.

In fact, people often look at this (group decision making capacity) as a powerful intangible element of organizations. This intangible gets various names: corporate culture, soul of the organization, core values, intellectual capital, etc. And it is considered a very strong element within these organizations. People (consultants) make their careers trying to divine this information for client organizations. Many people feel that if they can get a hold of this intangible that they can understand how their organization is going to act. But often times it is very difficult to get this information accurately and reliably.

This "intangible" piece is the human relational element or organizational values. All too often, organizational interventions or organizational consultants employ methods that have little or no focus on the values operating in organizations. It is this lack of understanding of the human relations context that makes organizations appear inherently complex. (Conversely, those that have come at it purely from the human relations perspective have not had the organizational or "big picture" view.) It is basically a psychological view versus a socio-organizational point of view. Most people view through the lens of one or the other, they need to look through both lenses to get the whole picture.

Part of this is because organizational values are non-linearly dynamic (there is not an A causes B dynamic). But it has been generally thought that values are not something that could be easily or properly utilized in an organization. More important, however, there has been no way to measure or compare accurately this values information. But if one introduces values measurement techniques such as suggested above then this issue can be addressed.

Chaos is basically the study of non-linear dynamics. To some people it may appear that individuals make irrational decisions when in fact they make decisions that use a different set of values. Why would this matter? Values are core to human decision making. A conflict in values orientation is going to make other decisions not seem logical. A common refrain that seems to bear this out is, "We always make decisions that way, this way makes no sense." Differing values orientations create this conflict. Keeping the values orientation congruent is key. Congruent values orientation make the organization more effective and efficient because it is not fighting the conflict of decisions based on differing values orientations. Consequently, values are the key to understanding how an organization makes decisions and moreover how they maintain effectiveness and overall performance.

People like to make decisions with others that are congruent to their way of thinking. This natural tendency towards congruent thinking and decision making is a form of self organization. In an organization, the complexity of how individuals make decisions is distilled down because people will tend to focus on decisions that are closer to or congruent with others in the organization. Core values develop and patterns of decision making become apparent. Values, specifically organizational values, are emergent properties that appear in and allow the organization to organize around certain issues or ideas. This focus is what many people would term, "core values".

Organizational values are the patterns that emerge by measuring the values of the individuals in the organization. Determining core values is not a black or white proposition. The core values simply identifies a tendency towards a type of group decision making behavior. It has at its core: desired cultural behavior, consequences, communications, processes, norms and goals.

The values interaction in organizations is non-linear in that it does not provide the "answers" for what the organization will do, but allows a predictive power to be enabled to see how organizations tend to react given their organizational values orientation (or core values).

The patterns from chaos theory that values measurement allows us to take a peak at, only scratches the surface. In a sense, it only allows us to know, as Sally Goerner suggests, the "spirit" of the organization [Goerner, 1994]. Matching the values with goals and objectives goals and objectives makes for very powerful, effective and efficient organizations.

A unique element of organizational values seems to be the way in which it lives in individuals in the organization (values of the organization are contained within individuals). The organization can spontaneously reorganize while maintaining a piece of the values orientation. The organization always manages to maintain its identity. Again, it is this is this idea of maintaining the spirit or soul of the organization.

Systems thinking and values measurement can work together with organizational management techniques to act as an analytical tool for organizational development and improving performance. As these techniques develop, I think that we will find that we will be able to harness the power of organizations. Maintaining their efficiency and effectiveness without removing any of the characteristics that make the organization unique.

This chapter is part of on-going research in the area of systems and values in organizations.

5. REFERENCES

Beck, D. E., and Cowan, C. C. (1996). Spiral Dynamics: Mastering Values, Leadership, and Change Exploring the New Science of Memetics, Blackwell, Cambridge.

Ellis, R. K., and Hall, M.L.W. (1994). Systems and values: A practical approach to organizational intervention. New Systems Thinking and Action for a New Century, Proceedings of the ISSS, Pacific Grove, CA.

Gleick, James 1987, Chaos: Making a New Science, Penguin Books: New York.

Goerner, Sally J. 1994, Chaos and the Evolving Ecological Universe, Gordon and Breach: USA.

Goldstein, Jeffrey 1994, The Unshackled Organization: Facing the Challenge of Unpredictability Through Spontaneous Reorganization, Productivity Press: Portland, OR.

Graves, C. W. (1974). Human Nature Prepares for a Momentous Leap. The Futurist 8, 72-87.

Hall, B. P. (1973). Values Clarification as a Learning Process: A guidebook of Learning Strategies, Paulist Press, New York.

Hall, B. P. (1976). The Development of Consciousness: A Confluent Theory of Values, Paulist Press, New York.

Hall, B. P. (1979). Developing Leadership in Stages: A Value-Based Approach to Executive Management, Manohar Publications, London.

Hall, B. P. (1986). The Genesis Effect: Personal and Organizational Transformation, Paulist Press, New York.

Hall, B. P. (1993). Values Power, Values Technology, Santa Cruz, CA.

Hall, B. P. (1995). Values Shift: A Guide to Personal and Organizational Transformation, Twin Lights, Rockport, MA.

Hall, B. P., and Joiner, C. W. (1992). Management by Values, Values Technology, Dayton, OH.

Hall, B. P., and Thompson, H. (1980). Leadership Through Values: An Approach to Personal and Organizational Development, Paulist Press, New York.

Hall, B. P., Harari, O., Ledig, B., and Tondow, M. (1986). Manual for the Hall-Tonna Inventory of Values, Paulist Press, New York.

Hall, M.L.W. (1994). Discerning Chaos in Organizations: Understanding Chaos and Complexity in Organizations Using Human Values. 1994 Chaos Network Conference, People Technologies, Denver, CO.

Hall, Martin L.W. 1997 Systems Thinking and Human Values: Towards A Practical Organizational Intervention Methodology. Ph.D. Dissertation, University of Lincolnshire and Humberside. Lincoln, UK.

Hofstede, G. (1991). Cultures and Organizations: Intercultural Cooperation and Its Importance for Survival, McGraw-Hill International, London.

Jackson, M. C. (1991). Systems Methodology for the Management Sciences, Plenum Press, New York.

Maslow, A. H. (1970). Motivation and Personality, Harper and Row, New York.

Maslow, A. H. (1970). Motivation and Personality, Harper and Row, New York.

Maslow, A. H. (1971). The Farther Reaches of Human Nature, Viking Press, New York.

Maslow, A. H. (1971). The Farther Reaches of Human Nature, Viking Press, New York.

Mason, R. O., and Mitroff, I. I. (1981). Challenging Strategic Planning Assumptions, Wiley, New York.

Robertshaw, J. E., Mecca, S. J., and Rerick, M. N. (1978). Problem Solving: A Systems Approach, Petrocelli Books, New York.

Rokeach, M. (1979). Understanding Human Values: Individual and Societal, New York, Free Press.

Chapter 2

The role of values in measuring performance of social systems

BRUCE BUCHANAN

1. INTRODUCTION

Our theme is the strategic role of values in the systematic guidance of behaviour. Our focus here will be the function of values, and the need for their systematic assessment in terms of relationships and results, rather than as fixed or arbitrary concepts. The light thrown on values, considered from a systems perspective as decision criteria through which needs are met, makes the challenge of reassessing the nature and role of values eminently worthwhile (Buchanan, 3,4).

In attempting to better understand values we face formidable problems. The subject is close to the hearts of many people, and a diversity of assumptions are defended with vigour and often much anxiety. One of the most eminent of philosophers, Michael Scriven (Scriven, 11), has even described a kind of "value phobia" as a feature of our time. Nevertheless, values are literally decisive in the shaping of purposes, steering behaviours and achieving results. They inform human affairs at all levels, and values that are covert or misleading can be a source of many problems. Governments and many large organisations are today being guided by values (e.g. social, political and economic indicators) which have not all been adequately appraised in what they purport to measure. (Some of these will be discussed in more detail below, in section 5. Values: Patterns and Specifics)

We may begin by considering the nature and role of values, in relation to system dynamics in general and the performance of social systems in particular. The distinction needs to be clearly made between (1) actual performances in the real world, which determines success or failure in life, and (2) performance *measures*, which reflect abstractions, always less than complete. While measurements are important, it is also vital to recognise the

The Performance of Social Systems
Edited by Francisco Parra-Luna, Kluwer Academic/Plenum Publishers, 2000

kinds of comparisons involved and their possible limitations. Measures of value require careful analysis, but must be justified by usefulness.

Power, it has been said (Newman, 8), is the production of intended effects. The factors involved in the exercise of power may be usefully described in terms of systems. A system is usually defined in terms of the specific factors which determine its outcomes. Thus the elements and relationships which comprise systems are identified for their contributions to the purposes and criteria, or values, which shape results. Inputs and components are controlled by feedback from criteria based upon governing values. (The environment is usually defined in terms of factors beyond the boundaries of the system and not under its control, but which may have important effects upon it.) Of course, power may also be described in terms of force and coercion (the implicit model being that of a machine), but the responsible exercise of social power must be transparent with respect to values and results if it is to be effective and efficient in the long run.

Social systems may pose problems of many kinds. Turning around a company that is losing its markets and is structurally impaired is difficult. It may be even more difficult to turn around a political entity damaged by its history, customs, and fears. It is especially difficult if such bodies are saddled with false certainties, convinced of the worth of faulty values. Finding creative directions for a world besieged by such problems may be hopeless without a vision based upon values which recognise needs for planning and change.

Peter Newman writes: "Inheritors often lose their nerve and are frightened of responsibility" (Newman, 8) Since fear and ignorance block communications, "telling the truth" to the power-that-be may not be possible, Finding a common ground and language is a prerequisite. But many basic assumptions may need to be challenged and revitalised, and no side has all the answers. At least since the times of Socrates and Jesus, this has been a dangerous kind of activity.

Of course, values do not describe an actual world "as it is". They reflect the use of abstract language, in terms of which alternative possibilities may be envisioned. Human evaluative processes also reflect biological and unconscious needs. To the extent that "universal" values may be described they are the product of the same kind of consensual validation as is involved in any scientific inquiry. Validation based upon shared experience is a more trustworthy foundation than imagined absolutes. Values are always provisional, i.e. subject to revision in the light of experience.

We live in schizoid societies, in which values, societal programs and daily activities are often in conflict, which may or may not be recognised. Leaders often do not "walk their talk". Thus, it is of the utmost importance that the values which govern performance in various areas be open to assessment. Effective values cannot be adopted prematurely and fixed; both the world and human needs change. Nor should one set of values be used for purposes of public relations, another set for self-promotion, and real responsibilities treated as an afterthought. Performance criteria should be able to distinguish such uses. Among the most important values which influence human lives and events we know that power looms very large. Some philosophers (e.g. Nietzsche) have seen the "will to power" as the primary drive. Other thinkers (e.g. Freud) and many professionals engaged in the arts of manipulating consumer choices, exploit human interests in sex. Certainly desires for power and sex are major motivating forces. And the desire to retain power far beyond security needs is perhaps a major obstacle to social change. However, it is beyond the scope of this paper to deal in any further detail with these values, important as they are. Our attention will be on the measurement of more pedestrian values.

What are the sources of values? They stem from the necessities of life and requirements for normal human development.All people have "rights" to life, to what they need to maintain their existence (Adler, 1). Such human rights have been described in terms of universal values which, in various forms and degrees, must be present in all enduring human societies.Yet even "universal" values can only be the product of consensual validation - as involved in any scientific inquiry. In addition, and relative to time and place, there are cultural values which reflect local and traditional habits and expectations, but are not as vital to life itself.

Parra-Luna's Axiological Systems Theory (Parra-Luna, 9), identifies nine "universal" values for consideration and possible measurement. (These will be discussed below - section 5. Values: Patterns and Specifics) In a hierarchy of values the final or ultimate terms depend upon process, and entail freedom of inquiry into new developments and possibilities rather than formal concepts - which depend more upon context, at lower levesls, Values are reflected in the functions of living systems and cannot be entirely captures in words and concepts. Indeed, the formalisation of any system of beliefs has the inherent disadvantage of closing the system to further consideration. In relation to human thought, language and values. it is essential to keep our ideas open to learning and new experience.

Our purpose here is to consider values in their potential role as measures of performance at various specified systems and levels (individual, group or institutional). We should also be clear about what may be seen as the major requirements for performance measures at any level. These include: 1) effectiveness of the system (whether an individual or social group) in doing what is required to meet the needs; 2) goals that validly reflect real human needs; 3) goals that take account of systemic limitations e.g. requirements for sustainability; 4) methods that take into account political factors involved in implementation; 5) aims that do not foreclose on novel and creative possibilities. It is not the purpose of this paper to provide specifics by which values may be measured. The first requirement is to consider an adequate intellectual framework for dealing with systems and action programs. In effect we are addressing the design requirements for a communication system with appropriate controls rather than specific programs; We are considering the design of a possible video-cassette recorder (VCR) to facilitate viewing, rather than characters and plots of programs to be watched.

2. NEEDS AND VALUES

Human needs are the primary source of our values. Both individuals and social group have essential requirements. These needs influence their perceptions of the realities within which they live. These real and pressing needs reflect the biological, psychological and social necessities of normal life and development. When these needs are not met, human life is impaired. There may be disagreements about where to draw precise lines, but basic necessities for life are recognised by all. Values and goals may, however, serve a variety of purposes, and may have more to do with wants or desires which have no natural end-points. The problems involved in defining and measuring such values have not all been well described, and certainly have not been solved. Efforts are required in this direction in the interests of transparency and perceived fairness. Needs and effects that are obvious may not be the most important, and what is easiest to measure may not be the most significant. Convenient measures of economic activity, such as the Gross Domestic Product (GDP) may be inadequate as measures of economic health. Within the totals of such an index there may be components which in reality are opposed in their effects e.g. commercial activities which may add to costs for health care, or armaments which negate rather than add to useful production. Societal ideologies such as "free enterprise" capitalism and democratic socialism are seen by their advocates as justified by their results. However, an ideological conviction is not evidence, and decision processes must be guided by facts. And it is less than honest to select only that evidence which proves a case, and to neglect facts which point to problems. In particular, it is vital to avoid the kind of restrictions which block feedback necessary for adapting to real world conditions. Systems must be open to be effective in the long run.

3. VALUES AND IDEOLOGIES

As used in this paper, values refer to decision criteria which steer behaviour, in particular towards the kinds of results intended to meet specified needs. Values are defined in terms of outcomes which can be described in operational terms, i.e. which can be clearly understood and experienced by others. Ideologies, in contrast, are abstractions which cannot always be described or tested in terms of experience and results. While there is a valid role for faith in unseen possibilities, the test of truth lies in evidence, and particularly in predicted fruits of action. Ideologies are often used as the basis for claims to authority and power which are usually shown in the event to be arbitrary, whatever the beliefs of their proponents. History provides many examples of such distortions, both well-intentioned and cynical. Claims or intellectual justifications on behalf of ideologies - e.g. capitalism, socialism, or fascism, etc. - tend to make selective use of evidence. From a systems perspective a valid or scientific approach must take into account all the relevant data. Elements of importance cannot be arbitrarily ruled out of consideration as "externalities", of no concern - as too often done by apologists for "free enterprise". Ideological positions, like other intellectual hypotheses, are only as good and useful and the evidence which supports them, and as the enhancements of knowledge that they bring. Many previous generations sought intellectual security in ideological absolutes. But genuine security requires responsibility - in examining evidence systematically, in making responsible evaluations. (Scriven, 11) and in the discriminating use of intelligence, no holds barred. This is the only process by which enduringly seful values can be established. It also means that in principle there are no arbitrary limits to be imposed on developments in human affairs.

4. VALUES AND PERFORMANCE: EFFECTIVENESS AND EFFICIENCY

The performance or results of an enterprise or social system must be measured against relevant criteria, which include values, objectives, goals and priorities. A variety of measurements can provide feedback guidance to various contributing operations, to modify processes as necessary to improve performance. (Drucker, 6) A measurement is a comparison, normatively to some standard. The standard may be be clear, but is often vague or implied. For example, rates of crime and/or imprisonment, in comparison with other countries, or as trends over time within one jurisdiction, may be offered as objective indices of social justice, or effective policing, or as support for some other possible value or objective. The validity of such claims must rest on empirical evidence which considers not merely various correlations of data but system dynamics e.g. policies, intended effects and actual results. It may be said that, from a systems perspective, the lowering of crime rates represent success for values which guide policy and actions. (Of course, all statistics must be open to scrutiny. The lowering of crime rates by falsification of reports and failure to register and prosecute crimes would be a perversion of the claimed values.)

Information which is readily available may or may not not be significant as a measurement. Statistical information can be meaningless and/or misleading. The validity and usefulness of the comparisons are empirical questions. The relevance of measurements may be illuminated by considering the relationships of values to actions and effects from a systems perspective. Not just any measures will serve, Values may be termed "objective" when they meet criteria which are agreed upon by different observers. Many measures in current use, e.g. statistics of crime, income levels, child mortality, etc., carry implied comparisons in relation to other jurisdictions and have been studied in many ways. Of course, values may be objective without being useful. The question of "subjective"values are more complex. This category may include whatever people feel or assert to be their values. These may be indeterminate for an observer. They may be used to justify actions regardless of outcomes e.g. on the basis of duty (deontological values.) But they may be important in their effects, and folk wisdom and scripture agree that "actions speak louder than words" and "by their fruits ye shall know them". We are here using the term value to describe criteria involved in the evaluation of actions and events. Thus, purported values which are held to be absolute, i.e. not subject to any further assessment, might also, and perhaps more clearly, be identified as Ideals or Principles. (These often reflect a Cartesian dualism which separates mind from body, thought and action, and is at odds with the non-dualistic views of integral systems being advanced here.) Of course values based on personal opinions can be assessed in various ways e.g. by questionnaires and interviews, etc. These may be important in their own ways., and may also be important as aggregate public opinion. However, it has been one of the tasks of social scientists, in attempting to describe such values, to bring the evidence for them into a realm of discourse which is consensual, that is, to objectify what is being described. Subjective values may be more strictly said to be incommunicable, not yet formalised for inclusion within the conceptual framework of scientific description and measurement. Thus, the burden of proof of the meaning of subjective values lies on the proponent, who has the obligation to demonstrate their meaning. Our consideration here will be limited to two kinds of "objective" comparisons which reflect the requirements of a systems perspective. Within the relationships which define any system there are two aspects of decisive importance. The first is the system EFFECTIVENESS in achieving the effects or outputs that are required to provide for its continuing existence - e.g. animals must obtain food and warmth, etc., social groups must perform the functions for which they are being supported, companies must provide services which meet market needs, and so forth. Effective performance may be described in terms of achieving intended goals and meeting real needs, the outcome(s) which it is the business or purpose of a system to produce. The second key consideration is the EFFICIENCY of energy and resource utilisation in achieving the intended outputs - e.g. the ratio of costs to presumed benefits will be influenced by waste, corruption, or diversion of efforts away from agreed intentions. This distinction has been recognised in the management literature (Drucker, 6)

Efficiency is improved as the ratio of output/input is increased. Efficiency involves doing better what is already being done. It may be noted that, if methods actually entail very different requirements and/or side-effects, simple comparisons may be misleading. For instance, apparent efficiencies which really depend upon external conditions, such as social changes for educational and transportation inputs, tax subsidies, waste disposal, etc., may require open and detailed accounting for a proper evaluation of efficiency and even effectiveness. In such cases simplistic comparisons may be of limited interest, and their advocacy may be suspect. The global description of a society in terms of all its inputs and outputs. including those internal to it, involves many complex paths and relations. Outputs of various components are in fact inputs to other components. Overall prosperity, i.e. effectiveness in meeting needs of people and markets, can only be judged in terms of a great variety of interests, as by electorates, where the key measures have to do first with effectiveness i.e. "doing the right things". Measures of effective performance involve the comparison of what is achieved to what is needed and intended. Performance is grounded in values and outcomes which meet human needs and requirements, including a healthy and sustainable environment. To establish such needs depends upon empirical inquiry. Effective performance cannot be assessed solely in terms of process - e.g. of rules of law, duties, roles, or functions alone. These are no more than means to end-values involved in meeting real needs. Constitutions and the role of law are not absolutes, but abstractions to be assessed ultimately by how well they serve the preponderant needs of a people.

5. VALUES: PATTERNS AND SPECIFICS

Parra-Luna has described A REFERENTIAL PATTERN OF VALUES (Parra-Luna, 9) which provides a useful focus for discussion here. Nine values described are: Health, Material Wealth, Security (Law and Order), Knowledge, Freedom, Justice, Environmental Conservation, Quality of Activities, and Prestige. According to Parra-Luna, we all pursue these end-values, but differ in the relative emphasis given to them. Such values are deeply embedded in our social fabric and growing in global importance. Precise definitions are subject to debate. Some values may lead to conflicts in practice (e.g. security vs freedom), and/or have been interpreted in ways that encourage behaviours which are unsustainable, (e.g. material wealth vs environmental conservation) . ordinary observation reveals other values - particularly power and sex - which frequently confound logical approaches. Moreover, the real values which determine actions are seldom to be known directly, and usually can only be inferred from the activities which they govern. To some degree such values are really products of analysis, and lie in the eye of the beholder. Ascribed values sometimes reveal as much about the observer as about what is being observed. In any case, any formalised schema or list of presumed values in the form of abstract concepts require assessment in relation to the phenomena to which they point. The identification of various specific indicators may be a useful way of pointing to unmet needs or to undesirable comparative standings e.g. in health, education, social justice or economic well-being. Few governments are happy to be seen as ranking low on such lists in comparison to other jurisdictions. Such indicators may be a kind of general spur to further analysis and political action. However, to translate such indicators into values, with criteria useful for management decision-making, requires that another step be taken.

This step is that of examining existential considerations - tests of experience which condition the meaning of abstract concepts and measures and provide for ongoing checks on such realities. The freedom to correct errors, and to assess and correct systems processes, to minimise the risks of future errors, must always be a premium value. An epistemological position is involved here. It is the case that any fixed list of values must, in principle, be incomplete. Specification of particular values in measurable terms can only be approximate and provisional, subject to endless improvements (Popper, 10). Such considerations bear upon measurement criteria, including those described by Parra-Luna, used to try to quantify values. The usefulness of convenient statistical indicators may be somewhat limited, and attention should be directed to performance measures which provide for feedback.

The following remarks are intended as points deserving of clarification, not as insuperable obstacles to progress

The values mentioned involve different dimensions, levels and relationships. The clarification of each is really a major undertaking. HEALTH as a concept is inherently ambiguous. It presupposes certain norms and expectations, and the absence of illness and disorder. Terms and measures are no more than approximate. Of course, health is relative to such factors as age, adaptive challenges and compensatory adjustments. Every organism is constantly learning and reorganising in various ways, such flexibility being a sign of health. Healthy stability at one stage may be maladaptive at another. Indicators which reveal and track needs for change may have special value, but no absolute measures are possible. MATERIAL WEALTH must also be seen as relative to expectations and conditions of existence. Economic criteria which take for granted ever increasing growth (e.g. Gross Domestic Product) and by implication present this as acceptable strategy, are sending misleading signals. It is likely to be just as important to identify and track changing conditions conditions which bear upon the creation of wealth, including natural resources and ecology, climate, and human resources. Measures which anticipate needs may be more useful than even current leading indicators. SECURITY (LAW AND ORDER) may be measured by global indices (e.g. by crime statistics, etc.) but effective management must really take into account the many conditions which shape human behaviours. Prevention as well as remedial actions, and insurance for damages, are required. In addition to law enforcement, social policies need to meet needs for education, support of family life, employment and other conditions found to be relevant. Societal values which may promote social pathology require a systems strategy for their management. Many such factors are well known, and their relevance needs to be made more explicit.

KNOWLEDGE, in the widest and most important sense, involves information which is reliable and relevant, Specific knowledge is the basis for sound judgments in all areas of human life. The validity and relevance of knowledge is not self-evident but requires value judgments. High performance in any field requires the continuing pursuit of improved knowledge and skill. A notion of knowledge per se as providing a fixed and reliable basis for ongoing measurements may be misleading. Once again, there are no absolutes.

FREEDOM has many meanings (the term is often used simply to obfuscate and avoid unwanted restrictions). In practice, when used as a signal or slogan to elicit unthinking response, appeals to "freedom" may be in effect opposed to responsible freedom. Freedom as a value stands for an increase in options for choice. Freedom is a kind of superordinate value, related to responsibilities, desicions and results, The exercise of such freedom contributes to conditions which help to realise other values e.g. health and security. Careful specifications are required.

JUSTICE depends upon knowledge, along with rules of law and due process, freedom of enquiry, and the enabling conditions of a healthy society. In addition to such formal protections, social justice also requires educational and work opportunities and political freedom. Here again, specifications for measures are complex and cannot be given in absolutes.

ENVIRONMENTAL CONSERVATION also involves values of knowledge and justice. The appropriate utilisation of natural resources is based upon science. Despoilation of the commons which is not the result of ignorance may be criminal. The knowledge, rules of law and measures required for long term sustainability present problems of the greatest ethical importance. Many values such as material wealth and human rights cannot be measured in the absence of this larger context i.e. an adequate "ecological" and social systems perspective..

The QUALITY OF ACTIVITIES, as these contribute to the potential for human development, are both influenced by, and contribute to, other values, such as knowledge, or individual or environmental health and well being. Like issues of prestige, they are closely tied to cultural expectations which may be involved in proposed measures.

PRESTIGE, as a proxy for values of respect and influence, may also include power, whether of individuals or states. The values by which prestige may be assessed are of great systemic importance in reinforcing other values.. And to be measured, and found wanting or even undeserving, is not without consequences. It is also desirable to distinguish the prestige of demagoguery from the statesmanship of a Nelson Mandela. Here again the operational implications of the value are of the essence.To summarise, the most useful reference pattern of values should be seen as providing a structure and opportunity to assess the individual and social criteria by which human lives are being guided., and to reevaluate these on a continuing basis. These are assessments of great comlexity, and any attempt to find simpler measures may run counter to Ashby's Principle of Requisite Variety, and to that extent be dangerously misleading. A first principle is still "Do no harm"

6. CREATIVE POSSIBILITIES AND VISION

Another value is that of CREATIVITY, particularly in solving problems, A deep understanding of the actual operations of an existing system is a necessary preliminary for considering how elements may with advantage be rearranged. Points of strategic leverage include the consideration of more effective goals as well as more alternatives from which to choose. At the dawn of history in Mesopotamia, royal courts were conceived as systems, patterned on the bodies and cycles of the universe - the life-giving sun, the supporting moon and planets, with priests and others through rituals transmitting the powers to govern life (Campbell, 5). The systems had utility in societies based on agriculture. Mythologies gave structure and support to societal functions, and concepts systematised to provide unity and purpose.

But a system of social unity cannot simply be imposed, Many values take on their meaning and power only from the "bottom-up", in terms of real and effective contributions to others. At a minimum what is required is an integrating vision supported by undistorted feedback concerning the business of the whole. A major obstacle to the serious consideration of values is a kind of phobia (Scriven 11), which may indeed reflect realistic fears of vulnerability to arbitrary powers. And hisotry teaches that any serious strategy for change must deal with values of aggressive self-promotion that maintain arbitrary powers in place.

An adequate strategy requires both conceptual and empirical methods. A systems approach can provide language and models of relationships which can be examined for effectiveness. Clear description of an actual system in all of its known relationships may provide clues. There may be faulty relationships at the levels responsible for integration - such as discrepancies between real intentions, stated goals, and actual results. There may be lack of information, or a lack of the specific feedback necessary to correct errors in direction. Disagreements may arise from different assumptions and mental maps, as shaped by values. A redesign may be possible, but would require much wisdom and commitment (Beer, 2) The final terms in the hierarchy of values which can be known to human beings have more to do with processes of free inquiry than with formal or conceptual truths. Formalisation of any system of beliefs has the inherent disadvantage of closing the system to possible further knowledge. These limitations are ever present for human thought and social life..

With respect to the possibilities for creating really adequate governmental structures we are perhaps in the position of mediaeval physicians ignorant of germs and physiology. While knowledge of management and systems sciences is growing, we still have much to learn about political power, the blindness induced in most of those who seek such power, and integration of distributed powers of governance. We need societal arrangements which better provide for mutual coercion, mutually agreed upon. (Hardin, 12) Above all we need to combine intelligence and responsibility in the service of values which reflect the conditions and requirements of the universe which supports us. To meet these requirements we need a much more precise and useful understanding of all the terms of this challenge. We need improvements in accounting, but must be ever mindful of the oversimplifications inherent in all abstractions.

In summary, this paper makes a plea, and sets out grounds, for considering values not as fixed or absolute, but as provisional concepts, subject to validation by observation and trial, intended to both reflect outcomes and guide action. The tasks and problems of measuring values in this sense can only be a cooperative enterprise guided by principles of a systems science (second order cybernetics) which takes human observers into account.

7. REFERENCES

(1) Adler, M. (1985) Ten Philosophical Mistakes. New York: Macmillan. (2) Beer, Stafford (1994), Platform for Change, John Wiley & Sons (3) Buchanan, B.(1997a) "Assessing Human Values", Kybernetes, Sept/97 - also see:
http://www.ceptualinstitute.com/genre/buchanan/homepageBB.htm

(4) Buchanan, B.(1997b) "Values, Systems, and Consciousness", The Noetic Journal, Vol 1 No. 1 June, 1997 also see:
http://www.ceptualinstitute.com/genre/buchanan/homepageBB.htm

(5) Campbell, Joseph (1959) The Masks of God: Primitive Mythology. Viking Press (New York)

(6) Drucker, Peter F. (1973) Management: Tasks - Responsibilities - Practices. Harper & Row (New York)

(7) Mesthene,Emmanuel, G.(1968), 'How Technology Will Shape the Future' in Purposive Systems: First Annual Symposium of the American Society for Cybernetics, p.67 ff. Spartan Books.

(8) Newman, Peter C. (1998), Titans. Viking/Penguin

(9) Parra-Luna, F., Axiological Systems Theory: The role of Values in the Evaluation of Social Systems (Universidad Computense do Madrid)

(10) Popper, K. (1992) The Open Universe: An Argument for Indeterminism. London: Routledge

(11) Scriven, Michael (1987), Theory and Practice of Evaluation.

(12) Hardin, G. "Tragedy of the Commons.", Science, and many reprints:
The role of values in measuring.

Chapter 3

Towards an axiological systems theory

FRANCISCO PARRA-LUNA
Facultad de Ciencias Políticas y Sociología. Universidad Complutense de Madrid

1. INTRODUCTION

Why do we live in Society? Why, with the exception of the occasional hermit, do we need others to survive? What are we seeking and what do we find? What process turns our needs into satisfaction, or more precisely degrees of need into degrees of satisfaction? Can what we need and what we get be measured and compared? Can we understand the "effectiveness" of a given society in terms of these two concepts? Would this provide us with a possible approach to appraisal of society?

These are some of the questions for which I will be attempting to anticipate an answers.

But mere contemplation of animal life provides us with insight into how imperative social relations are. With the exception of a few species that prefer to live in isolation, removed from their fellow creatures, (the cricket, robin, leopard, etc.) or other animals who live together for reasons other than social interaction per se (earth worms, spiders and crustaceans that live under stones for protection), there is quite an impressive typology of degrees of sociability and cross-species dependence of some individuals on others. The cases of parasitism (the sacculina that lives off of the crab) or feeding companions (the co-operation between the black-backed courser and the crocodile when the former "brushes" the latter's teeth or warns it of possible danger) are mere expressions of real sociability in other species. The social organisation of termites with classes and castes, nests of caterpillars or cliff swallows, the "settlements" built by beavers, flocks of sparlings, butterfly migrations, bee colonies, seal harems, packs of wolves, ape clans, etc. What

The Performance of Social Systems
Edited by Francisco Parra-Luna, Kluwer Academic/Plenum Publishers, 2000

instinct, what biological need brings them together, accepting strict discipline and rank?

Researchers have found, rather simply, that individual survival depends upon such behaviour. The group imposes its rule, competence and hierarchy (usually on the basis of sheer brute strength) because it affords safety and enhances survival. A pack of hyenas is thus able to steal a hungry lion's prey despite the latter's greater individual strength, and when a predator attacks a bird known as the rook in flight, the rest of the flock turn against the predator until it releases its prey. And in general, a socially integrated specimen benefits from systems of communication that warn it of danger, of the availability of food or water, and even the distance to and exact place where they may be found.

The individual animal in contrast finds life more difficult; such is the case of the solitary male cheetah when attacked by a pack of other cheetahs, or old or sick elephants that voluntarily stray from the herd seeking death. The optimisation of chances of individual survival, just that and nothing more, is what animal societies would seem to provide their members.

And what about human societies? What does society afford individuals that makes it impossible for them to live outside it? I will be discussing this question in the following chapters, but we might surmise that it provides basically the same thing as animal societies, except that for human beings "survival" means more than getting enough to eat and being able to reproduce; rather, they have considerably more complex, difficult and changing demands. Mankind lives not only on bread, but on good health, security as regards the future, degrees of freedom, prestige, wealth, justice, etc. This range of values is what human beings would receive in one degree or another from living in society. It seems then clear that societal problems can be defined as a kind of axiological imbalance since values can be considered to be the raw material which founds social phenomena (Buchanan, 1998, Hall, 1998, Parra-Luna, 1983). If "values" is the degree of usefulness or suitability of things to satisfy necessities", or to use more sociological terms, if "Value" is an element of a shared symbolic system which serves as the criterion for choice between indicative alternatives, then it can be extrapolated that everything, or almost everything in the field of human relations, can be considered as an enormous and complex framework of necessities and interests which can only be satisfied through the achievement of values. A central hypothesis in this book is to support that any systemic theory of society could not avoid to deal with values as the basic raw material for defining and measuring its performance and for achieving the best possible solution of societal problems. The "naked ape" as Morris and some sociologists would call us, would simply raise the level of expectations in what would seem to be persistent human, eternally dissatisfied behaviour: as needs are met they tend to be considered of a "lower" order. This is what "progress" would seem to be all about and this is the course chosen by what we call developed countries.

If this widespread phenomenon, sociability, is characterised by the synergies (and corresponding added value) exchanged in (by the way) closed groups, zealous enforcers of their territorial boundaries and wary of outsiders, the question arises as to whether a pure theory of mutual inter-influence is applicable, and in particular, whether a theory of synergies deriving from the very existence of the society or group would be useful. I shall begin this essay, then, by asking whether society is a "system" and whether the application of the corresponding theory has anything to add to social analysis.

2. IS SOCIETY A SYSTEM?

This is a question that even now, at this point in the development of System Theories in Sociology, is still relevant. Because if society is a system, it makes sense to treat it as such and apply the rules and principles deriving from the corresponding theory in order to put an end, once and for all, to anti-systemic fundamentalism; and if it is not, it would make even better sense to set such theories aside, at least until the two concepts (Society and System) evolve towards a closer or better established mutual ontological or isomorphic match.

Initially at least, I would dare to affirm that the world, the countries and human beings that populate the planet Earth, taken as a whole, and more and more often considered a global system due mainly to the advances in communication (P. Drucker's "global village"), cannot be considered to be a system strictly speaking. The reason for this is that at least one of the terms of the definition is missing (unity, as we shall see in chapter 7), as evidenced by the fact that two thirds of Humanity is starving, while the other third has food to spare. If the world operated like a "system", i.e., if it were governed by a single legitimate political authority, aided and controlled by the remaining participatory and critical institutions that characterise all the other social organisations, the Nation-State among them, it would be possible to prevent this dramatic axiological imbalance that makes contemplating hunger in the world unbearable. The greater, then, the radical division brought on by economic, religious, cultural or other kinds of particularisation, the smaller will be the degree of systemisation in the world. As long as there are sovereign territories mistrustful of their neighbours, no politically valid world-wide system will be able to make the correction of economic inequalities compatible with a respect for cultural particularities.

But in order to answer the question that heads this chapter with any accuracy, consideration must be given to what is meant by society and by system. Let us define the latter in a rather elementary and introductory way: a system is said to be a set of elements interrelated in such a way that if one of them undergoes change the set as a whole is changed. We shall see below how this basic notion (Von Bertalanffy) is used to develop the notion of transforming agent (Piaget, Easton) as well as the concepts control or regulation (Wiener). A system, then, is an engine composed of interrelated parts with specific functions: the human body made up of interrelated organs, muscles and bones or even a plant consisting of roots, sap, stem, branches, leaves, etc. in which each one of the parts mentioned contributes something to the whole. Buckley's (1968) well-known definition emphasises the effort of complex socio-cultural systems to adapt to the changing environment. J. Miller's well-known classification of existing systems (cell, organ, organism, group, organisation, community, society and supranational system) provides an idea of the degree to which the concept system has come to be universally applied. Bailey (1992), in turn, points out that all systems share the following features: 1) They are composed of basic elements, variables, attributes, etc. 2) Such elements are related via what may be called connections, correlations, interrelationships, communications, etc. 3) Such relations generally follow non-random patterns, which gives rise to a structure that abides by a certain order with a degree of entropy that is below the maximum possible level (or negaentropy). And 4) the system exists within a physical space although the boundaries of such a space may not be fully defined. Bunge (1977) expressed the idea formally with the expression: S = [R (a,b, ..., n)], where R is the set of relationships linking elements a, b, ..., n. It is less important to understand and classify the elements of a system than to understand the relationships "R" by which they are governed, although the former may be a pre-requisite for the latter. We might add as well that all systems are ontologically open, since there seems to be no system totally closed off from outside and with no exchange with the Environment surrounding the system. Systems are closed only as logical units to better understand and deal with them. This is why the specification of the system environment or boundaries is subject to the ends and means of each specific analysis or study.

And what do we mean by society? In the laxest sense, it is used to mean any more or less durable link between two or more persons through which they pursue the same ends. Wherever two or more persons are (explicitly or implicitly) involved in a project entailing coexistence in general or co-operation in specific areas, that is a society; and such project is characterised by the interrelationships, exchange or communications established among the persons involved along more or less patterned lines. Clear examples of societies in this lax sense are couples in love; families; clubs of friends;

community organisations; companies; large institutions such as the Church, the Army, etc., and so on until we come to the classic concept of the Nation-State, international organisations or even the planet itself and all its inhabitants. One feature that all these examples of society have in common is that the natural or legal persons of which they are composed think, consciously or unconsciously, that they have some individual "benefit" to gain from mutual co-operation. Otherwise, it would not be worth their while to belong to society, to be "with others" and depend on relationships that are often problematic and difficult. It would seem then obvious that the origin of the concept of society lies in the expectation of some benefit on the part of its members. Infants cannot survive without their mothers, who in turn do not feel fulfilled unless they take care of their offspring. Friends seeking one another's company tend to share pastimes that are inaccessible on an individual basis, and the members of associations or institutions seek some advantage that only membership can afford them. As soon as a social link fails to provide the "benefits" expected of it, it disappears. Such expectation is behind the chain of events that culminate in social action, following the inevitable and classic process Communication/Perception/ Memorisation/Decision/Action. It may be added that there are natural societies (the family, the tribe, or the couple in love) and artificial societies (marriage, a formal club of friends, a company, or the State), although this distinction is not particularly relevant from our standpoint since what we are interested in is the degree of individual satisfaction that belonging to a society provides. These expectations constitute the "cement" that holds individuals together in both natural and artificial societies, and if perceived needs are not felt to be met, the social entity begins to deteriorate such as, for instance, in pre-divorce situations in a marriage or temporary receivership or bankruptcy in businesses. In general, then, any society is a group of interrelated persons, where whatever happens to any one of them inevitably has an effect on the rest. Thus, for example, a father's unemployment, or a child's serious illness, are detrimental to the family's financial situation and to the satisfaction of its standard needs; other examples might be a corrupt politician whose conduct is detrimental to his/her party, a Nobel prize winner who lends prestige to the institution in which he/she works or our realisation that monetary inflation depends on interest rates which in turn affect investment, unemployment and economic growth. All of these variables are interrelated via mutual feedbacks, making each individual variable a powerful modifier of the whole. Any individual or group act usually leads to a series of consequences within the society in which it occurs; it does not take many examples to see this clearly. It has often been said that the assassination of Archduke Francis Ferdinand in Sarajevo in 1914 started the First World War. And various degrees of conflict may arise in a marriage, family, group or business when a given occurrence is detrimental to the interests of the whole. Lobbies, social movements, and in general any kind of individual or collective action must bear in mind not only the characteristics of the society in which it is taken, but the chain of events it may set off, the social responses and the reaction that those affected may adopt to respond adaptively to the new circumstances. Any social action, via such interrelationships, becomes a potential stimulus for social consequences or feedbacks of many kinds. Can it still be argued that Society is not a system (or does not behave like one) in view of the notions outlined here? It would not seem to be good for sociology to deny the obvious or cling to positions which, while they should be encouraged in so far as they afford criticism, diverge radically from reality as evidenced by the multiple empirical correlations that are observed among the variables. I feel, for these reasons, that we can conclude by assuming that all societies are systems and that therefore those who reject the concept of system in social analysis are mistaken. It would seem logical to infer that if society is or behaves like a system, it should be considered and dealt with accordingly, even though the purpose be to criticise, modify, or improve it or even to cause it to shed its systemic nature where such systemic attitudes begin to limit individual freedom, or more strictly, upset the axiological equilibrium that all societies pursue, as we hope will become clear in the pages that follow.

2.1 Systems - Unified, Semi-unified and Disperse or pseudosystems

Although all human systems without exception pursue certain objectives, one important distinction is the difference between unified and disperse systems, in terms of their internal hierarchy and authority. By unified system I mean one where: 1) the individuals of which it is composed choose (or admit) a single authority that represents them and assumes the dynamic management of the system; 2) such authority extends to the major systemic dimensions and is competent in respect of all objectives pursued. There are many examples of this kind of system: the head of family, the leader of a group, the gang boss, the manager of an organisation, the director of a company, or the president or prime minister of a State, are considered responsible for ensuring the proper operation of the system to the benefit of the largest possible number of its members. These systems are closed from the organisational standpoint, their objectives are well defined and there is awareness of what can and cannot be accomplished in relation to other neighbouring systems. The major advantage to this kind of system is that their membership is aware of the unitary and solidary nature of the system and therefore feels that it should have and indeed does demand a share in the benefits deriving from the system. Where personal, regional or other kind of identity clash with the solidarity among the members of the system, the latter usually prevails.

Semi-unified systems are those whose membership comprises sovereign systems with their own economic and cultural peculiarities that feel it advisable to integrate interests on a higher order, and consequently develop common unitary and compensatory institutions. This does not mean, however, that the original systems do not ultimately maintain nearly all their sovereignty. Where personal or national identity comes into conflict with the solidarity between systems, the former usually prevails.

In contrast, by disperse or pseudosystem I mean one in which: 1) there is no single authority in charge due to the sovereign independence of the systems of which it is composed; and 2) accordingly, the component subsystems attempt to produce their own benefits for their respective members only, despite growing pseudosystemic relationships or exchange. The planet Earth is currently representative of such a pseudosystem. In any clash between the above-mentioned identity and solidarity, the former inevitably prevails. It takes no comparison of the advantages of one kind of system over the other to perceive the gravity of the problems a diverse system creates on a universal scale, problems which do not necessarily exist as intensely on smaller scales or where the populations involved have a given standard of living. This would explain, on the one hand, nationalist tension in support of the formation of more disperse systems, represented by regionalist and independence movements; and on the other, international tendencies in support of the formation of more unified systems such as in the creation of large international zones. The two are the result of the natural tension between centrifugal (disintegrative) and centripetal (integrative) forces, which can be logically expected in any system. Systems Theory could definitely contribute to determining or defining the most fitting intermediate solution for each specific historic area and circumstance.

In short, taking the isomorphisms observed in the traditional definitions of society and system as a point of departure, it can be safely stated that societies (practically all kinds of existing society) are or behave like unified systems. One exception to this rule is the social globality of the planet Earth, which at best can be considered a disperse system or pseudosystem.

3. THE AXIO-OPERATIONAL APPROACH

Considering society to be a system, then, an obvious gap appears in social analysis that is not filled by any of the existing social sciences. Due to scientific-academic divisions, each one of the disciplines that study social phenomena ultimately review and analyse their own corresponding tree but none of them can see the forest in its entirety. I refer specifically to making the subject of research something that could be called "total productivity of a society in so far as it is an organisation, and which can be attributed solely to its existence as such". Such productivity may be expressed by the fundamental inequality: "Y" > sum of all "y", where "Y" measures the global production of society per se and "y" each one of its individual productions. This may also be translated to mean synergies, emergence or added value deriving exclusively from the fact that the society exists. Ignoring such concept constitutes a very serious omission considering that the forest that is not perceived is not a society of trees, but of men and women who have needs which they seek to satisfy via, and solely via, social interdependence. If such a state of affairs is allowed, the *raison d'être* of any society, i.e., the overall satisfaction - real and/or perceived - of its members, would never be the subject of analysis. The notorious short-sightedness of scientific specialities (now more than ever Ortega's "barbarian specialists" come to mind) slices up society in such a way that the result is similar to an analysis of the individual parts of an engine that has been taken apart. Its unity, its operation, the functionality of each part are ignored.

Hence, economists study production and resources with an emphasis, especially modern economists, on a few instrumental variables such as inflation, interest rates, investment, balance of payments and so forth. All of such indicators focus on supply and demand, centring concern, ultimately, on only one of the Values pursued by society: Material Wealth, which entails considerable bias in the assessment of such variables and their systemic effects on the whole. Unemployment, for example, an extremely important social variable, is considered by economists to be a by-product to be governed by the above economic variables. Other specialists, such as social anthropologists, study the customs, traditions and folklore of peoples and societies with more natural, less artificial connections. And although one of the major concepts of social anthropology is values, they are contemplated more from the standpoint of diversity (cultural relativism) than from the confluence of personal interests in a limited number of central values or End-Values which are the ones that justify and explain the existence of society. Social Psychology, in turn, studies individual behaviour of persons in society and, while it does take the individual as a set of objective and subjective needs, it focuses in particular on the social restrictions to individual behaviour but not, naturally, on collective behaviour. Human Ecology is concerned with the interaction and adaptation of societies to their physical surroundings, its core concepts being the "niche" or place occupied, density, domination, competitiveness, biological-physical symbiosis and so on. Historians are concerned with research into the past, focusing their analyses either on the behaviour of large social and political units or on the biographic study of celebrities, but not as much effort is put into comparing which systems of global end-values have been achieved by given societies in the past and what their axiological tendencies were. Philosophy (or what is left of it) seems to be devoting all its attention to formal logic, a topic some believe to be quite far removed from current concerns and events, an attitude which would seem to constitute a bit of a retreat with respect to this discipline's classic stand. Special mention must go to Political Science, perhaps the discipline that should define and broaden the organisational and social (political) answers to people's demands and needs. That branch of social science makes the mistake of considering itself, under the current academic curriculum, a subsystem of the social system that constitutes its subject matter; it as if it were analysing its own Environment. This merely academic and artificial separation, based perhaps on mimicry of specialisation in other sciences, is what brings authors with great prestige such as David Easton or G. Almond to consider the Outputs of the political system to be "Decisions" and "Action" (the former) and the three powers: legislative, executive and judicial (the latter). Not a word is said (Deutsch is an exception) about the specific needs of the public at large or the political solutions offered, matters which should be the major objectives of political and logical analysis. Political science reviews, rather, electoral processes, power, lobbies, political bureaucracy or international relations, but hardly ever the core political problem that consists of defining, measuring and comparing what a political system should do and does to meet the people's specific demands; this shortcoming is particularly exasperating considering that the people vote for or choose their politicians solely to do this job. In contrast, a discipline that is concerned with the "productivity" of a social system, is the theory of business organisation. From Taylor's and Fayol's classic studies, to those of Drucker, Perrow, Selznick, Simon, Tannembaum, Argiris, Barnard, Etzioni, Herzberg, Likert and many, many others as well as the efforts of institutions such as the Centre for Effective Organisations with headquarters in California, U.S.A., all such thinking points in this direction, but two considerable shortcomings are intrinsic to such attempts: first, no thought is given to epistemological globality, i.e., they stress economic profitability and tend to ignore other aspects that are essential to the operation of the company or organisation; and second, they do not express the system outputs in terms of End-Values, the only entities that can be matched directly to the corresponding needs of the individuals that make up the organisation. Finally, sociology, one of the oldest of the social sciences mentioned, would seem to be a kind of stronghold which has not yet been taken by specialisation. Unfortunately, however, with the exception of Marx in his early writings (and therefore his followers in that period, Agnes Heller, for instance) sociology has not been concerned with the study of human needs and especially of the extent to which bringing individuals together affords better satisfaction of such needs. It is true that in applied

Sociology, the discipline must necessarily address specific social problems, and hence the development of sociology of health, labour, unemployment, Science, economy, etc. But dividing the discipline into different specialities entails making the same mistake, ignoring "man", that whole made up of interrelated needs that lead him to establish a social system for the sole purpose of satisfying them. The history of sociological analysis reveals any number of subjects that have been treated as though they were the fundamental prime material for social analysis (role, status, class, communication, etc.), and this has made no small contribution to our current understanding of the social phenomenon. But these analyses have been partial, or covered partial aspects of society, either localised in terms of time and space (empiricism), or out of context and made extensive to any kind of society (ambitious theoretical or para-theoretical formulations). And while they may have interrelated concepts, they have been insufficient to justify societal phenomena in all possible processes and situations.

There seems to be, then, no science engaged in explaining, measuring and comparing the "benefits" that people obtain by co-operating with one another; or, in other words, there seems to be no science engaging in analysing "global productivity" or "added value" that any societal phenomenon entails. Such productivity must necessarily be expressed in terms of end-values since it is just that and only that, that justifies the existence of society per se. The pages that follow are an attempt to explore the possibility of filling this gap; I will try to consider this basic theory of human needs within the framework of the social phenomenon taken as a whole, pursuant to the principles of Systems Theory which specialises in the treatment of complex problems and units. Thus, I will refer to the axio-operational approach as the procedure in social analysis that combines the use of what is now classic Systems Theory (Von Bertalanffy, 1973; Buckley 1968)) and its Cybernetic derivations (Wiener, 1969; Maruyama, 1968) with modern contributions to the new theory of sociocybernetic systems (Luhmann ,1996; Bailey, 1992; Geyer 1996; Parra-Luna, 1999) and in particular with the part of axiology that studies those values that prove to be common to all individuals (Objective Theory of Values). This latter theory, I should emphasise at the risk of being repetitive, stresses the essential role of values in what I call the "societal phenomenon". This latter is understood to mean any co-operation between two or more persons intended to achieve certain ends, which may range from sexual gratification of the couple to meeting the needs deriving from a Nation-State, and includes the objectives to be reached by companies, organisations and groups. Now then, this societal phenomenon constitutes a whole which necessarily entails a globality of ends or values to be achieved, for which analysis is still outstanding. It is indeed ironic that Sociology, the science specialising in the study of such a phenomenon, is not concerned with analysing it as a global whole, which would seem to be indispensable to understanding and justifying the reasons, interests or motives that are behind two or more persons' initiating and maintaining a certain degree of co-operation over time.

This concept of social system as an entity comprising persons grouped in order to satisfy individual needs, makes it possible to formally define the *raison d'être* of any social system in terms of better fulfilment of such needs defined as values, and this is what might be called "axiological emergence of the social phenomenon". Indeed, if individuals a, b, n, need to satisfy needs N1, N2, ..., Nn and to do so they seek fulfilment of Values V1, V2, ..., Vn, from the time these individuals group together systemically, the obtaining (-->) of satisfactions (V) is governed by the following rule:

if a --> V1
and b --> V2
.
.
.
and n --> Vn

the system S = [R(a,b, ..., n)] --> "Y" where

$$Y > \sum_{i=1}^{n} v_i \qquad (1)$$

1

Which demonstrates the advantage to any association. The better ability to fulfil needs is what determines the existence of social systems.

Sometimes the inequality does not hold in that direction, so it may therefore be established that:

$$Y < \sum_{i=1}^{N} v_i \qquad (2)$$

if
>1, axiological emergence is positive
<1, axiological emergence is negative.

The second case only occurs in the event of social rupture (for instance, separation of a married couple or of industrial partners). The first or usual case is what justifies the existence of society and its myriad forms of association.

The inequality expressed in (1) and it alone justifies the societal phenomenon, so that Y is, or should be, barring some argument to the contrary, the sociologist's essential unit. The existence of society is insured for only as long as the inequality is significant: otherwise co-operation disappears, which is what usually happens when Y is no longer large enough to justify the association, or when the opposite of what is anticipated in expression (2) occurs. Bankruptcy, dissolution, separation or divorce are then the logical corollary to social failure, the result of Y's being too small or insufficient in relation to what the social actors expected. Y, which represents a set (system) of benefits (fulfilled values), as we shall see, and which must necessarily be treated as a complex whole or composite index of such values, is not currently analysed even by the Theory of Social Systems, which might be expected to engage primarily in the study not only of the social phenomenon, but in the analysis of globality. In such Systems Theory (in social applications of such theory), which has produced important developments, Y is usually ignored. Hence, the raison d'être of the social phenomenon as a system, the deep underlying motive that gives rise to social relationships, the cement that holds individuals together, the sole and unique reason that justifies the societal phenomenon as a system, is of no interest even to systemic sociologists who, instead of dealing with the concept "Y" (system outputs), specialise in the treatment of social processes (Buckley, 1967,1968)), communications (Luhmann, 1996), entropy (Bailey, 1992)) or Sociocybernetics (Geyer and van der Zouwen, 1978)). This represents a rather incomprehensible loss of systemic vision (of the overall social phenomenon, as Mauss would say), while sociologists stare at any given tree in the forest. Not even the formalising attempts made by Cortes, Przeworski and Sprague (1974) or Bailey's more recent attempts at integration put the necessary emphasis on outputs. The new Theory of Social Systems must then build on the foundations established by Cortes et al., Bailey, Luhmann and others, but going one step beyond such contributions via integration with the Values Theory.

What requirements should the new systemic theory of society meet if it is to visualise the entire forest? Undoubtedly the following, at the very least:

A. Sociologisation

Society should be considered as if it were analysed by all the individuals that make it up from the general perspective of the "man on the street", and not solely in university or scientific contexts. The social researcher must make the necessary effort to achieve an empathetic interpretation by putting him/herself in the place of the common person and analysing any social organisation from the standpoint of what individuals expect from it to satisfy their desires, interests or requirements. Systemic-social theorists cannot afford to make the same mistakes as other illustrious colleagues who have even defined socio-political system outputs or fulfilments to be "decisions" or "action" (Easton) or "setting rules", "implementing rules" and "monitoring it judicially" (Almond). Is it possible to criticise political systems using these concepts? Do not all, absolutely all systems, no exceptions made, both the effective and ineffective systems, make a wealth of decisions, take all kinds of action, set any number of rules, etc., without enhancing the well-being of the population at large? Is it possible to appraise or compare the governance of a political system or a company or any other organisation on the basis of such general mediative concepts? Again, what the man in the street expects of political systems is "benefits" that satisfy his needs such as salaries, good highways, enforcement of law and order, political freedom, environmental conservation, etc. I shall refer to these expectations as End-Values which will be arranged in a Reference Pattern of Values (PRV), discussed in chapter 6. This, in my mind inescapable, operation of interpreting the outputs of a social system from the standpoint of the needs of the man in the street, is what I call "Sociologisation".

B Axiologisation

In Axiological Theory values are considered to be the other side of the coin of need, and since the social system is formed solely to better satisfy individual needs, the outputs or achievements of systems can only logically be expressed in terms of values. Martín López (1997) addresses three types of Values in social systems: a) values whose nature makes them applicable to any social situation and which constitute the primary principles or foundations of the system, such as Justice, Equality, Freedom, etc.; b) those referred to some aspect of social life, such as professional or home life; and c) those associated with specific social positions or roles. For the intents and purposes pursued here, however, values must be separated into two large categories, i.e.: 1) the specific values peculiar to each culture or era (cultural relativism) and 2) general or universal values that are common to nearly all individuals and peoples regardless of other spatial or temporal considerations (Objective Theory of Values). In Part II I shall propose their use in any social system according to a Reference Pattern of Values that consists of the following nine values: Health, Security or Law and Order, Knowledge, Freedom, Justice, Material Wealth, Environmental Conservation, Prestige and Quality of Activity. The failure to refer the achievements (outputs) of social systems to specific systems of values that correspond to specific needs of specific people entails a more or less veiled falsification of Social Theory and unsuspectingly misleads the persons who make up society and uphold its scientific and academic institutions.

C. Cybernetisation

If Cybernetics is "the science of communication and control" (Wiener, 1969)) or "the science that renders action effective"

(Couffignal, 1972), by cybernetisation we mean the monitoring action taken by those interested in ensuring that the outputs or final states of the system abide by what was previously stipulated or anticipated. Fortunately business theory has developed Management Control procedures (Plan/Implementation/ Deviance) that are suitable for application to other social systems. Sociocybernetics today is, then, one of the branches of Social Systems Theory most likely to be applied. This is the least that might be expected of those in charge of any kind of management: the ability to monitor achievements to rectify them in time if necessary.

D. Self-appraisal

The cybernetisation of the System makes allowance for a self-appraisal of its behaviour, not only via the typical analyses of deviance provided by management control, but also, as we shall see in chapter 6, because all of the values and indicators that represent them are expressed in a common interval or standard ranging from 0 (minimum, poor) to 100 (maximum, optimum), and the differences between the actual and optimum (or projected) scores are what will inevitably involve radical criticism of what the system could have done and did not do with respect to each objective or indicator.

The most modern social-systemic theory (emerging from the latest developments in concepts such as autopoyesis, reflexibility, entropy, second order cybernetics, third cybernetics and others supplemented by the demands of the four requirements discussed above) is what I call the Axio-Operational Approach, a method that pursues mainly one thing: to attempt to draw up a rigorous interpretation of the principle of the complexity of social systems and their derivations, cautiousness and relevance to separate the primary (universal "end-values")from the secondary ("cultural" values), what is essential from what is not, what people do and do not seek in society.

4. SOCIETY AS A TRANSFORMING AGENT

All societies pursue certain objectives and attempt to reach them using a series of means: i.e., society can be defined to be a system for transforming means into ends, an operation that has been represented on various occasions (e.g., Easton 1965) by the process Input (X) Transformation (T) Output (Y), illustrated in Figure 1.

Indeed, any society, natural or artificial, large or small, is implicitly designed from the outset to convert something not directly "consumable" by the individuals that compose the said society, to something susceptible of being "consumed" by them. It can be compared to the classic scheme of manufacture whereby raw materials (money, effort and means) or, as the old-time handbooks said ("nature, labour and capital") were used to obtain certain finished tangible or intangible products intended to satisfy individual and social needs. Consequently, society arises to effect this transformation and would cease to exist should such transformation no longer takes place or becomes ineffectual. There can, then, be no doubt that the definition of what input and output are (the latter in particular) is essential to understanding the nature and origin of the social phenomenon, which is nothing more than a "transformation unit".

Let us begin with the output as it represents the primary motivation for social action and, as mentioned above, for the very existence of society. By way of illustration, let us take a question that the reader could conceivably pose, for instance: why am I reading this book? Obviously, because I want to w what it says, o because I seek to enhance my understanding; therefore, I as a reader, even as an "individual system" have used certain means (cost of book, reading time, etc.) to satisfy a need for knowledge. I am motivated by one major value, knowledge, but perhaps as well by the prestige that knowledge itself affords or perhaps even by the economic advantage I may obtain (consider a book bought to prepare a competitive exam for a position with the State, for instance). I am then inevitably "transforming" means into ends, input into output. I am therefore a transforming system. But purchasing and reading this book may also have an impact on other values that I also pursue, consciously or unconsciously. For example, I cannot read so much that it would be detrimental to my health, or what I gain in freedom - through enlightenment - may constitute an attempt on the principle of Social Justice if by buying the book I am depriving one of my children of essential medication. Such a decision would also affect my moral prestige as a parent (another of the values in play), or even have an impact on the value "environmental conservation" if, angry because the book did not live up to my expectations, I should throw it into the river to dispose of it, thus adding to pollution. My action inevitably "produces" and brings into play a whole system of values:

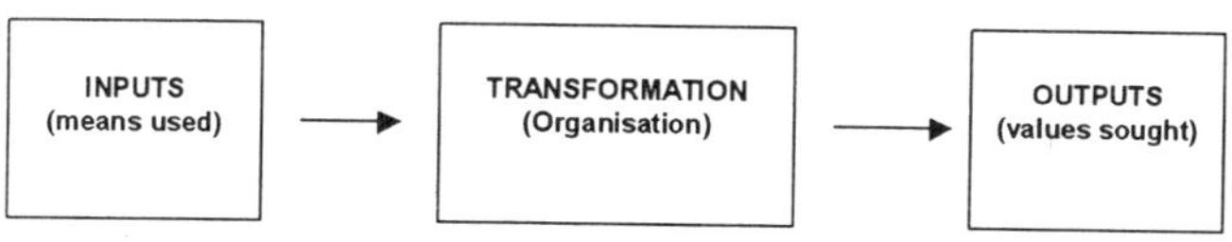

Figure 1. The transformation process

As a rational being I cannot avoid striking a balance between, on the one hand, the achievement of a system of values, and on the other, the consumption of a series of means to achieve it. The system of values to be achieved and the means to be used would certainly different from one individual to the next; hence the saying that each person is a world unto him/herself. But what brings us to live in society, apart from primary biological needs, is the empirical conviction that "grouped together" we will come closer to reaching the values pursued: from Physical Health to Material Wealth, including, inter alia, Knowledge or Moral Prestige. We shall see (chapter 9) that essentially we all pursue the same End-Values, the differences lying merely in the emphasis we place on one or another. Thus, we represent individual configurations of values that I call "axiological profiles" in the discussion that follows. Human beings are simply "bipedal axiological profiles" seeking the fulfilment of our maximum potential.

And the same may be said of businesses or any other organised social group. What, for example, are businessmen looking for? In the first place, to make a certain return on the capital invested, or more commonly, "to make money". But in order to do so they must at the same time achieve a series of inevitable and desirable values. They attempt to ensure the security and the future of the business by implementing a policy for depreciation and establishing reserve funds. They are concerned with the value knowledge in that they establish the qualifications and requirements their employees must meet and/or organise training courses; with the value Health in so far as they comply with rules and regulations on the subject or even go above and beyond such requirements in the firm's own interest; in the value Social Justice since they must pay salaries commensurate with the (quantity and quality) of the work done, or run the risk of creating a conflictive working atmosphere; with the value Environmental Conservation to the extent that they prevent pollution deriving from their businesses as far as possible; with the value quality of activity as they provide their employees with opportunities to reach the highest possible level of self-fulfilment so they feel they are a part of the company and thus perform more effectively; for the same reason they show a concern for the value freedom since it is only by allowing their employees freedom on the job that they will be able to propose suggestions, improvements and savings. And finally they are interested in the value prestige since business success calls for a reputation for reliability, quality, customer service, etc., and this should also be transmitted to employees so they feel proud of belonging to the company.

Such axiological principles are likewise applicable to institutions such as the university. What objectives should universities pursue? Firstly, to reach the highest possible level of Knowledge, by doing research and keeping professors and students abreast of the latest advances in sciences and the arts. But in order to do so they must also consider the value freedom of criticism and study; they should also allow for the maximum personal self-fulfilment of all their staff (teaching and otherwise) to ensure that the institution functions properly; they should also take an interest in the value Material Wealth, paying the faculty and service personnel appropriate salaries and spending as little as possible; the value Social Justice is a concern in so far as emoluments are established in terms of quantity and quality of work done; they should also be interested in the value Health, offering appropriate working conditions on campus, etc. etc. Universities, like any other organisation or institution (Church, Army, Family) cannot avoid pursuing and achieving a given system of values. This is the sole and inevitable purpose of social organisations.

Finally, the Nation-State itself studied in this light has no choice but to achieve values that satisfy the needs of its people. What are the concerns of the State? In an attempt to understand those of its people, it tries first of all to ensure that the economy behaves as it should, since all of the other values depend largely upon a healthy economy; its typical concern about economic growth (GDP), inflation, exports, balance of payments, etc. are mere expressions of concern for Material Wealth or standard of living; but at the same time it must take an interest in the value Social Justice and attempt to ensure that the national product is distributed as justly (or equitably) as possible; in the value Freedom by making provision for political, civil, economic etc. liberties; in the value Health by establishing hospitals, medical centres, etc. as public services or allowing private initiative to do so, thus enhancing the value freedom although, it is only fair to say, possibly at the expense of the Social Justice. The State likewise shows concern for the value Knowledge by financing research and training as public services; for the value Security (civil, functional, from outside threats, etc.) by protecting its citizens with police service and preventing riots, demonstrations, etc. as far as possible and maintaining an army to prevent invasion of its borders. It is concerned with the value Environmental Conservation in that it protects flora and fauna and bans pollution or cleans up polluted areas; and it is concerned, finally, with the value Prestige to the extent that it participates in international forums or provides aid to countries in need. In general, therefore, any society imaginable in space and time seeks to reach these specific End-Values; and since such end-values are all dialectically inter-related, mostly positively but sometimes negatively, what societies ultimately achieve, consciously or unconsciously, is a complex system of values.

What is unusual about the transforming process is that societies could not do otherwise even if they tried, since all their actions are either referred to the people's needs and interests or to those of the ruling elite, but they always are intended to satisfy the motivations of individuals. The basic problem in ideal political management consists then, of "producing" a system of values that best suits or matches the population at large.

If the production of Values (in so far as they are the other side of the coin of needs) is both inevitable and essential, a new formulation is in order for what may be the principal element of analysis in sociology and in particular in political science as a science understood not to be restricted to the analysis of the subsystem of political and judicial formalities, but rather engaged in the study of the maximum possible satisfaction of the needs of the people. The political subsystem must be understood to be a mere tool, a means.

This review of the Transformation process cannot, of course, ignore the Input or means used to produce the system of values pursued. But while important, input cannot be compared to output nor to the transformation process itself in sociological and theoretical significance. There is any number of examples of rather inefficient social systems which nonetheless have high input levels. Others, in contrast, have high output levels despite low input: Japan, Austria, Israel, Switzerland ..., countries practically bereft of natural resources which are nonetheless among the most developed in the world. Others, however, swollen with oil and raw materials, fertile land and a mild climate, huge plains and plentiful forests, are still relatively underdeveloped. The only thing that distinguishes one kind of society from the other is the quality of their respective "T"'s, the efficiency of their transforming body. In the function in which Output (dependent variable) is made to depend on Input and the Transformation process (independent variables), the truly explanatory variable is Transformation, understood to be social organisation or the combination of systemic elements whose aim it is to produce more from a lower level of Input. Hence, when we discuss the multiplicative relationship Y = X.T, "T" will be equal to the relationship Y/X, indicative therefore of the overall performance of the system due in all probability more to its internal organisation, to the position, role and functions attributed to each of the persons involved than to the means deployed; this makes it one of the expressions of greatest sociological value, in so far as it "translates" the suitability of social organisation.

We are going to add yet another thermodynamically relevant notion. As Bailey (1992) holds in his theory of the Social Entropy System, the greater the value resulting from "Y" (Output), the greater will be the production of internal negaentropy within the system; further, as a result of the application of the Second Law of Thermodynamics to these open systems, greater will be the entropy produced in the social system's Environment, from which it takes the amounts of energy/material needed for the negaentropic transformation process. Thus, what is good for the social system will be bad for its Environment. The fact is, however, that transformation "T" may consume a large or small amount of energy to produce a given level of output. This would explain the growing interest in rationalisation and optimisation of transforming processes so as to extract the minimum possible amount of energy from the Environment. But in conclusion, two are the aspects that should be stressed here: First, whether the system works well, poorly or indifferently, or uses small or large amounts of energy, inevitably transformation always takes place, since it is not possible to produce any level of Output (Y) unless some amount of Input (X) is consumed. And secondly, society is either the transforming social system that attempts to comply with the basic inequality given in (1) or it ceases to exist as such. These two principles of the new Systemic Theory would seem to be practically non-transgressable.

But, given the complexity of the subject, focusing on this transformation rate (efficiency, synergies or added value that internal organisation affords), given the complexity of the subject, will mean facing three challenges (Integrative, Modelling and Operational). By Integrative Challenge I mean the need to bear in mind the largest possible number of theories and approaches, sociological or otherwise, but which may complete the inevitably subjective view we take of social reality. Such integration would cover not only theoretical approaches but methodological procedures and techniques to be used. I understand the Modelling Challenge to mean the need to develop models, graphic or mathematical as far as possible and as isomorphic as possible with the reality they are intended to represent. Sight should not be lost of the fact that such subjective reality is a process, not a state or condition; a model should be built, the, for a film, not for a still photograph. Finally, I understand Operational Challenge to mean the need for each concept considered in the model to be dealt with as a variable subject not only to mere stenographic formalisation, but to being quantified via the corresponding indicators and empirical data which may come from secondary analysis (existing statistics) or primary (our own surveys or developments) sources. Bailey's (1992) work is a particular valuable reference with respect to these challenges, and in particular the integrative challenge.

5. DEVELOPMENT OF THE AXIOLOGICAL APPROACH

When the social system is seen as a producer of "universal values" and these values are operationalised in a standard or comparable manner (see chapter 6), it is absolutely feasible to consider the social system to comprise a series of sub-systems, each one of which conducts it own "system of values". The socio-political system, for instance, may be sub-divided into the following value systems: a) the sub-system pursued by the population as a whole, in keeping with its needs; b) the one pursued by each political party in keeping with its ideological interpretation of such needs; c) the one that the government in office tries to build in keeping with the alleged integration of interests that it pursues; d) the one that is finally achieved, objectively speaking; and e) the one that is achieved subjectively speaking (as perceived by the people). Although the overall system might be sub-divided into many more, these differences are indicative of the possibilities of applying a new concept of DEVIATION on the basis of computing the differences among all the axiological profiles (system of values) in which each complex socio-political system can be represented. A deviational analysis of the 35 deviations resulting from considering all the possible combinations of 11 profiles taken two at a time is given in Parra-Luna (1983. pp 413-424).

From this viewpoint, the concept of CHANGE might also be quantified by considering it to be merely a "deviation" over time and comparing the axiological profiles at times "t" and "t+1". The notions of PROGRESS and SOCIAL REGRESSION could be inferred from this, the first by summing all the positive deviations and the second by summing all the negative ones. The ratio between the two would yield a globalised Progress/Regression rate for the system as a whole, one of the most significant expressions of its performance.

The notion of SOCIALISATION may similarly be quantified by comparing the value systems of elderly people to those of young adults (on the grounds of appropriate surveys) or comparing the systems of values (or axiological profiles) pursued by the elite in power to those pursued by the population at large, which is something that can be viewed dynamically. For instance, do young people's value systems tend to converge on the systems advocated by the elite, or rather on the systems of values held by the population as a whole? Such comparisons would point to valuable quantified influences in socialisation processes.

A first attempt at quantifying the notion of ETHICAL BEHAVIOUR in social or political systems might even be reached in this way. For instance, in principle it would suffice to know the axiological profile that should or could have been achieved (the "what ought to be") in a given social system, to compare it to the profile actually reached (the "is"), after applying a coefficient to the latter to take account of possible contingencies not attributable to the system managers. This coefficient could be readily computed by a series of expert evaluators using Delphy type methodologies.

Another important application of the "Axiological Profiles" can be seen in the operational description of the concept "Corporate culture", one of the strategic notions in vogue in business management and "management theory". The idea of a business or corporate culture undoubtedly expresses systems of values whose components could hardly be other than those referred in the Referential Pattern of Values described above. See, for instance, the papers published in Spain by J.A. Garmendia (1994).

Finally, mention should be made of the extremely interesting possibilities open for the further development of certain social sciences, in particular from a critical perspective in which the aim is primarily to ascertain (in an operational and quantified manner) WHAT IS DONE rather than WHAT IS

NOT DONE to benefit the population. The methodology needed for this endeavour is explored in Parra-Luna, 1983.

6. REFERENCES

Bailey, K. Sociology and the New Systems Theory, State Univ. of N.Y. Press, 1992
Bunge, M. Finding Philosophy in Social Science, Yale Univ. Press, New Haven, 1998
Buchanan, B. "Information Requirements for a Viable Society", Paper presented at the XIV World Congress of Sociology, Montreal, 26 July-1 August, 1998.
Buckley, W. Sociology and Modern Systems Theory, N.J. Prentice Hall, Inc. 1967
Buckley, W., Society as a Complex Adaptive System, Chicago, Aldine Publishing co. 1968
Cortes et al., Systems Analysis for Social Scientists. N.Y., John Wiley and Sons, 1974
Couffignal, L. La Cybernetique, Press Univ. de France, 1972
Garmendia, J.A. "Tres culturas: organización y recursos humanos"; ESIC, Madrid, 1994.
Geyer F. and Van der Zouwen, J. Sociocybernetics, Martinus Nijhoff. Leiden, 1978
Geyer, F. "
Hall, Martin. L., System Thinking and Human Values: Towards Understanding Chaos in Organisation, IV World Congress of Sociology, Montreal, July 26. August 1, 1998
Luhmann, N., Introducción a la Teoría de Sistemas. Anthropos, Barcelona, 1996
Martín López, E, "La sociedad global" Fundación para la Formación de Actos Profesorales, Madrid, 1997
Parra-Luna, F. Elementos para una Teoría Formal del Sistema Social, Univ. Complutense, Madrid. 1983
Parra-Luna F. "The Role of Second-order Cybernetics in Sociological Research: Fashion or Progress ?, paper prepared for the Conference on "Sociocybernetics Bridges between Pas, Present and Future...", Kolimbari, Crete, May 26-31, 1999
Parra-Luna, F. "Towards an Axiological Systems Theory". Mimeo. 1998
Wiener, N. "Cibernética y Sociedad, De. Suramericana, B. Aires, 1969

Part II

Some Models for Defining Social Performance

Chapter 4

Some problems in the observation of performance

GERARD DE ZEEUW
Lincoln Research Centre, 11 Steep Hill, Lincoln LN2 1LT, Uk, University of Amsterdam

INTRODUCTION

An important characteristic of human consciousness is its ability to distinguish between two kinds of realities. The first is the reality of daily life. This is what one talks or communicates about and uses as a referent in activities. It includes writing cheques, taking a bus, being part of a family, a community, a society. The second reality refers to what is *not* recognisable as the first reality. It may be shared by many, as when it takes the form of a 'grand narrative', for example the bible-based narrative of medieval times, or the narrative of the Enlightenment (Lyotard, 1984). Or it may be personal and local, as in recent times where it also refers to the narrative of personal development.

The relation between first and second realities is complex and many faceted (Glanville, 1987). Realities of the second kind are used to explain events in the former and often derive from it. This holds for realities such as that of atoms, genes, the expanding universe and the 'invisible hand'. They also may give meaning to the first reality or provide ethical standards for people's activities, as when an 'original creator' or a 'personal philosophy' is introduced. From these examples it appears that realities of the second kind are mutually independent. They interact, overlap, or are in conflict only through being associated with the reality of the first kind. Together they constitute what is experienced as *social reality*.

Realities of the second kind are characterised by their 'authority', to use a generic term indicating preference—including qualities such as beauty, interest, and organisational power. The 'authority' of the reality of atoms for example depends on arguments in an observational and rational discourse.

The Performance of Social Systems
Edited by Francisco Parra-Luna, Kluwer Academic/Plenum Publishers, 2000

The 'authority' of a culture or philosophy depends on the possibility of sharing and contributing to their definition. The 'authority' of the reality of an original creator depends on the meaning it provides. It also sometimes depends on excesses of organisational power—while the 'authority' of the reality of legal contracts tends to depend rather on a balancing of such power.

There is a clear dynamic to such realities. New ones are produced, old ones disappear or adapt to new needs. Changes are not without pattern, however. There seems to be a preference for realities with higher 'authority' than previous ones. The major criterion used is that such realities do not change when used to improve on activities in daily life. This is labelled the criterion of *predictability*. There is a complement to this criterion, in case prediction proves impossible. It is less well explored, and presumably weaker. It applies when events in daily life combine into realities that remain unperturbed by events outside the combinations. This is labelled the criterion of *identity-maintenance*.

The criterion of predictability refers to realities that specify restrictions on reports by any observer, given reports by a subset of such observers. The restrictions may take the form of precise predictions, for example when future observations of a star's position are derived from present observations. Predictions also may be loose, as when observations of the behaviour of organisms derive from observations of other organisms. The criterion of identity-maintenance refers to realities in which combinations, or 'patterns', of observational reports are specified that remain the same even when there are reports of other, simultaneous, changes.

Although predictability and identity-maintenance refer to realities of the second kind, both are important to daily life. The larger the number of realities with either of these qualities, the more democratic daily life is. This can be argued as follows. Observations given predictable realities provide the same 'service' to any of their users, as these are not influenced by the (observable) ends the latter wish to achieve. Moreover, anyone maintaining a strong identity will be able to claim democratic rights. He or she will not lose pattern to those whose 'authority' depends on changing the patterns of others. Imposing a pattern is not uncommon for a central government, or for religious organisations.

Slow but steady changes towards realities showing predictability and/or identity-maintenance, and hence toward more democratic forms of daily life, seem undeniable. There also is another change. If organisations are able to achieve identity, observation of their performance (e.g. through public audits) should not change them, or their use. This implies that realities develop that satisfy predictability. One should expect therefore not only an increase in daily life of democratic forms, but also an increasing level of 'honesty'. These changes are not total, of course. In many societies people still are maltreated. Levels of education clearly are increasing however, as well as social responsiveness.

Another way of saying the same is that there still are many realities of the second kind that in a sense are 'in between': they tend to include non-democratic forms of 'authority', but also are continuously scrutinised in terms of possibilities to find realities that allow for prediction or identity. Managing organisations to perform better constitutes a prime example of this 'in betweenness' (Peters and Waterman, 1983). Much systematic effort is spent to move such management towards the use of realities that allow for prediction and identity-maintenance. These efforts usually are summarised as constituting research. The focus of this paper will be on the *design of research* to speed up the development of such realities.

The reason for this focus is that it may provide an answer to a question posed by the editor, reflecting his "critical position about some exaggerated views of constructivism". Quoting from his e-mail, suppose "we (the observers) arrive previously to an intersubjective agreement about the operational definition and quantification of the concepts Output and Input", where "Outputs" include "objective" as well as "subjective" statements. If we "work out the Performance (T) of a given system as Output (75.2)/Input (65.0) = 1.15", are we then able to "say that, although always "constructed", this measure of 1.15 represents the reality, even if it is contingent and provisional?"

The paper proceeds as follows. In the next section a basic aspect of research as a systematic effort to increase 'authority' is identified. It concerns the notion of an 'observer'. In section 3 collectives of observers are explored. In section 4 consequences of the use of collectives are discussed, in particular how they combine to produce practical forms of research. The paper ends with the conclusion that research is able to contribute to the development of representations or *models of* daily life, if it is based on explicit definitions of what observers are expected to do (De Zeeuw, 1988). Although such models contribute to the 'construction' of daily life, they do not require a constructivist reality (Butts and Brown, 1989), Von Glasersfeld, 1991; Schmidt, 1987).

1. THE SINGLE OBSERVER

Whoever answers 'yes' to the question posed has society as well as history on his or her side. All societies are rich in intersubjectively agreed upon definitions of 'outputs' and 'inputs'. An example is a company's balance sheet, using terms like debit and credit rather than input and output. Different societies may have different definitions. What seems especially important is whether the numbers involved are considered to be part of the first kind of reality, or to define a reality of the second kind. In the first case 'input' and 'output' simply refer to what has developed historically (Baecker, 1997). In the second case some effort may be required to find realities with higher 'authority'.

The second interpretation seems to be relevant as soon as there is some doubt as to the applicability of the existing notions of input and output (or of the derived notion of performance). Such doubt implies a reality of the second kind (with low 'authority'). The question that is raised is what reality should replace it. In some realities there will be an interest in small numbers, in others in large numbers. Either kind of reality may have an 'authority' that allows it to change dependent on what use is made of it. This suggests turning to research as a way to achieve high 'authority', traditionally using the criterion of predictability.

Traditional research relies on observations reported by individuals (given sufficient physiological apparatus). The aim is to 'improve' on them (Churchman, 1971; Luce e.a. 1963). This involves the *selection* of observations to be compared, and the preparation of a *summary* of the results with high 'authority'. Both activities define an 'observer', that is a reality that restricts the observations to be made in daily life. In an example suggested by the editor, one may report 'the number of women in Parliament', or 'the opinion the population of the system has about the number of women in Parliaments'. In the first case the selection of observations is identified as 'Parliament', in the second as a subset of observers.

There already are many ways to select, that is recognise what sufficiently belongs together to be compared. One is to make use of what is detachable in daily life, in the sense that it has a name and maybe a location. An example would be a company, observations on which may be summarised as yearly results, net products, plans, history; or an action such as filling out a cheque. Another form of selection is to follow the advice of someone with a high social status, or a 'confidence man'. Each may try to convince one that something undesirable (e.g. worthless shares) belongs to (is comparable with) what is desirable (valuable shares), and temporarily can be bought cheaply.

Although referents of either kind are proposed quite frequently, they obviously do not achieve the 'authority' research aims for. To identify a reality that satisfies the criterion of predictability it is necessary to combine the selection and the summary such that, a) the selection includes only what the summary may refer to, and b) all elements of the selection are summarised in the same way. Only if this kind of 'observational closure' is achieved will it be possible to guarantee that the reality defined by the selection satisfies the criterion of predictability. In this case the selection will be observable in the same way the summarised observations are.

The strictness of this definition suggests that it often will be difficult to identify realities that satisfy the criterion of predictability. This clearly holds for any reality concerning the management of a company, that is of its performance. Companies appear to be detachable only, as defined above. This seems to be at the basis of all attempts to identify companies as a 'system', that is as a set of coherent operations or relations (Klir, 1993). What is looked for thus will not satisfy the criterion of 'observational closure', although it may satisfy 'operational closure'—as the selection or recognition of the system remains independent of the summary one is looking for.

2. A COLLECTIVE OF OBSERVERS

The conclusion of the previous section is that the answer to the question posed by the editor must be 'no' rather than 'yes' (see section 2), at least as long one accepts the traditional notion of an observer. There is no way to create a reality that satisfies the criterion of predictability if one restricts oneself to detaching some system to measure its performance, for example in terms of inputs and outputs. This suggests at least three alternatives. One is to develop additional definitions of an observer that help to overcome the difficulty. The other is to search for realities that only satisfy the criterion of identity-maintenance. It also may happen that what one is searching for is a combination of both.

The main difficulty appears to be that it sometimes is impossible to select or recognise what observations to compare directly dependent on the summary one looks for. This suggests following the first strategy, and looking for an intermediate selector. The obvious candidates are activities. They select or define what belongs together in terms of what supports them. But the latter may produce precisely those observations that support the activity of getting the observations to be selected in the first place. This intermediary thus makes it possible to again achieve 'observational closure', but now through an intermediary that is not independent of the summary looked for.

One may thus define a new type of observer: anyone who is selecting observations through some activity (Axelrod, 1984; Luhmann, 1984; Von Foerster, 1970, 1991). To achieve 'observational closure' one needs a collective of at least two such observers—one to help 'stabilise' the selection of the other. Any observer in such a collective creates a reality that satisfies the criterion of identity-maintenance. None is disturbed by reports outside that reality—not by reports from within the collective nor by reports from outside. The latter is the case as the collective creates a reality that also satisfies the criterion of predictability. When used in outside activities, none of its elements will be perturbed therefore (Maturana and Varela, 1980; Rosen, 1993).

This type of observer does indeed help to find performance measures that have high 'authority'. It requires that an organisation be created, for example a company, the members of which behave like the observers defined above. This will imply that it will have the properties indicated, as well as other properties that derive from them. For example, they will maintain identity whoever of its employees is replaced (new employees have to adapt to the organisation). The organisation also will induce honest or non-corrupt behaviour (see section 1), both on the level of its employees as well on its own level—as dishonesty would imply a defection, which would destroy the 'observational closure'.

The effect of introducing this new type of observer appears two-fold. Firstly, what is gained is the ability to create organisations (the collectives) with desired properties (e.g. 'honesty', but also properties that depend on the situation in which the organisation is created). Secondly, the new observer makes it possible to find a referent that is created by the performance desired as well as vice versa (analogous to Spencer Brown's notion of distinction: 1969). This implies that observations will not be selected once and for all: if the collective has to adapt to changes in daily life, it can do so by changing the actions that determine the selections as well as the summaries that depend on the selections (Kline, 1979).

The idea of research designed on the lines of the new observer may be generalised by allowing for additional combinations. In the above an intermediary was introduced to extend the notion of an observer. In this way it became possible to satisfy the criterion of predictability from the point of view of non-participants, acting as outside observers. It seems reasonable therefore to consider situations where the latter also are part of some collective. This implies levels of collectives. What research should search for is what allows such hierarchies to develop. The reason for such a search is that it reduces the effort or cost of maintaining each collective (Simon, 1960).

3. RESEARCH DESIGNS

Suppose one notices that water is coming out of the ground in the garden. This observation starts a process of repair, but basically one of negotiation with the water company. Its first bid is that it has nothing to do with the water. It is coming from a well, as the garden is on a hill. Maybe, but there is loss of pressure inside the house. Then it is admitted that the water is from a broken pipe, but the leak is more than 60 feet from the public supply, so the owner has to pay. No it is not, only 30 feet. So a search for the main duct is started. Eventually the pipe is repaired, the water pressure on the house's outlets is re-established, and the water company pays.

This clearly is a story from daily life. Similar stories abound, with variations in different countries. Rather than with water they deal with journal subscriptions, with the supply of electricity, with unsolicited books, even with *not* having been at the place of a crime. Sometimes the outcomes are relatively innocent, sometimes they deal with life and death. In all cases what is at stake is the determination of what observations should be compared and summarised as an interpretation, diagnosis, or resource. In the example a light form of negotiation was used to determine in what reality the (lack of) performance could be identified, and what its 'authority' might be.

The negotiation did not prove too difficult as water companies in developed countries usually have a legal obligation to provide (household) water. The reality to be added to daily life therefore was relatively easy to define. It also is easy to get confused however. In the example the water company tried to 'impose' the reality it preferred, substituting its own authority as a national company for the 'authority' of that reality. In other countries it might simply declare the distance to be 60 feet, with no recourse. This raises the question how research might contribute to resist such a claim, if there would not be any possibility to do so in daily life.

The traditional observer would try to detach the company as a reality, given which its (lack of) performance might be determined. This reality would fail to satisfy the criterion of predictability however, as it the frame of comparison would be determined separately from the summary to be derived from that comparison. The obvious strategy for the company thus would be to continue to claim other realities, for example that of water running down the hill. In contrast, research based on the new observer (see previous section) would initiate a search for internal forms of communication that would lead to a reorganisation of the company such that 'observational closure' would be achieved.

What is important is that this approach does not take the leak as the reality of the first kind, to which a reality of the second kind has to be added. It starts its search for what may create a collective that helps to stop leaks, with minimum costs to clients as well as to the collective itself (see section 3). This might imply setting up a new company, or changing the existing company through an (experimental) change in internal communication. This should help adapt the realities used by the members of the company, and allow for an increase of its democratic accessibility, that is its 'authority'.

Negotiations thus would not be with clients, but inside the company, although possibly triggered by them.

Research of this kind suggests that to improve on the management of companies, the focus should not be on the development of performance measures, as implied by the question of the editor (see section 1). The emphasis should be on what helps to create collectives with desired properties, including performance measures in realities that satisfy the criterion of identity-maintenance and possibly of predictability as well. In other words, rather than derive a performance level T from inputs X and outputs Y given some organisation, one should try to identify what observations X may be constrained by actions Y in some collective, in order to achieve a performance T (Carse, 1986; De Zeeuw, 1986, 1998).

To design the research that leads to this kind of result, use should be made of the new type of observer defined in section 3. Various such designs have been experimented with, and implemented successfully (Vahl, 1995, 1997; Vahl and De Zeeuw, 1999; De Zeeuw, 1990, 1997, 1999). Two are summarised here:

- Suppose an organisation is not achieving a desired performance T. First one identifies what exchanges of the form 'if observation X, then action Y' are implemented. Secondly one allows members of the collective to mutually negotiate changes in X and Y, to improve on T, until the realities they create satisfy the criteria of identity-maintenance and possibly of predictability. Eventually the organisation should be able to maintain the desired T, given a wide range of changes in X.
- Suppose an organisation intends to develop a new activity, involving new clients. First one identifies what collectives exist among the clients. Then one establishes what forms of communication allow them to recreate these collectives as realities that satisfy the criteria of identity-maintenance and possibly of predictability. The collectives should show as a desired property that they are able to interact with the organisation to define what new activity is needed.

4. CONCLUSION

The starting point for this paper was a question about the measurement of performance. To answer the question various issues were considered. Firstly, measures of performance are widely used and quite useful in daily life—in the management of companies, in the determination of taxes, in the regulation of many activities on a societal level. Secondly, these measures often are imposed still, and hence partly resisted—leading to tax evasion, free rider-ism, defection on employee-ism. Thirdly, scientific observation seems misdirected towards preventing such resistance, by policing and punishment. It may be redesigned to allow for re-organisation of local observations to achieve desired performances.

This redesign involves the use of action as an intermediary to constrain or frame what observations to compare and to summarise as improved observations. To satisfy the criteria of identity-maintenance and, when possible, of predictability, it thus becomes necessary to consider collectives, and the way its members or elements interact. The interactions should allow for some level of variation: rather than considering observations, it will be necessary to consider classes of observation, each different from all others. These classes imply that members of a collective maintain a 'viewpoint', that is impose precisely those constraints on what they observe that help maintain the collective.

It has been emphasised that organisations (and other 'systems') in general do not allow for performance measures that have high 'authority', in the democratic and honest sense identified in section 1. This becomes possible only after they have been changed to the point where the organisations and their performance achieve 'observational closure'. There is thus a clear distinction between organisations to which realities are attributed with high 'authority' and those for which this is not the case. The former obviously have to be 'constructed', something which requires systematic effort, while the latter are the (unintended) result of changes in daily life.

5. REFERENCES

Axelrod, R., (1984), *The evolution of cooperation.* New York, Basic Books

Baecker, D. (1997), The form and culture of a social system. *Systemica* 11/pp11-29

Butts, R.E., J.R. Brown, eds. (1989), *Constructivism and science*. Kluwer, Dordrecht

Carse, J.P. (1986) *Finite and infinite games.* Penguin Books, Middlesex

Churchman, C.W. (1971) *The design of inquiring systems*. Basic Books, New York

Foerster, H. von, (1970) Molecular Ethology. An immodest proposal for semantic clarification in molecular mechnisms in memory and learning. In: G. Ungar (ed.), Plenum Press, New York

Foerster, H. von, (1991) Through the eyes of the other. In: Steier, F., ed., *Research and reflexivity*. Sage, London

Glanville, R.. (1987) *Objekte*. Merve Verlag, Berlin

Kline, E. (1979) Planning side-effects. In: Zeeuw, G. de, & Eeden, P. van den, eds., *Problems of context.* Amsterdam, Neth.: Free University Press

Klir, G.J. (1993) Systems science: a guided tour. *J. Biological Systems* 1/1/pp27-59

Luce, R.D., Bush, R.R., Galanter, E. (1963) *Handbook of mathematical psychology*. Wiley, New York

Luhmann, Niklas (1984) *Soziale Systeme. Grundriss einer allgemeinen Theorie*. Suhrkamp, Frankfurt a. M

Lyotard, J-F. (1979) *La Condition postmoderne - rapport sur le savoir*. Paris 1979. Eng. transl.: J-F. Lyotard. The Postmodern Condition: A Report on Knowledge. Manchester, 1984

Maturana, H.R., and Varela, F., (1980) *Autopoiesis and Cognition: the Realization of the Living*, Reidel, Boston

Peters, Th.J., & R.H. Waterman, R.H. (1983) *In search of excellence. Lessons from America's Best-Run Companies*. New York: Harper & Row, New York

Rosen, R. (1993) Some random thoughts about chaos and some chaotic thoughts about randomness. *J. Biol. Syst.* 1/1/pp19-27

Schmidt, S.J., ed. (1987) *Der Diskurs des Radikalen Konstruktivismus*. Suhrkamp, Frankfurt

Simon, H.A. (1960) *The new science of management decision*. Harper en Row, New York

Spencer-Brown, G. (1969) *Laws of Form*. Published by G. Spencer Brown, London

Vahl, M. (1995) Theories that create coherence in practice and other stories. In: Ellis, K. B. Young, M. and Ragsdell, G., eds., *Critical Issues in Systems Theory and Practice*. Plenum.

Vahl, M. (1997) Doing research in the social domain: concepts & criteria, In: Stowell, F., McRobb, S., Lander, R., Ison, R., Holloway, J. & R. Amstrong, eds., *Systems for Sustainability: People, Organisations, and Environments*. New York, Plenum, pp147-152

Vahl, M. and Zeeuw, G. de (1999) On redistributing values: a management methodology. Accepted for UKSS conference 1999

Zeeuw, G. de (1988) Modelling for applied social science. In: Trappl, R., ed., *Cybernetics and systems 1988*. Dordrecht, Kluwer, pp287-293

Zeeuw, G. de (1998) Improving on differences among viewpoints. In: Boog, B, Coenen, H., Keune, L. and Lammers, R., eds., *The Complexity of Relationships in Action Research*. Tilburg, Tilburg University Press, pp 153-175

Zeeuw, G. de (1999) Interaction of Actors Theory. *International Journal of Human-Computer Studies* (forthcoming)

Zeeuw, G. de (1986) Replacing models by actors: some ideas for metadesign. In: Gigh, J.P. van *Decision making about decision making: Metamodels and metasystems. Cambridge*: Abacus Press, Ch. 15

Zeeuw, G. de (1990) Coordinated cooperation and increasing competence. In: W.F. Schut, W.F. and Lohuizen, C.W.W. van, eds., *Managing knowledge for design, planning and decision making*. Delft University Press.

Chapter 5

The resolution of dilemmas and conflicts in approaching the performance of social systems through representation, evaluation and quantification

JOHN P. VAN GIGCH

1. MEANING OF METAINQUIRY

Modeling or **to model** implies that the modeler "abstracts" properties from **things** and/or **actions** in order to obtain a representation of the physical world.

It is easy to conceptualize that **the model** stands at one level of abstraction higher than **the things** and/or **actions** from which the properties are obtained.

The process of abstraction can be applied to **Modeling** itself, to obtain "a model of the modeling process" which we call a "metamodel."

The **metamodel** embodies the properties which are abstracted from all models. (There is more than "one model.") By **metamodeling**, we imply a process of design which is carried out at the metalevel and by which we define HOW the process of **modeling** (at the lower level of abstraction) is to be carried out.

Metamodeling defines the epistemology or design foundations of **modeling**. Design foundations consist of reasoning processes, guarantees of truth, proofs, axioms of validity, or any other logic which underlies a methodology.

Any discussion of **modeling** must, of necessity, involve **metamodeling** from which the former evolved. One cannot be considered without the other. They constitute a dyad.

Metamodeling is to **modeling** what the **theory of design** is to **design** (Simon, 1969/1981), what **learning to learn** is to **learning** (Bateson, 1972), or **decision-making ABOUT decision-making** is to **decision-making** (van Gigch, 1987). In this last label, the emphasis is on the word "ABOUT"

The Performance of Social Systems
Edited by Francisco Parra-Luna, Kluwer Academic/Plenum Publishers, 2000

which implies a distinction similar to the one described by de Zeeuw (1985) between **models OF the world** and **models ABOUT the world.**

Models OF the world are descriptions of the world from the perspective of an inquiring system which includes this world, whereas **models ABOUT the world** are descriptions of the world from the perspective of **the metaworld**, i.e., of the metalevel inquiring system, outside or beyond the world in which we are in.

A dialectic relationship exists between the two elements of each dyad where each element is said to originate in inquiring systems of different levels of abstraction or of logic.

Thus, by **Metamodeling** we imply the application of a conceptual process by which we postulate the existence of a metalevel or inquiring system at a level above that of **Modeling**. In brief, to metamodel implies to metadesign or to model at the metalevel.

See also van Gigch (1991a & 1997).

The modeling/metamodeling framework is applied to the resolution of dilemmas/conflicts which are considered "problems" found in the social science realm.

2. PROBLEM SOLVING AS A "THEORY OF PROBLEMS." THE EPISTEMOLOGICAL APPROACH.

According to van Gigch's (1997) **Theory Of Problems**, defining a PROBLEM involves a process, whose output is a statement outlining the discrepancies, (or lack of congruence), between WHAT IS and WHAT SHOULD BE.

- As the process to define a problem unfolds, a PROBLEM DEFINITION states, implicitly or explicitly, how to erase the discrepancies and achieve the expected outcome(s).
- A PROBLEM is an unwanted "mutation" of a system design.
- A PROBLEM exhibits *unanticipated consequences* of previous actions.
- To diagnose a PROBLEM entails *the acquisition of knowledge* which can be obtained by utilizing a hierarchy of cognitive functions .
- To solve a PROBLEM produces a series of statements of ever increasing COMPLEXITY involving DESCRIPTION, EXPLANATION, THEORY, HYPOTHESES, THEORY-VERIFICATION, etc. to finally arrive at PREDICTION.

For more details on this subject refer to van Gigch (1997).

2.1 WHAT IS A PROBLEM?

The Problem Is A Question.

It is advisable to pose THE PROBLEM in the form of *a question* in order to invite *solutions* .

To pose a question, narrows the field of possible solutions. To pose a question invites a pointed answer, instead of a rambling essay.

Answering The Question May Lead To More Problems

The solution must be *congruent with* (i. e. must fit) the question posed. The solution must *answer* the question posed and provide realistic and implementable solutions.

A PROBLEM should not be constrained by organizational boundaries.

As the definition "expands" to encompass more and more organizations outside the initial set, an impasse arises between the immediate goals of practicality and feasibility versus the ideal target of completeness and reason.

It is highly unlikely that the PROBLEM DEFINITION (and the corresponding proposed SOLUTION) can be limited within the narrow confines of the initial system(s) from which the problem originates.

In this case, deciding what to do, leads to more questions, a new problem definition and so on.

As we try to unravel the "cause" of every problem, we find that it leads to another question, and to more problems.

A stop-rule must be adopted to avoid infinite regress.

Consensus is Needed

When facing a PROBLEM, it is difficult to obtain agreement of what is the APPROPRIATE QUESTION. It is imperative that a consensus process to reach an agreed upon definition be adopted.

The Problem Definition Is Emergent.

A **problem definition**, as well as its concomitant **solutions**, are always "**emergent**" concepts.

The PROBLEM FORMULATION is a process without a definite answer. As the process unfolds a PROBLEM DEFINITION emerges, but keeps changing as more knowledge is acquired.

Neither a problem definition nor the corresponding solution(s) remain "fixed" for long. Both--definition and proposed solution(s) change, as more information is obtained. WHICH IS SPECIALLY THE CASE IN THE DEFINITION OF SOCIAL PERFORMANCE WHERE OUTPUTS AND INPUTS HAVE TO BE PREVIOUSLY DEFINED/MEASURED.

3. QUANTIFICATION, MATHEMATIZATION AND MEASUREMENT.

To be solved, problems need to be defined, represented and evaluated through some form of quantification.

- **Quantification** has several meanings. In its most general sense it includes ***mathematization***, ***measurement*** as well as the use of ***quantifiers*** .

In its strictest sense, the term "***quantification*** " originates in modal logic and predicate calculus and refers only to the use of ***quantifiers*** .

- **Mathematization** refers to the application of mathematical symbolism and mathematical methods to a problem, in order to solve it.

A scientific discipline "*improves its image as a science* " to the extent that its problems can be *formalized* , preferably through the use of mathematics--a process which is called "***mathematization***."

(van Gigch, 1997). Unfortunately, some disciplines have taken this motivation much too seriously and are driving mathematization to absurd extremes.

- **Measurement** refers to the assignment of numerals and numbers to represent attributes and properties in order to make choices and practical decisions.

The distinction between ***quantitative*** and ***qualitative*** measurement depends on the "*strength*" of the measurement scale which is used to evaluate the attribute or the property of a domain.

4. QUANTIFICATION MEANS REPRESENTATION AND EVALUATION

The Quantification OF SOCIAL PERFORMANCE AS A Problem refers then to an assessment of the extent to which mathematical and other quantitative approaches are used to solve a particular problem, in light of the hard vs. soft nature of its domain.

Quantification can be considered a form of ***representation*** involving ***evaluation*** .

The "***evaluation*** " portion entails:

- A determination of ***the appropriate form of quantification***, given the domain of the properties in question
- The choice of ***the suitable scale of measurement***

- The ***choice between qualitative-type and quantitative-type*** evaluations
- The ***selection of the correct methodology*** by which alternatives for decision making can be evaluated and weighed
- ***The assignment of value*** to attributes whose properties may change, depending on the logic level (of the inquiring system) where the problem is being considered.

5. REPRESENTATION THE HIERARCHY OF INQUIRING SYSTEMS

Problems have logic levels i.e. they must be represented as taking place at three levels of inquiring systems, namely:

a) a real-world inquiring system,

b) a modeling inquiring system, and,

c) a metamodeling inquiring system.

An inquiring system is a system dedicated to the acquisition and production of knowledge. The concept was pioneered by C. W. Churchman (1971).

The three inquiring system constitute a hierarchy of inquiring systems.

The real-world inquiring system, (at the logic level of lowest abstraction) takes place in the real-world, where practitioners apply the knowledge acquired in the other two inquiring system to real-world problems.

The modeling inquiring system, (at the logic level of middle abstraction) where scientists, engineers, technicians, middle managers work to solve problems and determine the specifications and technical details which are used by practitioners in their daily work.

The metamodeling inquiring system, (at the metalevel or logic level of highest abstraction), is devoted to solving the "extraordinary problems" and to unraveling logical "anomalies." It is in this inquiring system that the design of the modeling and real-world inquiring system take and where the so-called "constellation of beliefs" of the organization members, including their values and value system are forged.

6. AN ORGANIZATION'S VALUE SYSTEM A SYSTEM OF VALUES WITH THEIR CORRESPONDING IMPERATIVES FOR THE MEASUREMENT OF SOCIAL PERFORMANCE.

Through acculturation and the socializing process, the members of an organization adopt and accept its "value system." Organization values are shaped in the metamodeling inquiring system or in another meta-inquiring system of the highest abstraction.

According to Kant (1790), ethical and other commandments must be formulated in an imperative language.

An imperative is a pragmatic rule "which expresses the objective necessity of an action in such a way that the action would inevitably take place if the will were to be entirely determined by reason."

(Ferrater Mora, 1979: II,1633).

In practical terms, an imperative can be defined as a demand or an obligation to satisfy a particular value.

For our purposes, we postulate the existence of the following system of imperatives with their corresponding values:

1. The aesthetic imperative (of highest abstraction) which is explained by the desire to admire and enjoy beautiful objects or unique events which induce a feeling of pleasure and or satisfaction on the beholder.

Preponderant value and demand: To admire and protect what is beautiful.

2. The ethical imperative (in descending order of abstraction) which is justified by a the motivation to act morally.

Preponderant value and demand: To behave ethically i.e. follow moral principles.

3. The epistemological imperative which is explained by the motivation to ensure the logic of the system components and its coherence with other systems.

Preponderant value and demand: To ensure logical system coherence.

4. The political imperative which demands that conflicts within and outside the organization be brokered to the satisfaction of all concerned parties.

Preponderant value and demand: To ensure peace among organization constituents and with outside stakeholders.

5. The legal imperative which is imposed by the obligation of respecting the laws of the community and of the social system.

Preponderant value and demand: To know and obey the law.

6. The scientific imperative which imposes scientific and technological conditions on the design of the system. Preponderant value and demand: To abide by technical and scientific requirements.

7. The economic imperative which is exhibited by individuals who are driven to satisfy an economic goal.

Preponderant value and demand: To balance benefits and costs to ensure economic survival.

8. The management imperative (of lowest abstraction) which requires that managers resolve conflicts that arise from lower-level imperatives through efficient (practical) tradeoffs.

Preponderant value and demand: To keep the organization viable AND TO IMPROVE ITS PERFORMANCE.

Although, the order in which the imperatives are listed is not fixed, the nature of the aesthetic, ethical and epistemological imperatives is much more abstract than that of the scientific, economic or management imperatives.

Imperatives constitute a hierarchy, where the management imperative is at the bottom of the hierarchy (has more concrete attributes) and the aesthetic imperative at the top (has more abstract attributes). More than one imperative may be invoked at any level of the hierarchy.

Whereas the economic imperative occupies a preponderant role in the formulation of many issues, it should not be considered as the only imperative at work, nor should it be considered the most important.

Details on how the tradeoffs among the imperatives operate is discussed elsewhere (van Gigch, 1997).

The hierarchy of imperatives and their associated values is used to flesh the link between the beautiful and the pragmatic as well as what is morally desirable.

Rising abstraction is one of the main impediments to proper evaluation of the imperatives. For these purposes, the pragmatic" refers to the lowest imperatives and their associated values. We definitely know how to assign economic value. We also can find a way to define scientific value. However, our difficulties begin to mount when we attempt to evaluate attributes which pertain to higher level imperatives such as ethical, epistemological and aesthetic imperatives.

In what follows, the representation and evaluation of imperatives and their corresponding demands are outlined.

7. THE 'METAETHICS-NORMATIVE ETHICS-MORALITY' FRAMEWORK

A definition of what is "good" FOR THE MEASUREMENT OF SOCIAL PERFORMANCE CAN BE APPROACHED in a framework which uses the hierarchy of inquiring systems. It incorporates metaethics, normative ethics and morality as follows (van Gigch, 1991a; van Gigch, 1991b; van Gigch & Rosvall, 1991)

a) The inquiring system at the metalevel is concerned with metaethics where the definition of values is made. This inquiring system is also concerned with the cognitive functions involved in producing ethical judgments.

Example: An action is "good" if it never causes any physical or emotional harm to an individual person and if it never causes the individual to be deprived of his/her rights (We assume that "rights" are socially bound).

b) The inquiring system at the middle level is concerned with normative ethics where norms are set. A "norm" is a standard which is used for comparison purposes. At this level, laws and rules and procedures which will be used at the lower inquiring level are set.

Example: The action of battering a spouse causes physical and emotional harm and therefore is considered "bad."

c) The inquiring system at the lower level is concerned with morality i. e. the determination whether a certain action or a certain behavior deviates from the norm established by the inquiring system at the normative ethics level.

Example: Given the standard of behavior for a spouse which was established earlier, the actual behavior of a particular individual can now be judged to be ethical or not.

8. REPRESENTING THE REASONING PROCESS BY WHICH CONTRADICTORY IMPERATIVES ARE SETTLED

The metaethics framework can be used to represent how people reason when faced with contradictions among imperatives . Reasoning takes place at three levels of inquiry.

METALEVEL (Questions of **method and meaning)**

Here, the processes by which the ethical and other dilemmas will be settled are designed. In addition, the meaning of the terms used in settling the choice of alternatives are defined.

- **Method**: Establish the process that will be followed to make decisions
- **Method**: Establish rules of reasoning with imperatives
- **Meaning**: Define imperatives which will motivate behavior
- **Meaning**: Determine factors which are "considerable" (i.e. worth considering) in the situation
- **Meaning**: Establish mindscapes and inner beliefs

NORMATIVE LEVEL. Given **method and meaning** (from metalevel) what **principles will rule behavior** (at real-world level)?

At this level, we proceed to:

a) Choose the "principles" which will guide "the ethical commitment" (Elliot, 1991)

b) Choose the factors which are "morally considerable."

(Goodpaster, 1978) as well as their relative weights.

c) Define imperatives by which the choices are to be made.

d) Choose the appropriate scale of measurement for each factor after a determining the extent to which each of the factors can be quantified.

e) Formulate a decision to select a particular course of action according to a pre-established calculus.

- Principles: Establish principles or rules of reasoning that rule behavior.

REAL-WORLD LEVEL (Judge **morality** of individual to determine whether behavior conforms with **principles** obtained from normative level)

At this level, a judgment of individual behavior according to the principles and rules of behavior exacted from previous level must be made.

The resulting framework encompasses what is called "**morality**." **Morality** is sometimes explained in terms of the relative weight which an individual places on contradictory imperatives.

• **Morality**: Behavior that follows from normative rules.

CASE # 1: It is March 1996. California has a initiative on the ballot to repeal a ban on the killing of mountain lions. Mountain lions have been protected by a ban because its species is on the endangered list. However, in the last few years, mountain lions have been sighted close to urban areas and at least two incidents of human encounters with mountain lions have been reported. THAT MEANS ENTER IN A DISCUSSION ON SOME OF THE THEORETICAL DIMENSIONS OF THE CONCEPT OF OVERALL PERFORMANCE OF THE AMERICAN SYSTEM OF VALUES. FIRST, THE VALUE OF "CONSERVATION OF NATURE" IS CONFRONTED WITH THE VALUE OF "FREEDOM". SECOND, THE BEST PERFORMANCE OF THE SYSTEM REQUIRES TO EQUILIBRATE OR BALANCE THE LEVELS OF REALIZATION OF BOTH VALUES. How should I vote?

METALEVEL

Conceptual analysis

- Meaning: Determine mindscapes (or world views) underlying inner beliefs of main participants
- Meaning: Define imperatives and their proxies which will motivate behavior
- Meaning: Determine entities/factors which are "considerable."
- Method
- What method(s) will be followed to make decisions?
- Reasoning: Establish rules of reasoning with imperatives.

- Do not contradict truth.
- Do not contradict inner beliefs.
- Follow mandates of imperatives.
- Dilemmas: Establish rules to settle conflicts or contradictions and determine tradeoffs. Effect tradeoffs among alternatives on the basis of weighing relative importance of imperatives, actions or anticipated outcomes. Weigh the relative importance of imperatives
- NORMATIVE LEVEL
- Mindscapes or world views. Can be determined in each individual case.
- Imperatives.
- Economic imperative: represented by the value of sheep that have been mauled by mountain lions
- Ecological imperative: mountain lions are on the endangered species list
- Scientific imperative: represented by the statistics about the number of mountain lions which exist in the wild.
- Considerable
- Considerable (i.e. what is worthy of consideration): Economic reality i.e. value of cattle killed by mountain lions
- Considerable: Scientific truth i.e. statistics about the number of mountain lions which exist in the wild and their reproductive patterns
- Considerable: Ecological data: mountain lions are on the endangered species. On the ecological scale, humans come first, mountain lions second and sheep third. It is conceivable and therefore "considerable" that certain people value sheep more than mountain lions.
- Considerable: Safety of humans and safety of sheep, in that order.
- Dilemmas
- Management imperative: requires that demands be weighed relatively to each other.
- REAL-WORLD LEVEL
- Follow the demands of imperatives
- Economic imperative demands that the value of sheep which have been mauled by mountain lions be taken into account.
- Scientific imperative demands that I recognize the [scientific] "truth," of statistics about the number of mountain lions which still exist in the wild as well as the number of sheep that are known to have been killed.
- Ecological imperative demands that I recognize that mountain lions are on the endangered species list and therefore worthy of special protection.
- Outcome: Management imperative demands that the ecological imperative be weighed relatively more than the economic imperative. The number of sheep that have been killed by mountain lions does not justify that I vote to repeal the ban against killing of an endangered species.

EVALUATION

In the above TWO cases, outcomes were assumed by comparing the demands of competing imperatives. However, in order to compare alternatives they must be evaluated. Choices and dilemmas pose the typical problem of evaluation in the PERFORMANCE OF ANY SOCIAL SYSTEM. The evaluation problem can be stated as follows:

One the most intractable epistemological problem social sciences, is the formulation of a decision calculus to evaluate alternatives and determine the best course of action.

TRADITIONAL APPROACHES TO THE EVALUATION PROBLEM.

THE THEORY OF MEASUREMENT

According to Campbell (1957), measurement or quantification is: "The assignment of numerals [and numbers] to *represent* properties."

[Our emphasis].

The assignment of *numerals* leads to measurement with the nominal scale. *Numbers* are used for other scales such as the ordinal, the interval and the ratio scales.

Case # 2. Quantification in the Context of Measurement Theory

I can observe that watching the flight of ***a [particular] condor*** *is* ***very "beautiful".*** *Further, personally, I can vouch that making this observation gives me a* ***unique feeling*** *of pleasure which I call aesthetic.*

With reference to the definition of what is ***good/bad*** *I can state: "The elimination of an* ***endangered species*** *such as the condor, causes in me an emotional stress which I consider a* ***quite harmful****. To reach consensus, I can further buttress my assertion by alleging that* ***"a thousand*** *other* ***judges*** *think* ***the same way"****, and claim* ***universality****."*

Thereafter, the link from the aesthetic to the ethical can be made. Furthermore, I can set a standard that declares that killing ***condors [in general]*** *is unlawful.*

Finally, society can decide to punish ***[all] individuals*** *who are found flaunting a law which establishes this prohibition.*

The words implying plurality (or singularity) have been emphasized and the respective scale in which they are measured is noted:

- *a [particular] condor* NOMINAL SCALE
- *very beautiful* ORDINAL OR INTERVAL
- *unique feeling* NOMINAL OR ORDINAL
- *good/bad* ORDINAL OR INTERVAL
- *endangered species* NOMINAL
- *quite harmful* ORDINAL OR INTERVAL
- *a thousand [judges]* RATIO
- *same way* NOMINAL

- *universality* NOMINAL
- *condors [in general]* NOMINAL
- *[all] individuals* NOMINAL

The sentence *"good/bad"* implies the use of an ***ordinal or an interval scale of measurement***.

Some quantifiers are adjectives like *"very"* in *"very beautiful"* and *"unique"* in *"unique feeling."*

There exists several degrees of *"feeling,"* where the one referred to in this passage is *"unique."*

An ordinal or interval scale with different gradations of *"beautiful"* can be created.

Contrast the sentence *"[a particular] condor"* which refers to only *"one"* condor, to the sentence *"condors [in general]"* which refers to the whole family of *"condors."*

The words *"a thousand judges."* make use of ***the ratio scale***, where *"a thousand"* is a ***cardinal number*** which expresses an ***ascending order***.

9. THE APPLICATION OF QUALITATIVE AND QUANTITATIVE REASONING

Kant (1785) reminded us about an important distinction between ***quantitative and qualitative judgments***

- ***Quantitative judgments*** are related to the apprehension of the physical attributes of reality.

They represent *objective evaluations which are related to extensive magnitudes* which only admit the interval and ratio type scales of measurement.

- ***Qualitative judgments*** are related to the apprehension of *the influence* which these attributes have on our senses.

They represent *subjective evaluations which are associated with intensive magnitudes and* which admit the nominal and ordinal scales of measurement.

In the last few years, more attention has paid to a study of qualitative reasoning in the context of physical and social systems.

In the past, physical systems were the base for most mathematical and quantitative models in existence.

Nowadays, they yield clues for "softer" methods which have been fruitfully applied to problems in social systems.

In ***qualitative reasoning*** , quantification is restricted to the formulation of equations which merely reflect causal relationships. Variables are expressed in the nominal and ordinal scales of measurement, instead of equations with exact variables expressed in the interval and ratio type scales of measurement.

The most common error is committed when the inappropriate scale of measurement is used

If the variables only admit the nominal and ordinal scales of measurement, it is pointless to pretend to use stronger scales (such as the interval or ratio-type scales).

The dilemma is not "to pit" quantitative statements against qualitative statements.

Both quantitative and qualitative statements are acceptable forms of representations.

Instead, the argument centers on the differences among various forms of representation and their respective (suitable) domains of application.

10. A SIMPLE CALCULUS OF QUANTIFIERS

IMPLICIT QUANTIFIERS provide a "language" to verbalize the dilemmas which arise with obligations over IMPERATIVES.

In the following example, a procedure to articulate transitions among imperatives is illustrated by applying it to the illustrative case already discussed--the case concerning the value of an endangered species such as the mountain lions WHITHIN THE AMERICAN SYSTEM OF VALUES. See Case # 1 above.

10.1 METHOD OF IMPLICIT QUANTIFIERS APPLIED TO CASE # 1.The Killing Of Mountain Lions

- First, the dilemma imposed by each imperative is verbalized and recorded.
- In the resulting text, ***the implicit quantifiers*** are emphasized in **bold**.
- To repeat a statement made earlier: ***Implicit quantification*** exists whenever a word (or sequence of words) implies the idea of a scalar or a gradation by virtue of its context or meaning.
- Each imperative is taken in turn, to determine whether the imposed obligation produces results which are ***favorable/positive*** or ***unfavorable/negative*** to the enhancement of the preponderant value.
- The action is ***favorable*** (or ***positive***) when the value imposed by the imperative is enhanced.

- The action is ***unfavorable*** (or ***negative***) when the proposed action does not promote the value imposed by the imperative.

Illustration. (The implicit quantifiers are shown ***in bold***)

The aesthetic imperative demands that we **preserve** or **protect** what is **beautiful**. Given that I enjoy an **aesthetic pleasure** from watching the mountain lions in the wild, the action of **protecting** their habitat is a **favorable (positive)** action because it **promotes** the enjoyment of **beauty**.

On the other hand, **lifting** the ban to **protect** the animals is an **unfavorable (negative)** action with respect to the **enhancement** of my **aesthetic** experience.

Each of the imperatives is discussed in turn.

10.1.1 Aesthetic Imperative

We can observe that watching the running of a ***[particular]*** *mountain lion is* ***very "beautiful".*** *Further, personally, I can vouch that making this observation gives me a* ***unique feeling*** *of* ***pleasure*** *which I call* ***aesthetic****.*

- In brief, "It is **beautiful** to watch the mountain lions in the wild."
- This statement can be interpreted as:
- The action of protecting the mountain lions is a **favorable** (or **positive**) action, because it helps to promote the main value of **beauty**..

10.1.2 Ethical Imperative

With reference to the definition of what is ***good/bad*** *we can state: The elimination of an* ***endangered species*** *such as the mountain lion, causes in me an* ***emotional stress*** *which I consider a* ***quite harmful****. To obtain approval of my fellow citizens, I can further buttress my assertion by alleging that "****a thousand*** *other* ***judges*** *think* ***the same way****", and justify my claim of* ***universality****. Thereafter, the link from the aesthetic to the ethical can be made.*

- In brief: "To kill an ***endangered species*** is ***wrong*** and provokes ***emotional stress***. "
- The action of killing mountain lions is ***unfavorable (negative)*** or counter the ethical imperative which demands that I should refrain from ***killing*** [***needlessly***] an endangered species.

10.1.3 Epistemological Imperative

This imperative is ***more relevant*** *in this situation that may appear.*

In trying to decide ***whether*** *I would like to afford* ***further*** *protections for the mountain lions, I need to be* ***scrupulously fair****.* ***On the one hand,***

I want to obtain ***as many*** *facts* ***as possible*** *about the situation in order* ***to know "the truth." On the other hand,*** *I must allow people, may who* ***feel threatened*** *and* ***emotional*** *about* ***harming*** *mountain lions, to participate in the dialogue in order for the latter* ***to remain rational.***

- In brief, respecting the demands imposed by the epistemological imperative directs me to the following action : "***Encourage*** *efforts to* ***keep dialogue rational*** *in spite of any* ***emotional involvement.***"
- ***Favorable (positive)*** actions will keep the dialogue ***rational and fair***, while taking into account the opinions and emotions of all sides.

10.1.4 Political Imperative

Two or more *sides to the issue exists: Hunters and the gun lobby are* ***against any*** *restrictions placed on the use of weapons. Animal activists are dead* ***against further restrictions*** *and* ***threats*** *to the animal species of mountain lions.*

- In brief: To be political means to act as a broker to reach consensus.
- The main demands of this imperative are ***power*** and ***influence*** .
- It is impossible to know in advance which actions will ***help (or hinder)*** reaching consensus between the two sides--***those for*** and ***those against*** lifting the ban to kill mountain lions.
- The statement: "***Balance*** *the lobbying arguments from both sides of the issue* ," reflects the sort of ***favorable (or positive)*** action demanded by the political imperative.

10.1.5 Legal Imperative

*The community has approved a law [****a standard****] that declares that killing mountain lions* ***[in general]*** *is* ***unlawful****. Society can decide* ***to punish [all] individuals*** *who are found* ***flaunting*** *a law which establishes this prohibition.*

- In brief, "*To enact laws may be the only way to* ***protect*** *the endangered species* ."
- Enacting a law is a ***favorable action*** . It will promote a ***positive response*** to the demand of legal conduct imposed by the legal imperative.

10.1.6 Scientific Imperative

We have scientific evidence that mountain lions are an ***endangered*** *species. Mountain lions habitat is* ***restricted*** *to* ***uninhabited*** *regions of the foothills. Attack on humans are* ***very rare except when*** *the female* *lion feels that her cubs are* ***in danger****.* [***"very rare"*** & " ***"except when"*** are considered separate quantifiers].

- In brief, the scientific imperative demands ***rigor*** i.e. ***respect*** for scientific evidence such as that provided by information that mountain lions are an ***endangered*** species.
- Bowing to this evidence is *a* ***favorable (positive)*** *action* which promotes the scientific imperative.

10.1.7 Economic Imperative

Ranchers who own cattle ***near*** *the mountain lion habitats have testified to the Legislature that* ***stray*** *lions have killed* ***some*** *of their cattle,* ***in particular*** *in the winter, when their food supply is* ***scarcer****. As a result, ranchers have alleged an* ***economic loss****.*

- In brief, the economic imperative requires that ***costs*** and ***benefits*** be totted up, regardless of whether the result is ***for*** or ***against*** lifting the ban to kill mountain lions. The ***favorable*** action is that the calculation be done.

10.1.8 Management Imperative

After, listening to ***all*** *the evidence, I have decided to vote to* ***uphold*** *the initiative which* ***bans*** *the killing of mountain lions. My reasons are as follows. Hunters and ranchers are* ***a minority*** *compared to* ***the rest of us****. The gun lobby uses the issue about mountain lions to pursue their argument that owning a gun* ***is protected*** *by the US Constitution and that, therefore, killing endangered species is* ***acceptable****.* ***I do not buy*** *their argument.*

- In brief, the Management imperative requires that a decision be taken which promotes ***organizational viability FOR IMPROVING THE PERFORMANCE OF THE AMERICAN VALUE SYSTEM***. In the context of the present dilemma, ***organizational viability*** demands that I respect the opinion of the ***majority***.
- After listening to all the concerned stakeholders and adding up the ***favorable and unfavorable*** actions imposed by all the other imperatives, a decision must be made.

11. CONCLUSIONS

The decision to uphold the initiative which bans killing mountain lions is only illustrative of the process by which the demands of all imperatives can be taken into account. It does not reflect the actual result that may be obtained in reality. Furthermore, what is ***positive*** or ***negative*** depends on the personal opinions of each investigator. It is evident that these opinions may differ from the author's. THEREFORE, METHODS FOR ACHIEVING INTERSUBJECTIF AGREEMENT AMONG EXPERTS (F,I, DELPHY METHODS) HAVE TO BE USED.

We have taken the reader through a novel treatment of ethical and other dilemmas and conflicts which were cast in terms of imperatives and their corresponding demands. These dilemmas and conflicts are typical "problems" found in organizations AND THEY ARE SOME OF THE MAIN CONCERNS IN APPROACHING THE PERFORMANCE OF SOCIAL SYSTEMS. To be resolved they had to be defined, represented and evaluated. Because social science domains admit RATHER soft forms of quantification and measurement, new modes of evaluation need to be devised and introduced.

12. REFERENCES

Bateson, G., (1972), Steps to an Ecology of Mind, Ballantine Books, New York.

Campbell N.R.,(1957), Foundations of Science. The Philosophy of Theory and Experiment (Formerly Physics : The Elements), Dover, New York.

Churchman C. W.,(1971), The Design of Inquiring Systems, Basic Books, New York.

Elliot R., (1991), Environmental Ethics, in: A Companion to Ethics, Edited by Peter Singer, Oxford, Eng., Blackwell.

Ferrater Mora, J., (1979), Diccionario de Filosofía, Alianza Editorial, Madrid, Spain. In Spanish. Four Volumes, Vol. II, 1633.

Goodpaster, K. E., (1978), On Being Morally Considerable, Journal of Philosophy, 75: 308-325. See also Goodpaster, K. E., (1981), Morality as a System of Categorical Imperatives, Value Inquiry, Vol. 15: 179-197, 1981.

Kant Immanuel, (1785), Foundations of the Metaphysics of Morals, Königsberg, Prussia. Translation with an Introduction by Beck L. W.,(1985), Macmillan, New York.

Kant Immanuel, (1790). Critique of Judgment. See also Translation and Introduction by J. H. Bernard, (1951), Critique of Judgment by Immanuel Kant, Hafner, New York, 1951.

Simon, H. A., (1969/1981), The Sciences of the Artificial, 2nd Edition. Cambridge, Massachusetts: MIT Press.

van Gigch, J. P. (1987), Decision Making About Decision Making. Metamodels and Metasystems, Abacus/Gordon and Breach, London.

van Gigch J.P., (1991a), System Design Modeling and Metamodeling, Plenum, New York and Oxford, Eng.

van Gigch J.P., (1991b), The Importance of Metaethics, Human Systems Management, 10: 281-286.

van Gigch J.P., (1997), PROBLEMS HAVE LOGIC LEVELS. How to Improve Organizational Problem-Solving, Copyright © by John P. van Gigch, Sacramento, CA., 1997, 1998. To be published on the Web.

van Gigch J.P. & Rosvall J.,(1991), Multilevel Ethics of Conservation, Human Systems Management, 10-287-292.

de Zeeuw G., (1985), Proceedings of the Conference on Problems of Disappearing Knowledge, University of Amsterdam, Amsterdam. See also de Zeeuw (1987), in van Gigch(1987), Op. Cit.

Chapter 6

A model for measuring the performance of social systems

FRANCISCO PARRA-LUNA
Facultad de Ciencias Políticas y Sociología. Universidad Complutense de Madrid

1. INTRODUCTION

Modelling is one of the best methodological instruments to represent heuristically complex societal systems. We of course must try to model society in a way that is both so simple that it will afford a clear view of society and so thorough that it conceals nothing essential. Reaching such a balance will be no easy task and will certainly entail several attempts. Here goes a first such try. I am going to refer to a model on which I have been working for several years now (Parra Luna 1974, 1977, 1983) although the version discussed here is slightly modified. The starting point for this model, for which I am indebted to D. Easton and K.W. Deutsch, is the cybernetic-transformational approach that converts inputs into outputs via, as explained above, a Transforming Body usually called the "Black Box". To illustrate the model I will take as complex a society as possible, the current Nation-State, although I will also refer to other smaller social entities such as local governments, companies, groups or families. The model is intended to be applicable to them all. I realise that adopting the Nation-State as an example will make the reader even more wary, if possible, of the operational possibilities of this new systems theory that I am attempting to introduce. Nonetheless I will use this macrosociological dimension for three reasons: 1) because the model intends merely to be a sketch or outline for a wholly operative model and as such the relevance of any specific shortcomings that may arise in the description is limited; 2) because even if the model is only partially valid in relation to the Nation-State, the implication would be that it would automatically cover all other smaller systems (Institutions,

The Performance of Social Systems
Edited by Francisco Parra-Luna, Kluwer Academic/Plenum Publishers, 2000

Organisations, Groups, etc.); and 3) because the Nation-State, which represents an integrated macrosociological system, is meaningful for the set of end-values that make up what I will call the "Reference Pattern of Values". For example, the Nation-State illustrates an important consequence of this theoretical approach, i.e., the axiological differentiation between ideologies and political regimes, revealing their inner-most structure in relation to the interests, needs and motivation of their respective peoples or the ruling class that represents them, while at the same time shedding light on the explanation for much of the behaviour of the institutions, organisations, companies and groups that make it up. (Buchanan 1998 and 1999; Hall B.P., 1994; Hall Martin-Lewis, W, 1999). The first thing that must be explained about this model is what is meant by inputs and outputs, and the distinction between the two must be made very clear for the sake of understanding. By Outputs we mean anything that, with respect to Society, is simultaneously: 1) defined as "good" by a representative sample of experts; 2) wanted or desired by the people (because it tends to satisfy their needs); 3) obtainable onerously or through organised political action within the system. Two conceptual levels of Inputs may be distinguished: a) Theoretical Level: where inputs are considered to be the system's potential to produce outputs. This would mean including in a single expression (X) the Structural Factors used (FE), the level of people's Needs or Motivation (N), the level of Memory (M) or record of the recent past and finally the influence of the environment (ε); implicitly expressed: $X = f(FE, N, M, \varepsilon)$. And b) Operational Level, where to side-step the difficulties that the quantification of FE, N, M and ε entails, inputs are taken to be the system's overall budgeted expense plan expressed in monetary units, and this is the approach usually taken in all kinds of institutions, companies and organisations. Hence, on this simplified level, X would represent the budgeted expense plan.

In short, what enters into the transforming body is either the potential system production or the actual consumption in monetary units; what comes out are the goods and services consumed or made available to the society at large. Inputs are, then, the useable resources, or resources used; outputs are the ends achieved.

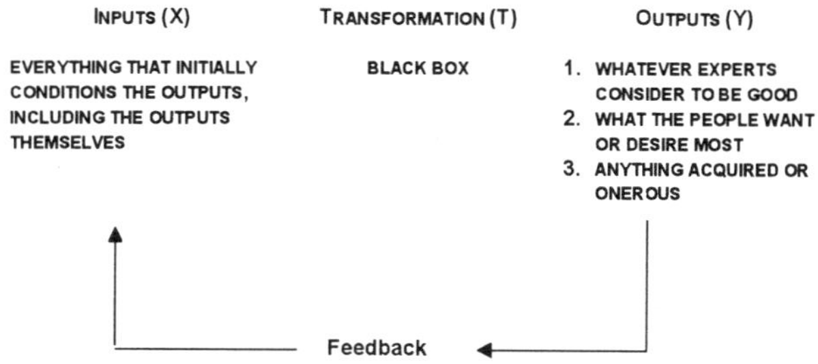

Figure 1. Distinction between "Inputs" and "Outputs" of social systems

Figure 1 intends to make this distinction very clear and at the same time enunciate the fundamental principle of transformation adopted in our model: NOT EVERYTHING THAT COMES OUT GOES IN, AND NOT EVERYTHING THAT GOES IN COMES OUT. In this latter case I will try, in due course, to justify the possible "reducing" role of the Black Box or transforming sub-system.

The operational model shown in Fig. 2, built on this basic model, is an attempt to explain the most relevant dimensions of the social system as a continuing process of transforming relationships.

Let us then see, first of all, what specific Inputs reach the Transforming sub-system in any social system. Initially, there are three kinds of inputs: the first consists of a series of Structural Factors, quite numerous and complex which, in the case in point, i.e., the Nation-State, would be: ecological, geographic, geological, cultural, religious, political, demographic, etc. factors. For example: Spain is a country with a series of natural resources such as an area of 504 000 km^2, nearly 6 000 km of coastline, with France, Portugal or even Morocco as neighbours, 39 million inhabitants with clearly defined physical and biological traits, a certain wealth in fishery resources, etc. All these are factors involving structure which are difficult to change, which remain more or less constant over time and for this reason may be referred to as "Structural Factors". They are represented as a pooled variable abbreviated as "FE" and constitute what I call the system's "Level of Potential". The use of these resources as inputs is likewise evident in smaller social systems. For example, in the Classroom as a system, inputs would be the professor's salary, the use of certain material resources (buildings, furnishings, etc.) and the time the students devote to class work. We know that with all that the outputs or system of values produced by the Classroom (stressing, of course, the value Knowledge) must be maximised, as we shall see.

Going back to the example of the Nation-State, as a result of the structural factors mentioned above, both individual and collective Needs arise. Sight should not be lost of the fact that the Structural Factors include the population factor which in the case of Spain consists of nearly 40 million inhabitants with a series of needs to be met. One way of seeing that needs are in fact also "coming in" to the system would be to consider that they represent the motivation or essential stimulus for system operation. In order to register these needs we will in due course refer to the underlying Values that correspond to them, and this will in turn call for the use of a general pattern of End-Values which theoretically form an exhaustive and closed axiological whole. Some of the Values that satisfy these needs are Security (Law and Order), Justice, Freedom, Health, Environmental Conservation, etc., in as much as they are the answer to specific needs felt by the 40 million inhabitants that make up the Spanish social system. They are represented as a pooled variable abbreviated by the letter "N" and constitute what we will call the system's "Level of Motivation".

The third category of inputs consists of what may be called system "Memory", since any organised society has a kind of historical "archive" (written or otherwise) which allows it to remember its specific history more or less forever. Using Spain as an example once again, the historical, cultural and linguistic ties with Latin America, its past Arabic influence, or even its recent civil war, may foster certain kinds of conduct or cause traumas and condition behaviour both in the population at large and in the ruling elite that may ultimately have a substantial impact on Spanish politics. It is, then, assumed that all societies have a memory that influences them and "comes in" to the "Black Box" to, together with the Structural Factors and Needs, be transformed into Outputs. This Memory is represented by a pooled variable abbreviated as "M".

Finally, the possible influence of the Environment represented by its level of Outputs or fulfilments is likewise an Input, although of an external nature. In the present example, the environment may consist of the other European Union member countries, computed to be the average of the national Output levels in such countries and symbolised as (ε).

But social systems differ from mechanical or biological systems in that the individuals making up the society are neither in agreement about the ends to be achieved to meet their needs (or the relative weight that each should carry) or about the resources to use or the way to organise the factors mentioned above in order to transform them into End-Values. This is a basic disagreement that affects the core of society; a nearly anthropological disagreement that is fundamental and specific to socio-cultural systems and one which generates a constant structural conflict in all societies. There is, then, an inescapable necessity to overcome such discord, a fundamental dimension of the system recognised not only by the Marxist tendency, but by previous traditional thinkers such as Polybius, Hobbes and others. The solution to this conflict entails an institutionalised struggle, more or less latent, more or less recognised, that is structurally present in all social systems. This struggle may be physical or simply dialectic, involve tanks or ballot boxes, an old-fashioned spanking or modern dialogue with one's children, but in any case it implies some sort of struggle as a way of solving the problem posed by axiological disagreement, which is practically inevitable. Such disagreement is represented by a pooled variable abbreviated as "Co" (Conflict) or "Level of Conflictiveness". Such a phenomenon is likewise present in other kinds of organisations: a company when there is a struggle for membership on the Board of Directors or in any other kind of organised group for a seat on the Steering or Executive Committee.

Such institutionalised conflict is usually more a real struggle than a friendly discussion with equitable distribution of capacity to define what the system should produce; in other words, the struggle normally generates "winners" and "losers" and as a result a hierarchy defined by the total power (economic, connections, positioning, physical, etc.) of individuals and groups, which hierarchical structure is usually more or less unchanging over time. We will see below how in this process involving struggle and resulting hierarchy it is the elite classes (economic, intellectual, military, labour, etc.) more than the population at large, that play the predominant role. For this reason the pyramid that the Inspiring Body (OI) represents in the system in Figure 2 is composed exclusively of the formal (explicit) and informal (implicit) elite that inspire society's workings. This body is represented by a pooled variable "OI" (Inspiring Body) and denotes its "Level of Hierarchy".

The elite shown on the upper part of this pyramid are the ones that decide what, when and how the system should produce. These ruling elite are the ones that determine the axiological profile that society should adopt and which in our figure is represented by the SIV or "Idealised System of Values" pyramid. According to the definition of this profile, the ruling or prevailing elite may consider, for example, that freedom is more important than justice (justice meant here in the sense of social justice), while in other axiological systems the contrary may be true. These are ideological options that either stand for more or less concealed personal interests, or derive from a deep axiological inclination. In any case the winners impose or try to impose certain principles for operation of the system which are summarised in what I have called the "Idealised System of Values": idealised, I repeat, not necessarily in the interests of the entire population, but in those of the ruling elite who are the ones with the power to define such a system and implement it in terms, moreover, of their "specific interpretation" of the people's needs. It is represented by a pooled variable abbreviated as "SIV" and denotes the "Level of ambition of the ruling elite".

This Idealised System of Values will drive the operational and transforming machinery of the system or the Transforming Sub-System which is subdivided, in turn, into two parts. The first of these is represented by the system of Values that governments usually programme, usually as a specific plan of action that consists of adjusting the scheme (or better, schemes) of Values idealised by the ruling elite (SIV), in the light of environmental contributions (ε) and the capacity of the structural factors (FE) of the system itself. This operation is represented by a pooled aggregate abbreviated "SVA" (Adjusted System of Values) and constitutes the "Level of governmental ambition". This SVA arises (or should arise) as the result of performing certain technical operations to optimise outputs in terms of SVI, N, FE, M and even "ε". In the mathematical sense, this optimisation operation is technically very complicated in complex systems, and for this reason is rarely performed. This would, nonetheless, represent the "Optimum level of possibilities" of the system. The second, unspecified part of the Transforming sub-system is of a material and organisational nature and comprises first of all the traditional powers: Representative, Legislative, Executive and Administrative. Moreover, it covers all public and private, institutional or individual entities, from huge ministerial bureaucracy to even the most insignificant of professional activities, which are what in practice implement (implicitly, through monetary, tax, social, labour, etc. policies) the directives of the Inspiring Body, transforming the incoming "raw materials" (Structural Factors, Needs, Memory, Environment) into "finished products" (Values) to be "consumed" by society as a whole. This residual part of the Black Box is represented by a pooled variable abbreviated "OT" and called "Transforming Body". In general, OT is the sub-system that comprises the process Communication-Perception-Memorisation-Decision-Action which are in turn the main prime materials for the transformation operation in as much as they are mere <u>means</u> or <u>tools</u> to reach certain levels of End-Values. The system begins to produce the goods and services expected of it, the outputs (Y) of the system, through this complex transforming process to meet the needs of the people. Two distinctions should be made, however, with respect to outputs: first, a difference should be drawn between objective and subjective outputs. The so-called "<u>objective</u>" outputs are verifiable with validated empirical data; for example, in the case of the Nation-State, kilometres of expressways or highways built, percentage of young adults that finish their schooling, degree of improvement in redistribution of income, enjoyment of freedom, etc., and for which there are, or should be, routine and reliable statistical data. <u>Subjective</u> outputs, on the other hand, are those where information about their existence depends on what people perceive or think about the system's Objective Outputs. It may well be that what people perceive as system outputs are not actually put out by the system at all, there being then a gap or kind of perceptive "discord", which may be positive or negative depending on whether the outputs are perceived to be better or worse than they objectively are. Strikes are a good example of discordant perception (justified or otherwise) with respect to the worth or suitability of objective outputs. Total system performance (Y) must then be the integrated sum of objective Y(o) and subjective Y(s) outputs. Second observation: theoretically there are two kinds of outputs being produced at two different times: before (dimensions 10 and 11 in Figure 2) and after (dimensions 14 and 15) regulation (No. 13). The system produces a primary group of outputs, both objective Y'(O) and subjective Y'(S), the integration of which is equal to "θ". But when these outputs are controlled, as we shall see below, and possibly modified by the regulating body R, they are expressed as Y(O) and Y(S) respectively, and their integration is equal to "Y". These are the final and definitive Outputs once cybernetic control has been performed. How does such control take place?

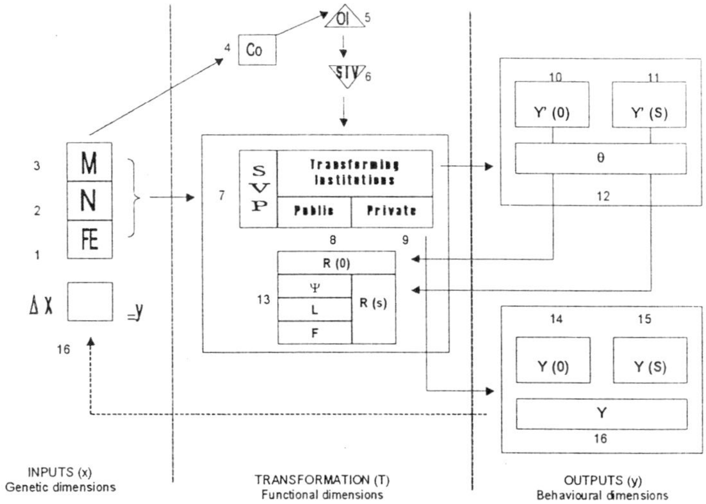

Figure 2. A general simplified model for social systems

The regulation (R) mechanism is explained as follows: in principle, whatever is put out, both objective and subjective, should automatically and unfailingly re-enter the system as new inputs. This sets off the following phenomena:

1. The primary objective outputs produced do not usually coincide exactly with the system of values programmed and optimised by the government or managing bodies (SVA) and even less with the ruling elite's Idealised System of Values (SIV). In view thereof, and given that such elite try to maintain the initial ideological scheme that best suits their interests or coincides with their own ideologies, they institutionalise the so-called Regulating Body (R) whose primary purpose is to modify objective outputs to adjust them as closely as possible to such initial schemes. Such

action is performed through what in Figure 2 is referred to as R(o), giving rise to the second group of objective outputs Y(O) or "final objective performance". The Social Security systems existing in the modern world usually constitute a foremost example of objective regulation. That is to say, since society is unable to prevent disease and accidents, it must make provision for medical and hospital care; since it cannot eliminate crime, it establishes a police and penitentiary system; or since it is unable to provide jobs for all those who want one, it must institutionalise economic compensation for the unemployed.

2. Even if there were no significant difference between programmed and actual outputs, it often happens that the population at large is not satisfied with what the system does. Since such dissatisfaction enters the system as feedback, this could cause a series of disturbances or difficulties that could be harmful to the axiology of the system inspired by the elite classes. What do the ruling elite usually do to avoid this "danger"? What they do and have been doing since the beginning of time, regardless of ideology or form of government, is to exercise a kind of psychocybernetic control through which they either modify the perception of the population at large, or they mitigate the impact of the disturbance produced. The use of primitive versions of political "servomechanisms" or "thermostats" (propaganda, manipulation, threat, etc.) before the servo theory appeared, is as old as humanity. Therefore, the Regulation (R) in Figure 2 should also register the subjective reaction of the system's population and on this basis send a communication intended to modify either the actual outputs or the people's perception of such outputs. The second function of the regulating body consists, then, of discovering when outputs are not accepted by the population; that is when the psycho-regulating mechanism is set off, whereby the impact of any feedback on the existing balance of power and therefore the axiological scheme assumed by the ruling elite is kept to a minimum; this action is performed through R(S). This second regulating function usually establishes, in turn, three kinds of control: the first is Psychological, consisting of messages conveyed by the ruling elite via all possible media (direct and indirect) systematically devoted to singing the praises of their successful management and supporting (even clearly contradicting actual facts) that what is being done is the best that can be done, that any other option would be less favourable for society as a whole and therefore that any other elite or group that might come into power would do a worse job than they. This pattern of psychological control is applicable to a greater or lesser extent to any social or political system. Although in the most extreme version it entails "brainwashing" the population, it is still perhaps the least harmful way of exercising subjective control. Governmental political propaganda, stressing its own success; "doctored" annual company reports; or a father's long-winded speech in reply to what he feels are his sons' or daughters'

excessive demands, all of this, justified or not, are simply examples of this kind of psychological control intended to uphold the hierarchical, family or axiological *status quo*, which the elite classes consider to be beneficial, even though the social situation does not necessarily favour a majority of the people.

But there are individuals or groups within the system who are aware, critical or intelligent enough to understand the nature of this kind of control and questioning the reliability of the messages and thereby endanger the effectiveness of such psycho-regulation. Naturally, the ruling elite make provision for the minorities which for one reason or another cannot be controlled psychologically; this sets off a second kind of control mechanism which we will refer to as <u>Legal</u> control. Such control consists of making laws through the corresponding legislative machinery to implicitly issue the following message, from a position of legal power: "You have a right not to believe what we say, but there is a law that says this or that, and if you do not obey that law you will have face the consequences (illegality, risk of fine, imprisonment, etc.)," all of which usually works. For example, it is legal to support the doctrine that the means of production should be State-owned, but any practical action in this regard would contradict legislation providing that private ownership of such means is a natural, inalienable or sacred right. Or it may be thought that is it unfair, to use the Spanish State by way of illustration, that Prince Felipe of Spain is heir to the throne instead of his elder sister Elena. But such sexist discrimination, hardly justifiable in today's world, is to be found in the Spanish Constitution itself. Whoever rises firmly against these principles must be prepared to accept the (implicit or explicit) consequences of such action.

And yet despite this second kind of control part of the population may still scorn the risk of illegality and be prepared to assume the consequences. Logically, the dominant elite are also aware that there is a minority of individuals or groups who are bold enough to break the law; the Executive Power then implements a third kind of control that may be called <u>Physical</u> or <u>material</u> whereby the recalcitrant rebels are marginalised, made to pay fines, imprisoned, tortured and, if necessary, killed, all quite legally, of course. Parental punishment, dismissal of workers or expulsion or a partner or member, are mere extensions of this kind of control.

In short, the System includes a Regulating Body with important duties, which I will represent operationally as follows: total regulation is "R"; regulation of objective outputs is "R(O)"; and regulation of subjective outputs is "R(S)"; this latter term can be subdivided into three variables, still pooled, represented by: Ψ for Psychological control; "L" for legal control; and "F" for Physical control. The effect of "R" on the primary outputs Y'(O) and Y'(S), the integration of which is represented by "θ", leads to the secondary or final outputs Y(O) and Y(S), the integration of which is represented by "Y". These total Outputs, at least partially "Purified", automatically feed back into the system as new inputs (Δ x) (body number 16) which are thus assumed to cause the least possible impact on the existing balance of powers. This would explain the densely homeostatic or conservative tendencies of all kinds of pyramidal social systems.

The direction of the arrows in Figure 2 illustrates society's most representative process, a never-ending and contingent (vicious/virtuous) circle. Society turns over and over on its initial Structural Factors, modifying them as well as the other initial conditions via, *inter alia*, the outputs produced by the operation of the system itself. The term "Autopoyesis", coined by the biologists Maturana and Varela to denote systems' self-transformation capacity could be correctly applied here provided that the final outputs entail significant change and can be explained exclusively by bodies numbers 4, 5, 6, 7, 8, 9 and 13 in Figure 2

Formally speaking and pursuant to the transformation principle, the outputs produced during period 0-1 depend upon the inputs "X" and the transformation operation "T" in such period; i.e.,

$$Y_{01} = f(X_{01}, T_{01}) \quad \textbf{(1)}$$

and due to feedback, the inputs in the following period "2" depend, in addition to another series of factors represented by I_{12}, upon Y_{12} or the outputs in period, "0,1" immediately preceding, where the subscripts "0", "1" and "2" denote the beginning and end of each period studied. Hence

$$Y_{01} = f(T_{01}, X_{01})$$

$$X_{12} = f(Y_{01}, I_{12}) \quad \textbf{(2)}$$

$$Y_{12} = f(T_{12}, X_{12})$$

the full cycle being, then, a mere continuation of the above functions.

The model shown in Figure 2 represents both a process and a structure: a <u>process</u> because it attempts to represent the flows as they appear in real life which - a look around suffices to confirm - is a series of events that can only be recorded as a motion picture, not as a photograph. But it also represents a <u>structure</u> because the dimensions and processes indicated in the model are repeated time and again, for better or for worse, with scant prospect of change. In other words, it is difficult to conceive of a society that is not, ontologically speaking, an Input/Output <u>transformation system</u>, encompassed by an <u>Environment</u> that influences it and is in turn influenced by it, and consisting of <u>Structural Factors</u>, <u>Needs</u>, <u>Memory</u>, <u>Control Bodies</u>, etc. It is, then, a process in as much as it is a living, adaptive and relatively changeable entity; and it is a structure in as much as it remains essentially unchanged over time and is applicable to any existing social system, ranging from mere couples to large supranational organisations.

2. DOES THE SYSTEM CHANGE ?

According to this description, where the final result ($Y = \Delta X$) appears to be perfectly regulated and controlled by the ruling elite, it may be concluded that the possibility of change is minimal. Two observations come immediately to mind in this regard: first, that any social system has a series of servomechanisms designed to minimise change, a property known as system homeostasis. History shows that, except for a few historical revolutions (the English, American, French, Russian, Chinese, Iranian, and other lesser revolutions), systems resist drastic and truly significant change. Even in these cases it is questionable whether or not such revolutionary change really affected the population at large in terms of values "consumed", since the changes in institutions on the political, legal or even the sociological level should not be mistaken for real change in levels of Health, Security (Law and Order), Material Wealth, etc., which are the ones that have a true impact on the people. In other words, any analysis of change must consider the concept in the light of the model in Figure 2 in order to determine whether change has come about in the Outputs (behavioural dimensions) or on what I have called the "Genetic" or "Transforming" level. These latter may be of great explanatory importance without entailing real change in the satisfaction of the needs of the population at large, which is what should be considered to be real change from the sociological standpoint. Whether a government has 10 ministries or 20, whether a State comprises one or two legislative houses, or is governed by a constitution that is new or old, means nothing if this does not affect the welfare of the people. The slight attention that Fundamentalism pays to aspects of change is due essentially to its abuse of this interpretation, drawing analogies, besides, between social and biological systems in respect of the tendency to reach equilibrium; but it is not at all mistaken in stressing the inevitably conservative nature of all social systems. However, the fact that such systems are ontologically conservative or follow an irresistible tendency to be so, by no means implies (contrary to erroneous interpretations sometimes put forward in this regard) that the theory that describes such situation is equally conservative.

But despite the above controls and the ruling classes' need to maintain the *status quo* (essentially with respect to the hierarchical structure of its

Inspiring Body OI), there is no question that there is a whole series of factors that prompt change in the system in one way or another. Hence, for instance, Structural Factors may change (e.g., the discovery of new minerals, a ban on fishing in traditional fishing banks, natural disasters, etc.). The people's Needs also change because systems of values change and evolve over time; historic Memory changes due to the accumulation of significant events or the appearance of new interpretations of history; the processes whereby disagreements and conflicts are solved and the system hierarchy established change; the nature of Environmental action changes, and with it the enormous influence it has on the system; the above example of the political change in Spain in 1976 is a clear illustration of changes taking place under the influence of the Environment and how psychological, legal and physical controls may, after all, be ineffective. In this regard events in Spain after November, 1975, deserve some comment. When General Franco pronounced his famous statement to the effect that Spain's future after his death was "all set, and well set" he did not or would not perceive that the Spanish political system (formal dictatorship) was already "upset", that the socio-cultural bases had changed and that this process had begun with the First Development Plan subsequent to the stabilisation process of 1960 that had prompted emigration, tourism, etc. Rapid development in the sixties had an impact on a series of cultural traits such as more and better schooling, contact with foreign tourists, the secularisation process, etc. A series of occult cultural transformations took place whereby the system, despite the Regime's typical psychological, legal and physical control, underwent change which the ruling elite seemed to be either unaware of or unable to counter. Moreover, the ideological clash among the elite themselves was causing profound change in the structure of the Inspiring Body through what might be called "internal axiological discord" or a power struggle among the elite, even without elections, in an attempt to influence legislative and governmental programmes. The Idealised System of Values (SIV) that the elite had proposed barely held up as designed, then, because pressure was brought to bear by all factors and from all directions. Whether or not these changes were significant in the short term or were good or bad for the population at large is another matter.

In general, this is the manner in which the ruling elite in social systems try to conserve their ideological schemes. Change becomes, then, a dialectic process deriving from a series of forces working for and against it. Although it is premature to go into detail at this stage, the difficulty lies in the definition of "significant change". Significant change is rare in social systems, at least over short periods of time. But as we know, the process of change is an unceasing and inherent aspect of such systems. We will discuss, in due time, how both the concepts of Change (homeostatic and neutral) and

Progress, Development, Social Regression and others can be operated with and quantified.

The model of social system presented here, simplified for the sake of comprehension, is rounded out with the description of its three basic dimensions: genetic, functional and behavioural as shown in Figure 2.

Table 1. Fundamental dimensions of the social system

1.	Genetic	
	1.	Structure-related (structural factors)
	2.	Shortage-related (people's needs)
	3.	Memory-related (historic archives
2.	Functional	
	4.	Conflict-related (axiological disagreement)
	5.	Hierarchy-related (socio-political stratification)
	6.	Teleology-related (pursuit of objectives)
	7.	Projection-related (order and optimisation of Ends/Means)
	8.	Transformation-related (conversion of means into ends)
3.	Behavioural	
	9.	Related to achievement of primary objectives (use of primary outputs)
	10.	Primary perception (perception of primary outputs)
	11.	Related to total primary performance
	12.	Regulatory or control (conservative tendency)
	13.	Related to final objective (use of secondary outputs)
	14.	Related to final perception (perception of secondary outputs)
	15.	Related to feedback (total final performance)
	16.	Adaptive (environmental impact)

These sixteen dynamically interconnected dimensions are initially assumed to be valid for the analysis of any kind of society or institution organised around the achievement of its ends, regardless of the nature thereof or the ideology that inspires it. Nation-states and tribes, multinational companies and small enterprises, families and groups of friends, schools, churches, ... all seem to more or less fit the model described. .

Finally, it must be admitted here that for the sake of clarity a highly simplified scheme is being used to represent complex socio-cultural systems. But the following must be understood: first, that each one of the dimensions presented may in turn be subdivided into any number of sub-dimensions, classes and types; second, that the arrows in the model proposed in Figure 2 refer only to the main direction of flow, but that given the complexity of the inter-relations among all the elements of the system, the arrows showing causal relationships, which are often mutual or recursive, may point in any imaginable direction. Modelling the social system calls for schematisation, but this by no means entails an over-simplification or loss of significance. It is a question, then, of stressing the model's Transforming (or inputs or

Outputs), Cybernetic (Control of Outputs), Axiological (in the search for End-Values) and Sociological (values considered from, by, and for the people) orientations.

The systemic approach, like any theoretical orientation, leads us "despite ourselves" in a certain direction in our research. But we must try to see if this direction is appropriate and if it contributes something or improves upon other sociological approaches. According to A. Regnier, at least, "drawing up a model is to bring scientific discourse to a level of rigour where the formal logic of the predicates is valid". My argument, moreover, is not that it improves upon other theories, but rather that it integrates them; such integration consists of assuming and using them selectively and complementarily in accordance with the specific problems faced in the pursuit of the end in question. But what should perhaps be emphasised again is that the systemic approach constitutes a theory of relevance and error, as Churchman says; what this kind of globalising modelling intends to stress is precisely those relatively relevant or essential aspects that have been forgotten or ignored; the description of the model is both a description of the errors introduced in our "scientific" projection, and of the possible epistemological adulterations and subjective and ideological manipulations that may arise in each individual case.

Once again, the importance of the four epistemological operations - Sociologisation, Systemisation of Needs, Systemisation of Resources and Teleologisation - must be emphasised. The core method and origin of the theorising process consists, then, of understanding the needs of the population and of taking the standpoint of the group as a whole, consisting of each and every one of the individuals that make it up as a starting point. In the proposed approach humans are at the hub of the system, are its *raison d'être*, for whom the system should operate and for whom CHANGES should have a positive meaning. Moreover, although the group is no more than a pool of persons, it is necessary to take the group viewpoint, to empathetically take the place of the man in the street to interpret "what the system needs". We shall see below that system needs thus interpreted call for other functions therein that are not political "action" or "decisions" (Easton), nor the division of powers (Almond) or even Parsons' famous AGIL scheme ("regulatory stability", "integration", "pursuit of ends" and "adaptation"), formulations which are inoperative not because they are too abstract, as is often said, but because they are sociologically out of focus. The sociological perspective calls for integration of the abstract (values) and the specific (Needs just as people perceive them). It demands higher salaries and lower prices, good public service, good highways, security (law and order), freedom, an unpolluted natural environment, etc., and on the basis of these concrete needs to build a system of Values that are sufficiently abstract and operative to withstand axiological critique, as well as measurement and appraisal. Being overly abstract is as bad as being overly specific and, naturally, as bad as the failure to interpret needs and their corresponding functions in terms of self (Luhmann), i.e., as perceived by the system's own population.

A discussion of the explicit ends shown in Figure 2 is now in order and may be summarised as follows: 1) to measure the System's resources; 2) To determine its Needs; 3) to determine who is to interpret what the system is to do; 4) to analyse the proposed system of values; 5) to analyse the implementation or transformation process, and 6) finally, to analyse the deviance between what the elite want, what the governments or bodies at their service achieve and what is finally perceived by the group as a whole. All of this seems to me to be by far the most interesting subject for debate in any social system, and the analysis and explanation thereof is what the axio-operational approach suggested here addresses. According to Churchman (1979), the systemic approach often generates questions that we would not otherwise have posed.

Finally, the ultimate consequences of this systemic modelling lie in proposing the End-Values as the centre of epistemological attention. Indeed, nearly all the concepts drawn up under this approach must relate in some way to the Values pursued and/or achieved by individuals or by the system as a whole. This is Van Gigch's (1978, p. 150) line of reasoning when he maintains that values should be assigned to objectives, to alternatives, and to the Outputs so as to be able to make rational decisions. Since the model, through the operation that we call "parsimonious selection", obliges us to focus on the most relevant and at the same time the most directly human aspects of the complex phenomenology, Values become the essential language of the different system expressions. And according to Van Gigch (p. 150) what is more important yet is that values, like any other attribute, may be measured; it is merely a question of defining the scale to be used.

It is the systematic use of what in the following chapter I will call Reference Pattern of Values that links these possible operational definitions of sociological concepts and calculations that in turn may make the axio-operational approach a more appropriate tool for social analysis.

3. NEEDS AND VALUES: THE REFERENCE PATTERN OF VALUES

Any social system inevitably revolves around two axial concepts. NEEDS AND PERFORMANCE: needs of the people and performance for the people, both concepts referred to a theoretical list of common objectives that I will call the Reference Pattern of Values (PRV). The reciprocity of this pattern is such that if the needs of a given society can be expressed in a model of, for instance, 97 empirical indicators (such as the example I will be using), Performance must necessarily be expressed in terms of such 97 indicators; and if 22 indicators suffice to validly represent the needs of a second society, Performance must refer to these 22 indicators. Satisfaction depends on need and vice-versa, and the relationship between the two is shown in Figure 3.

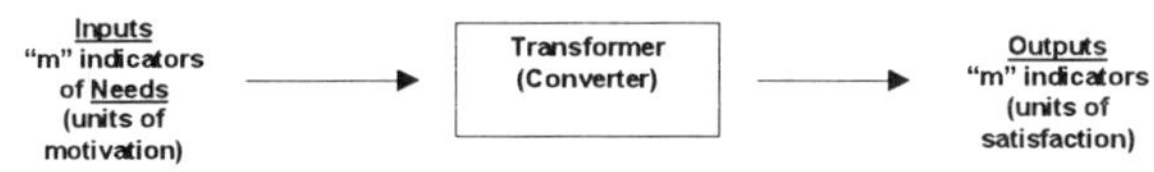

Figure 3. "Transformation" of needs into indicators of satisfaction

All the transformation operation does (or should do) is to satisfy the people's needs, and the PRV is what acts as a theoretical-conceptual bridge to make Needs and Performance, degrees of motivation and degrees of Satisfaction, Inputs and Outputs, means and ends comparable. The PRV is designed, then, as we will see in the chapters that follow, to attempt to measure - with an acceptable degree of accuracy - the overall Efficiency of the System by comparing the two magnitudes.

The Reference Pattern of Values (PRV) therefore is a nominative relationship of End-Values assumed to be desired, pursued or sought by the population at large. It is assumed to be valid to measure both the performance of social systems and important sociological concepts such as change, progress, social regression, socialisation, deviance and so on. It is postulated that most individuals, located in no-matter-what space and time pursue these values from cradle to grave and the only thing that differentiates them is the emphasis or weight that each individual places on each value in a given circumstance. By System of Values, in contrast, we mean the axiological configuration (or combination of the relative importance of the end-values) that each individual adopts as a living standard or pursues at any given time. The PRV is, then, a universal pattern while the System of Values ranges so widely that there are probably as many such systems as individuals in society. The author (Parra Luna, 1974, 1977, 1983) has been using a Reference Pattern of Values that seems to have been useful on several occasions. Such pattern is based on the adoption of nine essential Needs, which the System attempts to satisfy via the achievement of nine quasi-universal Functions and Values, as shown in Table 2.

The order in which the needs and values are shown has no particular significance. Nevertheless, Health and Economic needs (which correspond more or less to what Maslow called physiological needs) are generally considered to have priority over Protection against contingencies. The latter, in turn, is usually felt to be higher on the list of priorities than Knowledge and naturally than other more political needs such as Justice (social) or Freedom. Lowest on the list would be Power, Harmony with Nature, Self-fulfilment and Power and Prestige.

Table 2. Reference pattern of values

Needs (N)		Value pursued (Y)		
Kind	Symbol	Function	Kind	Symbol
1. Physical and mental well-being	N1	Health care	HEALTH	Y1
2. Material sufficiency	N1	Economic	MATERIAL WEALTH	Y2
3. Protection against contingencies	N3	Insurance	SECURITY (LAW AND ORDER)	Y3
4. Under-standing and command over Nature	N4	Research and Education	KNOWLEDGE	Y4
5. Freedom of movement and thought	N5	Liberating	FREEDOM	Y5
6. Equity	N6	Redistributive	JUSTICE	Y6
7. Harmony with Nature	N7	Naturalist	ENVIRONMENTAL CONSERVATION	Y7
8. Self-fulfilment	N8	Humanistic	QUALITY OF ACTIVITIES	Y8
9. Power and esteem for others	N9	Prestige-building	PRESTIGE	Y9

This nine-Value model is based on two premises mentioned above but which should be stressed here.

1. In the complex world of needs and culture, values can be classified under two major categories: variable and constant. The former are those that are characteristic of each specific culture, which will be considered secondary (dress, greeting, eating, celebrating important events, etc.). The latter are values that are common to all societies and the ones on which we have focused to establish the PRV (health, safety, justice, etc.). The former are culture-specific, the latter general and universal.

2. The degree of specificity and abstraction of the model must strike a balance between a small number of highly abstract system functions (as in Parsons) and a large number of specific indicators. Actually, we find that we must choose between a single Value, that consists of achieving General Well-being, and among all the aspects or sub-values that may define it operationally. Well-being seems to be the goal on which we all target and think that we can reach through multiple factors that may be represented by the nine values proposed in the model and the hundreds or even thousands of indicators that make them operative. But while a single concept of well-being would not be useful because it would entail concentrating information too highly, it would be equally unwise to use so many indicators that it becomes unmanageable. The Reference Pattern of Values to be used must, then, be a model comprising not more than ten essential values, to strike a happy medium both allowing for an understanding of the theoretically important dimensions of the system and providing a way of formulating them in a structure that is clear and meaningful to the population. It is a matter of balance between maximum specificity and minimum detail to ensure manageability, but also and above all, of adopting a sociological approach, that is to say, of ensuring that the meaning of PRV is directly perceived by and for the population.

Before going into a list of the indicators under which the nine Values in Table 5 would operate, certain comments about the specific indicators to be proposed are in order. First of all, as I mentioned above, the indicators are designed for application to a national system. This entails the advantage *mutatis mutandis* that it is easy to move from these complex social systems to simpler ones. The number of indicators proposed to define a given Value has no bearing on the relative importance of the value; a Value represented by only a few highly concentrated indicators may be considered to be more important than others with a larger number of indicators. The system of indicators is not limited to using existing information, but in fact empowers it by pointing out the need for additional information. Thus, for example, routine statistics on the "number of working days lost due to accidents" do not exist or are not available, but they would be very useful, and this suggests a possible indicator. Indicators are characterised by being as direct as possible, and therefore highly concentrated. When an attempt is made, for example, to see how well the value "(social) Justice" is working, we will not review the personal distribution of economic income, which is the usual procedure, but rather certain special mobility indicators that consider not only the distribution of such income, but also of other - cultural, political, social etc. - "incomes". If these mobility indices show that the situation is satisfactory, it must of course be concluded that economic income is appropriately distributed. The indicators proposed always intend to be the shortest route to the information, focusing more on ends than on means.

Finally, the indicators are as unambiguous as possible; in other words, they represent values desired by a majority of the people. Thus, for instance, a Religiosity index is not included because there is no clear consensus about whether being religious is "good" or "bad" as there is, for example, about being healthy or ill. There are, of course, some indicators on the suggested list that are somewhat ambiguous in this regard, but they have been kept to a minimum. Moreover, certain areas of concern may be missing; needless to say, this is an open and dynamic list that must be adapted to cover the information most suitable to each kind of system reviewed or specific research study. The list presented here intends merely to cover a number of generally applicable essential indicators.

In short, the Values and indicators that will be used must meet the three basic requirements discussed to be included as OUTPUTS in our model: 1) they must be defined to be good (we will see how this can be done by a weighted sample of "experts"); 2) they must be desired by the population at large; and 3) they must be obtainable onerously or must be of an "acquired" rather than an "ascribed" nature. GOODNESS, DESIRABILITY and ONEROSITY are then, the three basic traits that a value must have to be included in the Reference Pattern of Values. I will discuss this Pattern below, listing the indicators that provide the operational definition for each one of the nine values.

4. THE INDICATORS

4.1 Health

It is assumed that one of the fundamental objectives pursued by most individuals and peoples is to live as long as possible in the best possible health or physical condition. With respect to the indicators, it is suggested, as per Table 3, that this Value be quantified using the indicators traditionally employed by the World Health Organisation to define the state of health of the people in each nation.

Table 3. "Health": minimum indicators.

1. HEALTH		
1.1.	OBJECTIVES dimension	
	1.1.1.	Life expectancy
		1.1.1.1. Life expectancy of one-year-olds (Females)
		1.1.1.2. Proportional mortality rate at the age of 50
		1.1.1.3. Infant mortality at the age of 1 (Inverse index)

1. HEALTH		
	1.1.2	Quality of life
		1.1.2.1. Days not worked due to illness or accident (Inverse index)
1.2..	MEANS dimension	
	1.2.1.	Medical or paramedical staff/10 000 inhab.
	1.2.2.	Hospital beds/10 000 inhab.

4.2 Material wealth

It is assumed that most individuals and peoples wish to live as comfortably as possible.

Table 4 shows the indicators that refer in general to material standard of living, under two general headings, "standard of living" and "economic security". The indicators intend to focus on the actual standard of living of the population as a whole. The index on savings, not at all easy to obtain in full, provides information both on economic power and on the degree of future security and as such should be interpreted.

Table 4. "Material wealth": minimum indicators

2. MATERIAL WEALTH		
2.1.	Standard of living	
	2.1.1.	Homes with no running water (inverse index)
	2.1.2.	Homes without electricity (inverse index)
	2.1.3.	m² of housing and land *per capita*
	2.1.4.	% of GNP spent on tourism abroad
	2.1.5.	Number of telephones/1 000 inhabitants
	2.1.6.	Homes with automatic dishwashers/1 000 homes
	2.1.7.	*Per capita* Gross National Product
2.2.	Level of economic security	
	2.2.1.	Savings

4.3 Security and law and order

It is likewise assumed that individuals prefer, in general, to live peacefully and orderly in accordance with known rules and where contingencies and the need to rise to new circumstances do not entail too much discord or call for making inordinate efforts to adapt to them.

Table 8 includes two dimensions in an attempt to measure what we have called Public and Operational Safety, and assuming that the indicators suggested validly represent what is usually understood to be Safety in a modern Nation-State.

Table 5. "Security and law and order": minimum indicators.

3. LAW AND ORDER

- 3.1. Public safety
  - 3.1.1. Domestic
  - 3.1.1.1. Crime rate (inverse index)
  - 3.1.1.2. Number of deaths due to political unrest or revolutionary processes (inverse index)
  - 3.1.1.3. Total number of demonstrations (authorised + non-authorised) (inverse index)
  - 3.1.1.4. Number and power of nuclear power plants for peaceful purposes (inverse index)
  - 3.1.1.5. Foreign military bases (inverse index)
  - 3.1.1.6. Number of *coups d'état* (inverse index)
  - 3.1.1.7. Number of (proven) attempted and unsuccessful *coups d'état* (inverse index)
  - 3.1.2. Foreign
  - 3.1.2.1. Number of deaths due to conflicts abroad (average index)
  - 3.1.2.2. Military power
  - 3.1.2.3. Nuclear military risk (inverse index)
- 3.2. Operational
  - 3.2.1. Days not worked due to labour conflicts (inverse index)
  - 3.2.2. Number of accidents (job, traffic, etc.) (inverse index)

4.4 Knowledge

It is assumed that most individuals and peoples are curious to know about their surroundings.

Table 6 attempts to quantify the Value Knowledge in three dimensions: "educational level", "reading" and "research". Other indicators that have not been included, such as intermediate educational levels, the domestic consumption of paper for printing (deducting exports) or the number of patents registered, may also be used.

Table 6. "Knowledge": minimum indicators

4. KNOWLEDGE		
4.1.	Education level	
	4.1.1.	Illiteracy rate (inverse index)
	4.1.2.	Children not enrolled in school (inverse index)
	4.1.3.	% of university graduates
4.2	Reading level	
	4.2.1.	Press. Index
	4.2.2.	Number of books per home
4.3.	Research	
	4.3.1.	Royalties trade balance. Index
	4.3.2.	% GNP devoted to research

4.5 Freedom

The following working hypothesis is assumed: freedom (political, of expression, of movement, etc.) is desired and pursued by most individuals regardless of the kind of social organisation they live in.

Table 7. "Freedom": minimum indicators

5. FREEDOM		
5.1.	Political	
	5.1.1.	Number of political prisoners (or persons shut away in psychiatric hospitals, deported, etc.) (inverse index)
	5.1.2.	Rate of authorised/non-authorised demonstrations
	5.1.3.	Number of free elections, referendums, etc., per year.
	5.1.4.	Freedom of the press and editorial freedom
	5.1.5.	Mentions in Amnesty International (inverse index)
5.2	Social	
	5.2.1.	Foreign migration rate
	5.2.2.	Domestic migration rate
	5.2.3.	Freedom of coexistence
	5.2.4.	Freedom of reproduce
	5.2.5.	Freedom to commit suicide
5.3.	Religious	
	5.3.1.	Separation of Church and State
	5.3.2.	Freedom of worship
	5.3.3.	Freedom of propaganda

4.6 Justice

It is likewise assumed that equality among human beings, regardless of race, religion, ideology, sex, etc., is a generally accepted principle among civilised peoples.

Table 11 shows how well the "total national pie", i.e., economic, cultural, functional, etc., is distributed among the individual members of the social system. The 21 indicators used are somewhat different than the indices most commonly used to measure distribution of wealth. However, as mentioned above, the distribution of economic wealth is implicit in these indicators. The result of this set of indicators would undoubtedly be quite surprising, especially if actual percentages are compared. Thus, for example, it is maintained that if 51 per cent of a population are women, ideally (i.e., the target) 51 per cent of the executives running the economic, political, religious, etc. sub-systems should be women. If this is not the case, the implications are, aside from the effect of certain biological determinants in women's professional activities, that for whatever initial and historic reasons, the operation of the system is "structured" to keep women's influence, or at least their direct influence, to a minimum. Deeply entrenched educational principles and unconsciously sexist attitudes lead to situations in which there are, for instance, no female generals, bishops or union leaders in certain social systems. The general discriminatory principles that are at the root of this system conduct are likewise present in the case of other groups that are discriminated against, such as worker or underprivileged class children, blacks, etc. The ideal, of course, would be for all of the indices comparing % of positions held to % of population be as close as possible to one. Any deviation from that result would in principle mean undeniable *de facto* discrimination. Table 8, then, provides a very general summary of the actual level of "equal opportunities" existing in the society under study.

Table 8. "Social justice": minimum indicators

6.	SOCIAL JUSTICE		
	6.1.	By gender	
		6.1.1.	Percentage of women in Parliament and the Government
		6.1.2.	Percentage of female executive presidents of the 1 000 largest (public or private) companies in the country
		6.1.3.	Percentage of women generals in the armed forces
		6.1.4.	Percentage of women editors-in-chief of the 200 largest newspapers in the country
		6.1.5.	Percentage of women university presidents or deans
		6.1.6.	Percentage of women in the 200 highest positions in Labour Unions
		6.1.7.	Number of women bishops
	6.2	By race	

6.	SOCIAL JUSTICE		
		6.2.1.	Percentage of blacks, Jews, gypsies or other marginal minority in the Parliament and Government
	6.3.	By social background	
		6.3.1.	Percentage of members of Parliament or the Government who were born in the working class.
		6.3.7.	Percentage of bishops who were born in the working class

4.7 Environmental conservation

It is further postulated that one of the urgent needs of modern society and one of the targets of economically developed countries is conservation of the environment, to counter the detrimental effects of such development. Table 9 lists the indicators suggested for "Environmental Conservation". Sight should not be lost of the fact that most of the information required for this Values either does not exist or is still being compiled, so institutional efforts are called for in this regard.

Table 9. "Environmental conservation": minimum indicators

7.	ENVIRONMENTAL CONSERVATION		
	7.1.	Fauna	
		7.1.1.	Quantity – index
		7.1.2.	Diversity – index
	7.2.	Flora	
		7.2.1.	Quantity – index
		7.2.1.1	Hectares lost to fire (inverse index)
		7.2.2.	Diversity – index
	7.3.	Atmospheric pollution rate (inverse index)	
	7.4.	River pollution rate (inverse index)	
	7.5.	Sea pollution rate (inverse index)	
	7.6.	Percentage of park areas in cities > 100 000 inhabitants	

4.8 Quality of activities

Finally, it is assumed that all human beings should develop their full potential, which development should take place in all dimensions and aspects of life, from work to relaxation, from suffering to pleasure, from birth to death. Table 10 shows the indicators that represent this value called "Quality of Activities" which are an attempt to measure in some way the extent to which human beings participate, create, and reach self-fulfilment and self-development via the enhancement of their human capacities.

Table 10. "Quality of activities": minimum indicators

8. QUALITY OF ACTIVITIES

- 8.1. UNEMPLOYMENT dimension
  - 8.1.1. Unemployment rate (inverse index)
- 8.2 PARTICIPATION dimension (worker management and self-employment)
  - 8.2.1. Rate of worker participation
    - 8.2.1.1 Self-employed workers
    - 8.2.1.2 Workers in worker-managed firms
    - 8.2.1.3 Workers in co-operatives
    - 8.2.1.4 University professors and public school teachers
    - 8.2.1.5 Other self-employed professionals
  - 8.2.2. Participation in public life
    - 8.2.2.1 Time devoted to political militancy
    - 8.2.2.2 Time devoted to labour union militancy
    - 8.2.2.3 Time devoted to militancy in citizens' groups
- 8.3. LEISURE TIME AND ARTISTIC WORK dimension
  - 8.3.1. Hours worked per week (inverse index)
  - 8.3.2. Artistic production
    - 8.3.2.1. Number of painting exhibited
    - 8.3.2.2. Number of recordings
    - 8.3.2.3. Number of sculptures exhibited
    - 8.3.2.4. Number of play openings
    - 8.3.2.5. Number of art books published
    - 8.3.2.6. Number of motion pictures produced.

4.9 Prestige

Another working hypothesis on which the Reference Pattern of Values is based assumes that individuals need to give, do something valuable for others, and to seek love, admiration and respect.

Table 11 shows the indicators for Prestige, here understood in the international sense, since initially it is referred to the Nation-State. Application to other systems will of course call for the adaptation of such indicators.

Table 11. "Prestige": minimum indicators

9	PRESTIGE	
	9.1	Percentage of GNP devoted to donations abroad
	9.2	Number of international organisations with headquarters in the country
	9.3	Rate of foreign immigration
	9.4	Number of foreign military stationed in national territory (inverse index)
	9.5	Population growth rate
	9.6	Rate of success in international sports competition
	9.7	Rate of success in international art shows

This, then, is an initial list of indicators that are intended to quantify the proposed Reference Pattern of Values in an attempt to reflect both the needs of any socio-political system and therefore what should be done, regardless of the emphasis placed on each specific Value, dimension or indicator. The pattern, merely a nominal model of values to be pursued by any society, intends to be a unique tool, i.e., to serve as a common scheme for the analysis and comparison of societies. What will differ is the Systems of Values arising from the different emphasis or possibilities of practical implementation of the Values, and the indicators that represent them, which depend on the dimension and type of social system.

To make the distinction between "Reference Pattern of Values" (PRV) and "System of Values" perfectly clear, Figure 4 stresses the nominal nature of the former and the gradual nature of the latter.

The roster of needs is, then, one of the highlights of the theory of social systems, the core around which the operational theory proposed revolves. Given the central role that PRV plays in this theory, we feel that a more complete formulation of the pattern is one of the most pressing needs in current systems theory. The present discussion is simply a first step taken in the hope that work will continue to be undertaken in this important field by others, better qualified or more fortunate.

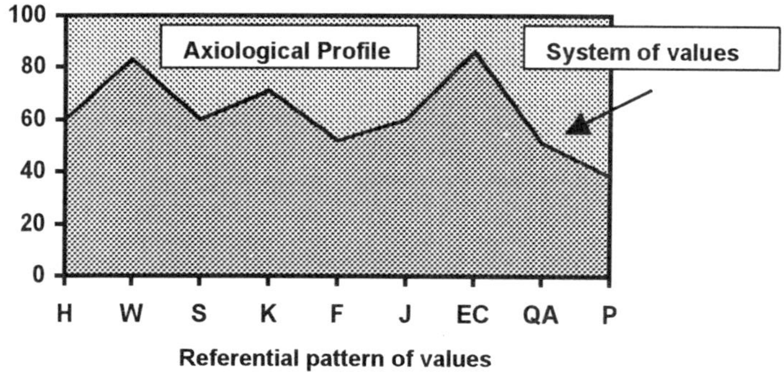

Figure 4. Differences between referential pattern of values and system of values

5. SYSTEM PERFORMANCE

After defining the outputs by means of the most appropriate indicators in each case, we have then to define the inputs. Operating with such inputs may be difficult in the case of complex national socio-political systems, since strictly speaking, on the one hand they would have to include the so-called "structural factors" (geographic, geological, orographic features, and so on), which form the basis of their economic systems, primarily; and on the other, the population's needs (system motivation), memory (historical archives) and even the surrounding area insofar as it comprises the series of countries that are economically, politically, geographically and culturally comparable to and interact to a significant extent with the system considered. Nonetheless such complexities can be readily avoided, in particular in respect of organisations and companies in general, whose INPUT is considered to be correctly represented by the total cost generated by their OUTPUTS. Any company's or organisation's economic-financial budget is the expression of its input (X), the sum of which can then be compared to its outputs (Y) to calculate its performance from the equation (T) = Y/X.

It should be clear, then, that Y is generally a complex vector of indicators (y_1, y_2 ... y_n), whose values are standardised to range between 0 and 100, where "0" represents the minimum and 100 the maximum possible level of fulfilment of each indicator; "X" (or the overall cost borne by the system), in turn, represents a single figure which is also standardised to range from 0 (minimum cost) to 100 (maximum cost).

Three possible circumstances may be inferred from the resulting value of T: if T is greater than 1, the system is "multiplicative", if T = 1 the system is neutral and if T is less than 1 system behaviour is "reductive".

I have decided not to go into an extensive discussion of why each indicator was included in the above nine tables so as not to make this paper excessively long. I trust that for the time being behind the assignment of each indicator to its respective Value is more or less transparent.

Having introduced PRV as the general measurement standard, the first methodological concern that arises before proceeding to quantify the model involves, as noted, homogenisation or standardisation of the indicators to make their results directly comparable, which is not difficult to perform in any of its varieties. This will be the next endeavour within the research process on the program Axiological Systems Theory.

Nonetheless it should be added that the main problems encountered in calculating the degree of Performance "T", have already been solved;

1. The basic relationship, Performance (T) = Y/X, where "Y" is outputs and "X" inputs, can be seen, for instance, in Cortes, Przeworski and Sprague, (1974). Today this relationship is generally accepted.

2. The outputs, "Y" are, inevitably in social systems, "COMPLEX SYSTEMS OF VALUES", as recognised by a large number of colleagues. Se, for instance, the latest publication that I know of on this subject (Hebel, 1999). According to this author, social systems cannot help but produce values.
3. The OPERATIONAL DEFINITIONS (through empirical indicators) of "Y" and "X" follow the well-known methodological procedure described by Lazarsfeld. The typical problems of summing up (integrating) heterogeneous indicators are, of course, statistically solved (See Ivanovic, 1970).
4. The numerous applications of this universal method (the world-wide SOCIAL INDICATORS movement) have proved its usefulness and heuristic validity in most cases. Even where sufficient good data is not always available, it is obvious that what we are considering is the date to which we have avail and nothing more.
5. The method for STANDARDISING indicators that I am applying (all indicators are expressed on a scale of from "0" – pessimum – to "100" – optimum – is taken from the work done by the UNRISD (United Nations Research Institute for Social Development) around 1973, to obtain "comparative profiles of indicators according to their relative degree of development". I call these profiles "axiological profiles", which are directly comparable to one another.
6. The concept of Performance, T = Y/X, is generally applied in entrepreneurial circles, where it is essential to know the PROFITABILITY of a specific line of business. And even when a final firm profit of, for instance, 15% is computed (which means that the surplus earned, "Y", divided by "used capital", "X", yields 0.15), the approach used actually involves applying the formal concept of Performance.

What I am adding to the discussion at this point is an attempt to generalise the method for social systems and to suggest some possible future lines of development for the concepts of Deviation, Socialisation, Ethical Behaviour and certain others, as a mere application of the systematic comparison between several of the AXIOLOGICAL PROFILES produced by social systems.

In this regard, my pretension is to do no more than take a first small step towards an AXIOLOGICAL approach to Systems Theory.

6. REFERENCES

Buchanan, B. "Information Requirements for a Viable Society", Paper delivered at the IV World Congress of Sociology, Montreal, July 1998.

Buchanan, B. "The Role of Values in Measuring Performance of Social Systems". Mimeo. 1999

Churchman W., "The System Approach and its Enemies", Eng. Cliffs. N.J. 1979

Cortes, .. *et al.* "Systems Analysis for Social Scientists", Wiley, 1974. Previously published in Oscar Lange (ed.), "Wholes and Parts: A General Theory of System Behaviour", Panstwowe, Warsaw, 1965.

Deutsch, K. W. "Politics and government: How people decide their fate". Boston, Houghton Mifflin Co. 1974.

Gigch van, J.P. "Applied General Systems Theory", Harper & Row, Pub. N.Y. 1978

Hall. B. P. "Values Shift", Twin Light Pub., Rockport, MA., 1994

Hall, Martin-Lewis, W. "Systems Thinking and Human Values;: Towards Understanding the Chaos in Organisations", Paper delivered at the XIV World Congress of Sociology, Montreal, July 1998.

Hebel, M. "World-views as the Emergent Property of Human Value Systems", "Systems Research and Behavioral Science", May-June 1999.

Ivanovic, B. "Problème de l'identification des pays les moins avancés parmi les pays en voit de développement", UNCTAD, U.N., Geneva, 1970.

Lazarsfeld, P. "Des concepts aux indices empiriques", in "Le vocabulaire des sciences sociales", Mouton, Paris, 1965.

Parra-Luna, F. "Towards Comparing National Social Performances", Univ. Lausanne, 1974

Parra-Luna, F. "Las Organizaciones y sus sistemas de valores", Boletin de Doc. Confed. Esp. Cajas de Ahorro, Vol. IX, Fasc. 3- Julio- Sepbre. 1977

Parra-Luna, F. "Elementos para una teoría formal del sistema social". Univ. Complutense, Madrid., 1983

Parsons, T. "El sistema social". Rev. de Occidente, 1976

Chapter 7

The potentialities of the sociocybernetic model in assessing political performance in a non-stable society

PHILIPPOS NICOLOPOULOS
University of Piraeus, Crete

1. THE CHANGING SOCIAL ENVIRONMENT, THE NEED FOR ORGANIZATION AND THE SECOND ORDER CYBERNETICS.

The modern society is a rationalized, a differentiated and mainly a quantitative and complex one. However, it is a kind of society which includes a lot of contradictions and fluxes which are not a positive factor for the establishment and maintaining of stable political situations, which are full of tensions. The increased degree of changes often are deriving from technological improvements or from pressures which are included in general social trends or from contradictory social elements, which lead to a dissolution of social units or to a sort of destructive social processes. In addition the chain of changes mean neither a change of the basic structure of society in each case of a transitional situation nor a stable course of developments towards a certain type of society, known to humans according to a deterministic way of thinking.

Therefore the instability in a social system neither is sure where may lead nor is susceptible to an analysis based on a process of exact prediction. The strict deterministic sociological theories could not convince us that a series of particular changes or non-stable situations in a time t necessarily will lead in a time t_1 to another type of society which is supposed to belong to another stage of social evolution, superior than the previous ones (Isaiah Berlin & Ramin Jahanbegloo, 1991:35-36). Of course, we can't exclude the approach of another type of society stable and more qualitative, in the stream of a

The Performance of Social Systems
Edited by Francisco Parra-Luna, Kluwer Academic/Plenum Publishers, 2000

general social trend, but we can speak neither about a predestination nor about a certain creation of a necessary improvement of sociopolitical structures. The whole process includes structural pressures and impacts, as well as individual initiatives and influences.

Thus we must be pragmatic about the final outcomes of the state of changes, non-stable situations or conflicts. The only thing about which we can be sure is our will as organizations or individuals to contribute to a process of progress and qualitative stability. Moreover we can be sure about our choice to organize our activities in this direction.

Just at that point the intervention of the sociocybernetic model and related approaches (general systems theory, functional theory, information theory, chaos theory etc.) might be useful (Felix Geyer, 1994:1). Especially, **the second-order cybernetic way of thinking** with its main concepts, as «self-organization», «self-reference», «self-steering», «autokatalysis», «autopoiesis», «anascopic approach» and «positive feedback loop», can contribute to **the organizational attempts facing problems of very complex social situations.** It can help them to find the appropriate balance between organizational initiatives, plans and goals and the empirical reality (Felix Geyer, 1994:7-16; Jonathan Turner, 1991:120-121; Felix Geyer & Jonathan van der Zouwen, 1978:1-3).

The second-order cybernetics is applied mainly in living systems with increased ability for meaning-based actions communication and consciousness (Felix Geyer 1994:13-14, Jonathan Turner 1991:100-102). These systems don't always intend to maintain a static «normal» state for themselves, but often they have plans for growth, expansion and new formations of themselves and environment (the question of morphogenesis). The social and political systems and their subsystems consist of living systems, that's why the second order cybernetics in their case is more proper scientific approach than the first-order one. The sociocybernetic model is just a model which applies in human systems and belongs exactly to the latter sort of cybernetics, which of course in turn was affected by the whole problematics about the features and reactions of human communities and personalities. The sociocybernetic model is an extension of cybernetics in order to embrace the so-called **flexible** and **negotiated systems**, as right John A. Busch and Gladys M. Busch define (1992:200). This sort of approach can accept both feedback loops, negative and positive, given that the flexible and negotiated systems often have alternate guidelines, i.e. sometimes they choose the maintaining of an internal steady state, sometimes they choose the differentiation and change of a previous steady state towards growth and expansion or creation of a new state with different structural elements (John A. Busch and Gladys N. Busch 1992:199).

2. NON-STABLE SOCIETY AND ENTROPY

Speaking about non-stable society we must at first define what we exactly mean. It is also required to define the meaning of its entropic tendencies, once the whole attempt of sociocybernetics is to reduce these tendencies as much as it is possible, based on elaborate information and organization (John A. Busch and Gladys M. Busch 1992:209-212, 230-237).

We may say that **a non-stable society is a society in which the basic political, economic, cultural and other social institutions are not stable, and there is a big enough discrepancy between them and the real needs, expectations, perceptions and beliefs of the people.** In addition there is often the sensation that something changes, that something dies and something is coming to emerge, but without always an exact perception and prediction of the new forms of institutions and social relationships which are coming to shape. So, in that social situation there is a problem of **legitimacy** of the existing institutions. In any case either an impetus for change is perceived or not, people realize that unshakable basic social axes (institutions, values, norms, public regulations etc.) don't exist or don't exist in the required degree, so the social and political relations are shaped with a very fluid way. The **lack of certainty about a stable concrete system of values and norms may generate** – and actually it generates – **strong anomic tendencies in society,** which in turn contribute to its disorganization (David Jary and Julia Jary 1991:21-22).

In a non-stable society, as we described it and defined it above, it is potential to emerge a political crisis, even if there are authoritarian state mechanisms. The **reason is the increasing lack of legitimacy of political institutions** and in addition the problem of the reduced efficacy and effectiveness. Between lack of legitimacy and low degree of efficacy and effectiveness there is correlation[1] (Juan Linz 1978:16-24). The discrepancy or much more the antithesis between beliefs of the people and the existing institutions, in the one hand, and, on the other, the instability of the latter create a context, in which legitimacy, efficacy and effectiveness of the political system gradually are decreasing. In that situation the rules of political game are unsteady, the power centres might be shifting and the governments might change quickly and irregularly without always correspondence with the real needs of ordinary people.

[1]. Between efficacy and effectiveness there is a slight difference. Efficacy has something more theoretical – to solve problems in a theoretical level, to know what should be done – , while effectiveness has practical character, i.e. to solve problems and to be successful in the field of action. Juan Linz adopts this distinction. Efficacy and effectiveness are included in the more comprehensive concept of performance.

Thus the social instability, either is connected with a transitional period or is not, create general tendencies for disorganization with corresponding political repercussions. Just in that situation the degree of **entropy** is increasing, i.e. the tendency for fall in increasingly lower levels of disorder and uncertainty, where the energy, included in the activities of actors (collective or individual), is being diffused and degraded. – Correspondence with the second law of thermodynamics (John A. Busch and Gladys M. Busch 1992:227) – Respectively the degree of **negative entropy** is decreasing, i.e. the ability of the system to organize its function and activities with order and certainty based on an effective information network.

3. POLITICAL LEADERSHIP, CHARACTERISTICS OF THE SOCIOCYBERNETIC MODEL AND THE QUESTION OF POLITICAL ALTERNATIVES

In a non-stable social situation the role of political leadership (I don't mean only the government but the leading mechanisms of all the basic political organizations) can prove very important, because that leadership can have initiatives and make decisions, by which the instability might be reduced and in contrast the emergence of a new social formation might take place. As it is known, the political system is the subsystem of the whole society which is connected with the final decision-process and the function of goal attainment according to Parsonian distinctions and classification (Talcott Parsons 1951, Talcott Parsons et al. 1953). The political leadership is the ultimate control subsystem of the political system, so it is connected with the final decisions and crucial initiatives par excellence (see figure 1).

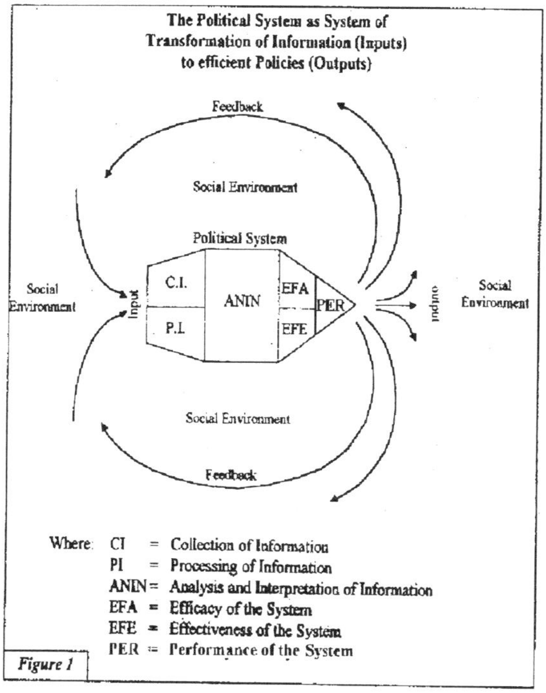

Figure 1

What, of course, was described in the foregoing about the potential functions of the political system, exists only if this system is not blocked by established classes or predominant social groups, which tend to use it as their instrument or the mechanism which can perpetuate their power. The game between political system and established classes or groups constantly is enacted, and which party finally will be the winner and which will be the loser is not predestined in every historical moment. The outcome is depending on many interrelated and interdependent factors which can confine or in contrast increase the relative autonomy of political system.

In any case we pose as presupposition for the potential benefit of the application of cybernetic model the relative autonomy of political system. Blocked political system means, less or more, undermining of the crucial cybernetic process «input-output-feedback». It means finally containment of the steering subsystems by social forces which put their self-interest above the common interest.

In a fluid social condition a political leadership is confronted with many, various and unforeseen difficulties. At first it must begin with three main guidelines:

1) How to have clear perception about its goals.
2) How to realize the exact situation of its environment based on empirical approach (collection, elaboration and analysis of empirical data), and

3) How to map a sensible policy which will give the chance to it to materialize its plans in combination with an effective adjustment to the existing situation. Adjusting to the environment, how to adapt parts of its to its goals (namely to the goals of the political leadership). In this process the very points is to find a convergence or a «gold mean» between adaptability and intervention in order to create a new formation in some social institutions or relationships or much more in the basic social structure (the question of morphogenesis, as we say in cybernetic terminology).

For that prospect the political leadership must be very flexible and avoid the retention on its already materialized plans. It should take measures in order to preexist **an empirical knowledge** (and not only information) about the total social milieu in which the collective or individuals actors will interfere. The difficult point in that process is not to insist on a dogmatic-deductive approach, if the empirical reality doesn't agree totally or partly with the initial propositions which are attached to that approach. At that point we would like to underline that in some political thinkers and social scientists the deductive-nomological explanation of sociopolitical phenomena is a temptation – temptation connected mainly with the acceptance of the historicosocial determinism (Karl Popper 1960:35-49).

We would like, of course, not to exclude the above type of sociological explanation, but to underline the need for **a sensible convergence of inductive with deductive method and for checking the deductive generalization by empirical verification.** In a complex contemporary situation, where the role of technology often is very important, the political leadership must map out its policies with scientific way. And scientific way of thinking and acting means endless pursuit for empirical checks and verifications. The deductive-nomological explanations (covering law model) in the social phenomena is not built up as easily as we think or as we would like to build up (Karl Popper 1960:105-1090; Vernon Pratt 1978:69-76).

The sociocybernetic model can just contribute to the increase of the scientific quality of the decision-making process of political system. It doesn't exclude the so-called problematics about the **substantial rationality**, in order to use the Weberian terminology, but insists on the necessity of an initial down-to-earth perception of the dynamics of the environment and the complexity of the interrelated systems.

The political leadership based on the sociocybernetic model must carry out and develop the following policies, abilities and work:

1) **Collecting exact data and information about the environment** (natural, social and political), as much as it is possible.
2) **Improving the problematics about the «absolutism» of the perspectives** (John A. Busch and Gladys M. Busch 1992:36-37). Every citizen, politician or leading group has the right for his and its

own perspective, but in order to collect exact information about his and its environment needs not to permit that perspective to be an obstacle in the collection and analysis of information. At that point the «absolutism» of the perspective is always a danger, because it may make one to see just what wants to see and to lose those parts of empirical reality, which one doesn't like to see and accept (John A. Busch and Gladys M. Busch 1992:33-60, 78-85; Israel Scheffler 1965:14-15, 75-91; Vernon Pratt 1978:90-106).

3) **Processing, analyzing and interpreting the data and information collected.** Through this way we pass from the stage of **information** to the stage of **knowledge**, which of course is something deeper and more systematic.

4) **Organizing the existing «matter» and «energy»** in the political system and in its environment, based on the elaborated and analyzed data and information.

5) **Developing a flexible and effective adaptation and response ability** in respect with its environment, based on the model «input-output-feedback». Inputs are all data and information which exist about the sociopolitical system, its natural environment (including its biological parameters), its potential dynamic trends and especially about the demands, claims, expectations and support of people which appear in various levels of social and political system. The outputs are the choices, decisions, initiatives and plans of the leadership. The outputs have an objective dimension (what really achieved) and a subjective one (what people perceive about the achievement of political leadership). Francisco Parra-Luna, on the basis of this distinction, speaks about objective, subjective and total outputs and presents some criteria according to which each category of outputs can be appraised (Francisco Parra-Luna 1995:105-113). According to him total outputs are the integrated actual and perceived ones (Francisco Parra-Luna 1995:111-113). The outputs don't need always to be simple corresponding responses to the former, as dependent variables, because on the contrary sometimes they need to be independent variables.

6) **Finding balance between adaptation to environment and goal attainment.** Especially the adaptation to the human environment is a complicated process and must take into consideration a lot of elements and parameters of human being. If the adaptive mechanisms of a leading political organization neglect these elements and parameters and view the human environment with an absolute positivist and a-historical way they are likely to fail. The political plans and goals can be materialized and be actually accepted by people, only if they find a sensible way to adapt their

content to some crucial cultural, psychological and historical dimensions of the concrete society of those people.

7) **Perceiving the change trend in the society which might generate new types of its structures and institutions.** Inability of the political leadership to perceive the exact tempo and content of that trend might lead to its collapse. The same thing may happen, if it is opposed to the tendency of change, because of its ideology or because is blocked by an established class or group, which tries to control the whole political system.

The leadership must constantly understand that doesn't govern a static society or a «flat» environment. The society is determined by deep historical trends and differentiations, as well as is full of conflicts among various interest groups. The class conflicts is one from these conflicts. If the decisions of the leadership are not right in the terms of those historical parameters and struggles between established and developing forces, there is a serious danger for it to be overthrown. In that case it will prove, in last analysis, an inflexible decision-making mechanism. Very often this weakness of the political leadership is brought about by the predominant social class or group which blocks the leading political mechanisms and attempts to use them as «instruments» subordinated to its plans and interests.

Thus **the sociocybernetic model is not connected necessarily with a static approach of the society.** The perception of change, and its promotion and its backing is included in the same approach. It is always a potential alternative, as the fall also of the inflexible leadership is another one.

Among the abilities and virtues that a political leadership, as subsystem, must have in order to implement the above mentioned guidelines, one can include the increased degree of its performance (PER). It is not an excess to say that a great part of the above guidelines is compressed, to an extent, in the notion of this performance (Francisco Parra-Luna 1995:58, 114-117). The performance leads to the satisfaction of people needs. With this aim in view the political leadership should systematically try to increase the quantity and quality of its work while in parallel it minimizes its cost[2]. So performance (PER) in last analysis is the ratio of work done by it to time and cost spent for that work, according to the following equation:

[2]. The concept of performance (PER) is similar to the concept of efficiency. Nevertheless the meaning of PER refers to the entire satisfaction of human claims and needs more than the corresponding meaning of efficiency (so, it is more obvious how the anthropocentric element complements the systemocentric one). That is why we prefer it. The latter has a meaning more quantitative, abstract and neutral regarding human needs, while PER has something more functional in social terms (Francisco Parra-Luna 1995: 58-62).

$$\mathbf{PER} = \frac{\mathbf{w}}{\mathbf{t} + \mathbf{c}}$$

Where, PER = Performance

w = work done by a system

t = time

c = cost

Equation 1

The more the leadership has the right information and what can be called **Practical Transformation Insight (TRIN)**, the more it can increase its total performance. The practical transformation insight means the capacity of transformation of a system of «immaterial» elements (aims, ideas, plans etc.) to a concrete successful policy, political action and generally to a concrete of «material» elements, which will constitute finally a part of the sociopolitical reality in which the leadership is called to intervene. The TRIN, in last analysis, is the capacity of materialization of the theoretical plans to a real world and is the crux of all grand radical policies. Thus we can represent the above relationship with the function shown in the box on the followng page.

The sociocybernetic model and the problematics of performance can also be related with the whole question of the less or more radical change of society. The choice of revolution or reformation must not be based on a theoretical deductive way of political thinking. The decision that in concrete historicosocial and political circumstances the necessary and sensible choice is, for example, revolution of reformation, is something depends on the right perception and analysis of the empirical reality in which the interested political actors are involved.

Thus, if the collection and analysis of the appropriate information from the concrete social and political environment point out that radical structural changes can be realized peacefully without the overthrow of the existing political institutions, then the active political organization pursuing those changes mustn't choose the way of a violent revolution. It can proceed with peaceful revolutionary steps or with an accumulation of reformations. Especially in a modern, quantitative and complex society a violent revolution or a violent generally overthrow of a political regime can increase the degree of social entropy in an unanticipated way.

$$PER = f(I+TRIN)$$

Where, PER = Performance

I = Information

TRIN = Practical Transformation Insight

Function 1

Generally speaking we should underline that every revolutionary process can strengthen the entropic tendencies of a social environment. Therefore it is not necessary to promote it, if there are other alternatives or a more peaceful way of changes. The spirit of sociocybernetic model always includes the pursuit for alternatives analogous to the possibilities emerged in the concrete environment in which the interested political organizations act. If of course the process of decision-making in the political system is blocked by an unfair socio-economic establishment and the structural changes are necessary, then the way of violent revolution probably is inevitable (Tom Baumgartner, Tom P. Burns and Philippe DeVillé 1978:46-47; John A. Busch and Gladys M. Busch 1992:199-200).

But even then the right decisions of the revolutionary leading organization must be based on the exact perception of the dynamics of the empirical reality. Furthermore the question of performance (continual assessment and comparison of the political results with the costs) must be taken into account independency of the initial need for revolutionary steps. A runaway violent tendency always is a factor for increasing the social entropy and must be confined by negative feedback loops (Jonathan H. Turner 1991:120-121).

4. CONCLUSION

The sociocybernetic model is actually useful in a non-stable society, if it is used by political leadership with sensibility and a sensation of historicity. Its essence is found in the conception of social life as a complex system in which among parts and relationships of theirs there are mutual influences (John A. Busch and Gladys M. Busch 1992:118, 154). Based on that perception the leading political units can organize matter and energy of the whole sociopolitical system, if and when use and elaborate the appropriate information. The model itself is not conservative or radical, but it depends on how it is used by political actors and how is combined by them with the messages and the demands of people and the understanding of the historical trends.

5. REFERENCES

Baumgartner, Tom, Tom R. Burns and Philippe DeVillé, «Actors, Games and Systems: The Dialectics of Social Action and System Structuring», pp. 27-54, in R.F. Geyer and J. Van der Zouwen (eds), *Sociocybernetics*, vol. 1, Martinus Nijhoff Social Sciences Division.

Berlin, Isaiah and Ramin Jahanbegloo 1991, *Conversations with Isaiah Berlin*, New York: Charles Scribner's Sons.

Busch, John A. and Gladys N. Busch 1992, *Sociocybernetics: A Perspective for Living in Complexity*, Social Systems Press.

Geyer, Felix 1994, *The Challenge of Sociocybernetics*, paper presented in 13th World Congress of Sociology, San Diego, CA: Sociological Abstracts Inc.

Geyer, Felix and Johannes van der Zouwen 1978, «Introduction», p.p. 1-14, in R.F. Geyer and J. van der Zouwen (eds), *Sociocybernetics*, Martinus Nijhoff Social Sciences Division.

Jary, David and Julia Jary 1991, *Sociology: The Harper Collins Dictionary*, Harper Perennial.

Linz, Juan 1978, «Crisis, Breakdown and Requilibration», pp. 3-97, in J. Linz and Alfred Stepan (eds), *Breakdown of Democratic Regimes*, Baltimore: John Hopkins University.

Parra-Luna, Francisco 1995, *Why Society? An Inquiry Into its Consequences Through Systemic Axiology* (Working Paper), Madrid.

Parsons, Talcott 1951, *The Social System*, New York: Free Press.

Parsons, Talcott, Robert F. Bales and Edward A. Shills 1953, *Working Papers in the Theory of Action*, Glencoe, IL:Free Press.

Popper, Karl R. 1960, *The Poverty of Historicism*, London & Henley: Routledge & Kegan Paul.

Pratt, Vernon 1978, *The Philosophy of the Social Sciences*, London & New York: Routledge.

Scheffler, Israel 1965, *Conditions of Knowledge*, Scott, Foresman & Company.

Turner, Jonathan H. 1991, *The Structure of Sociological Theory*, Belmont CA: Wadsworth Publishing Company.

Chapter 8

Criteria of systems performance from a macro-sociological viewpoint

Karl-Heinz Simon
Center for Environmental Systems Research, University of Kassel, Germany

1. INTRODUCTION: THE CONCEPT OF PERFORMANCE OF SOCIAL SYSTEMS

The approach introduced in this contribution tries to adopt the view of functionalism, in the sense, that tasks and functions of social systems are the starting point of our analysis. In order to evaluate the performance of these systems inputs and outputs have to be considered. Good performance, in this view, is given when a certain ratio of input and output characteristics is ascertained, at a given state of systems behavior or in principle.

Hence, for evaluating the performance of social systems there are two decisions to be made: (1) a selection of the criteria which are used to represent functioning of the systems, and (2) a definition of operational properties of these criteria, a determination of threshold values or scales.

But before these two types of information are gathered, there remains a very important aspect to be approached: the description of the object of assessment itself, that is a description of phenomena of society as those items, the performance of which are evaluated using the criteria and applying the formal machinery.

Without discussing this point in sufficient detail, we will be arguing later on in this paper from the standpoint of "naive systems theory". In other words, we take the existence of certain kinds of systems for granted, as well as the premise that systems can be analyzed with respect to tasks and functions and that inputs and outputs can be determined, that is, systems boundaries are observable or at least determinable in one way or another.

The Performance of Social Systems
Edited by Francisco Parra-Luna, Kluwer Academic/Plenum Publishers, 2000

But despite this primary decision some aspects of alternative views on analyzing societies are very briefly discussed in the following paragraphs.

In order to look at the structural decision in a more formal way, the proposition of Parra-Luna is adopted, namely to consider the relationship: X → T → Y. In that formula certain inputs X are transformed by a (social) system T into outputs Y (X and Y are vectors in the normal case). In the following, we argue for the possibility of combining the „performance approach" with the „orientors approach" and similar endeavors like social goal functions.

Without going into more detail here, some problems are listed which have to be solved in order to be able to apply the performance approach to social systems:

Firstly, the boundaries of the systems considered have to be stated explicitly. Therefore, X and Y as well as also T have to be described precisely with respect to membership in the system in focus.

Secondly, system functions have to be specified as explicitly as possible. Problems for analysis occur when functions are distributed all over the system and are not located at specific, definable positions.

Thirdly, a chance must be given that inputs are connectable to outcomes, that is, mechanisms within the system boundaries and with respect to specified functions have to be identified.

If the above-mentioned conditions are not fulfilled then, nevertheless, empirical research on systems and a kind of black-box approach are still available, but hardly transparency about systems performance can be achieved.

2. REMARKS ON A CONCEPT OF SOCIETY AND A BASIS FOR DESCRIBING PERFORMANCE OF SOCIAL SYSTEMS

In this paper a macro-sociological viewpoint is adopted, that is, we are not dealing with organizations or families or similar lower level social systems but with the society as a whole or with functional subsystems of society. And, as announced above, systems are described as concrete entities (as it is the case in Miller's analyses of Living Systems (1978)) with properties and attributes, for example, certain input and output characteristics. Certainly, this methodological preliminary decision is not accepted in general. On the contrary, sociologists formulated severe objections against such a view on social systems. What seems to justify that position, nevertheless, seems to be for us (i) the urgency of problems societies are confronted with (as in the case of sustainability) and (ii) that good hopes can be put forward that some more pragmatic steps (without proper backing in theory) can be taken in order to achieve progress in solving these urgent problems.

In our search for criteria to evaluate the performance of social systems we give priority to a look on Parsons' theory and after that on more recent proposals for analyzing social goal functions and „orientors" of system functioning and development.

2.1 The AGIL Scheme

In recent history of social theory building Parsons specified four essential tasks social systems have to fulfill on different analytical levels in order to be successful in functioning and surviving. These tasks are arranged in his well-known AGIL scheme and they are: Adaptation, Goal-attainment & Goal-selection, Integration, and Latent pattern maintenance. On different levels different subsystems are responsible for taking care of these tasks. In Parsons' architecture, for example, economy is responsible for adaptation at the level of the social system. As Münch (1982) pointed out, these tasks are correlated with fundamental system problems. Jackson calls them „the four imperatives that must be adequately fulfilled for a system by its subsystems if that system is to continue to exist" and explains:

> „The meaning of the terms that make up AGIL is as follows
> A. Adaptation - The system has to establish relationships between itself and its external environment.
> G. Goal-attainment - Goals have to be defined and resources mobilized and managed in pursuit of these goals.
> I. Integration - the system has to have a means of coordinating its efforts
> L. Latency (or Pattern maintenance) - the first three requisites for organizing survival have to be solved with the minimum of strain and tension by ensuring that organizational „actors" are motivated to act in the appropriate manner." (cit. after Francios 1997)

Having his suggestions in mind, it seems to be useful to develop criteria centered around the AGIL framework. But as we are convinced that more flexibility in the conceptual framework would be helpful in integrating different types of systems we are introducing an alternative approach, called the orientors approach (using an „EXFSAEFC-scheme" as explained below). Nevertheless, Parsons' impulse on considering functional properties of social systems should be kept in mind as a touchstone for recommendations of evaluation criteria of social systems.

2.2 The Orientors Approach

From a more general systems theoretical viewpoint Bossel developed the so-called orientors approach.[1] Though not completely identical with the definition of orientor in the important reader edited by Müller and Leipelt (1998), Bossel has introduced in numerous publications the concept of an assessment approach to value system states and system development since the seventies. We describe that approach in a very brief way, and we are going to discuss the potential it will have to bring more systematics into judgments about social system performance.

In his approach Bossel proposes six fundamental dimensions which have to be included every time an evaluation of the system's state and development has to be performed.[2] These dimensions are:

- EX. Existence - Attention to this indicator is necessary to insure the immediate survival and subsistence of the system in the normal environmental state.
- F. Freedom of Action - The system must have the ability to cope in various ways with the challenges posed by environmental variety.
- S. Security- The system must be able to protect itself from the detrimental effects of environmental variability, i.e. variable, fluctuating, and unpredictable conditions outside the normal environmental state.
- A. Adaptability- The system should be able to change its parameters and/or structure in order to generate more appropriate responses to challenges posed by environmental change.
- EF. Effectiveness- The system should on balance (over the long term) be effective (not necessarily efficient) in its efforts to secure scarce resources from, and to the extern influence on its environment.
- C. Coexistence - The system must be able to modify its behavior to account for behavior and orientors of other (actor) systems in its environment. (Bossel 1980, p. 80)

We do not have enough space here to discuss his proposal in proper detail. But some very brief comments are necessary to clarify the standpoint from that the following extensions are discussed.

Let us begin with those dimensions which are easy to describe. Take freedom of action for an example: It seems to be clear that the identity and autonomy of a system requires certain degrees of freedom in deciding and sufficient independence from external advice. Also the security dimensions seem to be not problematical to understand. Again a certain degree of stability and protection against alterations in the environment is absolutely necessary. Another well-known concept is efficiency, for example in discussions about technical systems or about means / aims relationships. Often forgotten, but nevertheless also quite well intelligible, is the adaptability dimension. The main idea is to take precautions because of the dynamic of the environment which cannot be predicted in its consequences for the system in the future.

Somewhat more complicated are the remaining dimensions *existence* and *coexistence*, and indeed discussions are going on about integrating them into the theoretical framework (Hornung 1998). One problem with existence is that a clear dividing line between indicators of the EX dimension and those of the other dimensions (perhaps excepting the C dimension) is extremely difficult and seems often to be a matter of subjective decision-making. And with respect to coexistence, the justification why it might be useful for systems to include this dimension in their decision-making is overdue.

There is a certain correspondence between the AGIL-scheme and the above list of orientors. This correspondence is stated here merely on the level of concepts without going into any detail concerning underlying social theory.

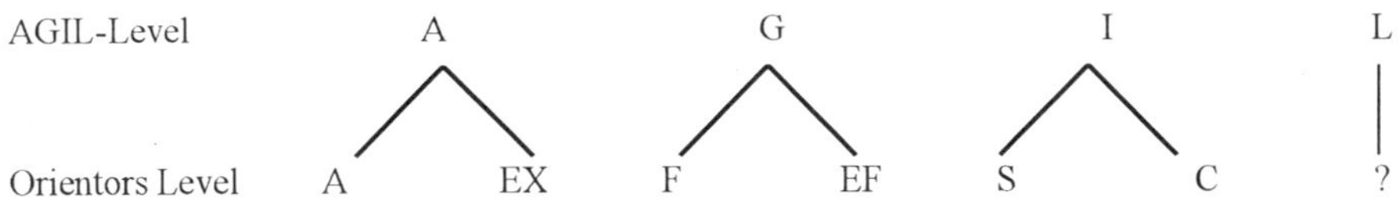

Figure. 1 A possible correspondence between AGIL and orientors

The A-box of the AGIL-scheme is corresponding with the A-dimension of the orientors. As also the supply of the system in question is integrated in Parsons' scheme, we have also related the EX-dimension to that box. In goal attainment aspects like freedom to choose proper behavioral options and the efficiency of measures are important; hence, we established a relationship there. A third box is Integration. Hereto, we associated the S-dimension and, at least in part, Coexistence. For the fourth box of the AGIL-scheme, namely Latency, we miss a clear correspondence with orientors dimensions. That points out that in social theory an issue like „continuity building“ is considered explicitly, whereas in the orientors approach that aim seems to be associated with several dimensions and is therefore treated implicitly.

After this short introduction of the dimensions, a very brief view on the justification of the dimensions is added. Bossel tries to derive the dimensions from his analysis of properties of system environments. That is the realm where he found the same concepts as those he decided to be important for systems performance. But that rises the crucial question what environments of social systems are constituted of? We will come back to this question later. A more general problem with such an attempt is, that it might not be possible at all (i) to define systems without their corresponding environments and (ii) to analyze environments without specifying whose environment is analyzed and with respect to which system problems and to which questions and hypotheses.

Table 1 Examples for indicators for social systems (source: Bossel 1996) (taken from a list of 220 indicators selected

	E - Existence
L02:	Net population growth rate
W14	Rate of change of social service capacity; net rate of change of social service capital per cap (i.e. investment depreciation)
W15:	Pct. security of funding, or of secure social processes, for next five years.
	P - Psychological needs
W26	Average size of cohabiting family unit.
W27	Avg. rate of intense family type social contacts per day.
W28	Avg. distance between living places of members of extended family
P15:	Avg. proximity of places of rest, beauty, spirituality, culture (e.g. forests, parks, temples, churches)
	E - Effectiveness
W23:	Fraction of working age population employed (paid or unpaid) in social service work
W22:	Social service unit cost: annual per cap cost of social service per services persons vs. avg. annual income
W24:	Ratio of volunteer services (hours) to paid services (hours)
B31:	Health cost of environmental pollution: pct. of lifetime lost to environmental pollution (illness, early death: see DALY index)
O12:	Percent of GDP going to graft, corruption, and political favors

In Table 1 we illustrate Bossel's approach with respect to the social subsystem. The social system is understood in this report merely as a system of social welfare and Social Security systems. In later versions of the report dealing with the problem of finding indicators for sustainability of social systems, he widens up the perspective. Within the social subsystem he then arranges aspects like population development, ethnic composition, income distribution and class structure, social security, medical care, old age provisions. The system is described as follows:

> The 'social system' is here understood as that vital fabric of relationships between humans that both supports and constrains individual human development. It provides the collective support, solidarity, and community necessary for human welfare, and societal self-organization and evolution. Many of its social functions supplement the functions of the economic system; very often they are necessary to rectify its deficits. The social system enables and enriches individual development. Its functions and performance depend on the infrastructure system. They are regulated and shaped by the government system. Much of what happens

in the social system is in response to impacts of the economic system. Value change, and social and political action to regulate the economic system and protect (or exploit) the environment and resource system are first generated in the social system. (Bossel 1998, p.167)

Table 1 Continued

	F - Freedom
W17:	Unemployment rate: pct. of working age adults who want to, but cannot find work
W22:	Social services unit cost: annual per cap cost of social service per serviced persons vs. avg. annual income
W13:	Percent of social needs effectively dealt with by the system
W18:	Rate of change of unemployment rate
	S - Security
W20:	Social support ratio: (children + old people + sick + unemployed) / (working population)
W21:	Rate of change of social support ratio
W16:	Probability of adequate financing for social support processes in twenty years
	A - Adaptability
O28:	Fraction of self-organizing (NGO) vs. total social activity
O15:	Level of institutional bureaucracy: bureaucrats and administrators in government and industry per working adult
Q07:	Avg. quality and level of education and skills: equivalent years of European education
Q17:	Percent of population reached by quality media information (newspapers, magazines, TV, radio, books)
	C - Coexistence
W01:	Degree of social inequity (e.g. percent of population under discriminatory conditions of gender, race, culture, religion, poverty, class)
W02:	Burden on future generations due to excessive demands: (nonrenewable resources use / need for sufficient lifestyle); (debt /cap)

For making the approach more operational, he proposes an Agenda for Change separated into problems, goals and tasks to be taken out (Table 2).

We will come back to the approach after having introduced the goal functions approach.

2.3 The Social Goal Functions Approach

The orientors approach has a counterpart in (social) goal functions, developed recently in ecosystems research. Müller and Fath introduce the idea of goal functions in the following way:

„The central idea of the orientor approach ... originates in modern systems analysis, complexity science and synergetic. It refers to the idea of self-organizing processes, that are able to build up gradients and macroscopic structures from the microscopic „disorder" of non-structured, homogenous element distributions in open systems, without receiving directing regulations from the outside. In such dissipative structures the self-organizing process sequences in principle generate comparable series of constellations that can be observed by certain emergent or collective systems features. Thus, similar changes of certain attributes can be observed in different environments. Utilizing these attributes, the development of the systems seems to be oriented toward specific points or areas in the state space ... The respective state variables which are used to elucidate these dynamics, are termed orientors. Their technical counterparts in modeling are called goal functions. They are variables or objects of optimization procedures, which are guiding the system's functions as well as their structural development ... (Müller/Fath 1998, p.15)

Table 2 Agenda for change - problems, goals and tasks for social systems (source: Bossel 1998, p.280)

No.	Problems	Goals	Tasks
S1	Population impact exceeds regional carrying capacity	Adapt population impact to carrying capacity	Change lifestyle, reduce birth rate
S2	Inequity splits society	More equitable society, and similar changes for all	Equal value of individual lifetime; no inherited privilege
S3	Social security and self-support ability eroding	Ensure adequate minimum of social support and individual contribution	Share social service burdens in intergenerational contract
S4	Loss of diversity and innovation threatens adaptability	Maintain critical and creative potential of participation	Recognize and support contribution of non-governmental organizations

Without going into too many details here - several aspects of the statement of Müller and Fath actually would require additional comments - we look only for connection points to social theory building. But it should be clear, that it is especially the mentioned aspect of optimization where connection points exist. Instead of commenting on the statement we list „potential social goal functions" like such having been collected by Jüdes (1998) to give a first impression of the variety and multitude of possible „candidates" for such functions. Each entry in the table would have to be discussed in sufficient detail before applying them for evaluation purposes.

Table 4 shows - in our opinion - a more concise assignment of the table's entries to categories. Only a part of the original entries are adopted and associated with „true" social goal functions. The selection carried out is surely not the only way to separate the functions, but it shows the problems with such a kind of listing, especially if no further justification is cited and no additional descriptions are given how to understand precisely the particular goal function. In addition, after re-formulating certain entries they would also fit into the category of genuine social functions. For example, if innovation is altered to social innovations, that means, development of new types of collective schemes of interaction for solving new social problems, then it could also be included in the group of true social goal functions.

Table 3 Listing of goal functions in different disciplinary fields (source: Jüdes 1998)

Ecological goal functions	Cultural goal functions	Social goal functions	Economic goal functions
complexity	perception	pluralism	benefits
retinity	rationality / reasoning	participation	Economic production
biodiversity	sense / meaning	communication	market
stability	self-formation	social integration	innovation
exergy storage	religion/myth/sanctification	prosocial attitudes/behavior	costs
carrying capacity	morality / value system	cooperation / partnership	efficiency
regeneration	self conception	manners	Economic cooperation
resilience	common good	harmony / peace	sufficiency
selfregulation	zeitgeist	collective identity	Circular economy
fitness	worldview	work	Economic risk management
ascendency	time perception	humanization of work	Social compatibility
cyclicity	enculturation	equality of opportunity	substitution
	family / cultural group	self-help	
	homeland	solidarity	
	cultural identity	community work	
	tradition	social security	
	innovation	social distance motility	
	differentiation	sufficiency	
	cultural diversity	structural nonviolence	
	adaptive lifestyle		
	freedom (of rational action)		
	institutional freedom		
	social freedom / liberty		

In order to make full use of the heuristics provided by the orientors approach, in a next step the remaining social functions are associated with orientors to assure completeness and „systematicity".

Remarkable is the absence of orientors representing effectiveness and adaptability, as well as coexistence (with other partner systems).

In the form presented, both approaches - social goal functions and orientors - suffer from severe shortcomings. They are formulated without redress to social theory as it is presented for example in the work of Parsons and Luhmann. In contrast to goal functions as they are derived from ecosystems theory, social goal functions are unfortunately largely ad-hoc constructions only, which may be useful in particular situations but which are highly questionable in their reliability and profoundness. That might reflect once again the differences in theory building between social and natural sciences. It is also possible that the approach is too new to get sufficient attention up to now.

Table 4 A more precise analysis of the candidates for social goal functions; re-distribution to different disciplinary fields (entries see Table 3)

	Social goal functions		
political	social	psychological	economical
pluralism	participation	prosocial attitudes/behavior	work
structural nonviolence	communication social integration cooperation / partnership equality of opportunity solidarity social security social distance motility	manners harmony / peace collective identity	Humanization of work Community work sufficiency

Table 5 Correspondence between goal functions and orientors.

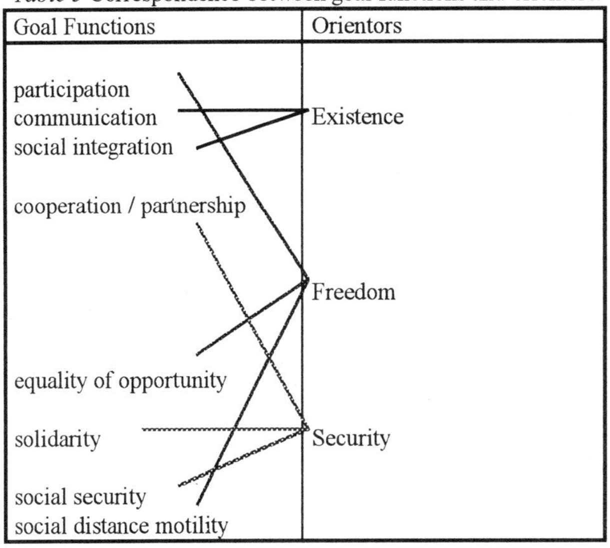

In the following, we try to elaborate on a kind of intermediate level between the more general level of the orientor dimensions mentioned and the manifold indicators that can be found in discussions on social systems performance. As we are convinced that the stage of theory development does not allow general solutions, we discuss some examples in order to illustrate the approach and present a platform for criticism and further improvements.

3. CRITERIA TO EVALUATE PERFORMANCE OF SOCIAL SYSTEMS

Referring to the discussion above it seems to be appropriate to (i) apply a common set of evaluation criteria, but (ii) to adjust these criteria carefully to the specific evaluation object by a process of differentiation.

Hence the questions are, in principle, the same for different types of social systems or different levels of analysis, but the guidelines for answering the questions and, therefore, the answers themselves can be completely different.

Up to now, theory building in social systems theory has not yet reached a level as in ecological systems theory with respect to „operationalization". In this research area a certain degree of theoretical backing or consistency of indicators has been reached within recent years. A similar effort like the „functions of nature" (deGroot 1992) would be necessary to take steps towards progress also in the field of social systems. Due to these gaps in theory we are going to introduce some examples only, showing how indicators could look like which are derived from some more general ideas about the status of social systems and their aims.

The common set of evaluation criteria is taken, in the following, from the „orientors approach". In this first attempt to apply them to different problem areas of social systems performance, we take - as mentioned above - social systems more or less as concrete entities and will completely ignore constructivistic positions. If such a „naive" view on social phenomena is really warrantable, however, remains an open question to us.

We are dealing with four examples covering different aspects of social systems. A first example is referring to communicational aspects of societies as a prominent issue in modern social theory. The second and the third example are dealing with some more pragmatic aspects of societies, namely social order and social security. The subject of the fourth and last example, eventually, is the material and energetic supply basis of societies.

It has to be kept in mind that with the separation into different particular views on social systems, it is not intended to neglect the concept of the whole system. Especially the first example, the treatment of autopoiesis and communication, refers to a conceptualization of societies where all other aspects should be subordinated if not even banished from social theory. For pragmatic reasons we are not trying to achieve such a comprehensiveness or purity.

3.1 Maintenance of the Autopoiesis of the Social System / Process of Communication

In all the examples following we take the EXFSAEFC-scheme as a framework for ordering several criteria which could provide a basis for evaluation. In modern social systems theory communication is regarded as the central process which maintains the functioning of the social system (Fuchs 1994, Luhmann 1991/1992). In our suggestion for a set of indicators we focus on the participation to communication and the quality of questions dealt with.

Table 6 Criteria on the level of the communication processes

EX	Continuity of communication processes and intelligibility (common language means)
F	Possibility to contribute to communication processes (or to keep oneself out of it) and to raise issues voluntarily
S	
A	Possibility to gain resonance for the „right“ or pressing questions (e.g. with respect to the „ecological question“)
EF	Organization of public discourses; time lags in responses
C	Equal access to communication processes; access to adequate technical means

Table 7 Criteria on the level of sustaining social order

EX	Establishment of a common scheme of values
F	Societal restrictions and pressures imposed on members of society (e.g. from the political system)
S	Frequency of involuntary involvement (of individuals or social groups) in dangerous situations; social entropy
A	Adequacy of existing structures to changing societal circumstances and societal evolutionary processes
EF	Comparison of measures and outcomes
C	Legality and legitimacy of restrictions and their distribution within society and treatment of immigrants

3.2 Stability of Social Order

One of the prominent themes in social systems theory is the stabilization of order. Theoreticians like Parsons started their theory building from that question (Parsons 1966). Some ideas for indicators can be found in Table 7.

3.3 Assuring Social Security

With respect to social systems one important question is that about security. In this context, it is not referring to stability of order but to old age provisions and other aspects of assuring life support for those members of a society who finished work age, who are handicapped etc. Much of the indicator's concept introduced here is related to monetary terms but there are also structural aspects of the problem which are to be mentioned.

Table 8 Criteria for Social Security systems

EX	Present people receive at least a minimum of support to have the opportunity to live a life free from burdens of care
F	Constraints for decisions for or against different security systems
S	Continuity of security systems over longer periods of time
A	Diversity of several functionally equivalent resources on which the security systems rely
EF	Ratio of investigations to outcomes
C	Guarantee of minimum support independently of performance principle

3.4 Coverage of Essential Energy and Material to Support Population

Several subsystems exist to fulfill the task of supporting people with bare necessities of life and with other means to allow a good quality of life. Raising the question without specifying which aspects of the support systems are in fact taken into account seems not to be very helpful, therefore further differentiation is needed. Hence some subsystems have to be evaluated first - like food supply, energy etc. - and in a second step the outcomes of the subsystem's assessment have to be integrated into a valuation of the superior general supply system. As an example, indicators for the energy supply systems are listed in Table 9.

The examples given are far from being complete and there is no guarantee that they are really applicable. We outlined first ideas on possible solutions of the evaluation problem only. In a next step more specific indicators must be defined below the more general level considered above, quantification should be carried out and the practicability of the approach must be substantiated (examples are given in Parra-Luna (1982, unpublished)).

Undoubtedly, the task to establish a close relationship between social theory and evaluation measures is not solved with the remarks above. One necessity to build up such a relationship is to have at least a minimum of transparency with respect to the viewpoint which gives a basis for evaluation. Crucial differences have to be taken into account whether the fundamental basis of social systems (e.g. communication processes) are analyzed, or the material basement (like energy supply structures) is considered. The proposals above are understood as an attempt to keep such different „spheres“ separated.

Table 9 Sketch of criteria for energy supply sub-system

EX	Availability of a minimum of energy demands for present consumers
F	Possibility to influence decisions relevant for future energy system
S	Safety of energy-conversion processes; Guarantee of access to necessary amounts of energy; Dependence on energy sources outside the realm of direct influence
A	Degree of centralization/decentralization of energy production systems, e.g. power generation plants
EF	Consumption of resources efficiency factor
C	Consideration of the demands of other social systems, e.g. on the basis of definition of the „environmental space“ on a global level

4. OUTLOOK

As mentioned before, the quantification of indicators is a crucial next step which often is not being taken. Up to now, in most cases only an intuitive association of values is undertaken, e.g. on the basis of marks like A (very good) to E (extremely bad) as in former experiments of Bossel in this field (Bossel 1996). Investigation of self-organization processes in virtual environments (Krebs/Bossel 1997) produced first results which are useful for further improvement of the approach. An example is shown below, where cultural theory is taken as a methodological starting point, and different „cultural types“ are evaluated with respect to the set of orientors dimensions described above. Several single indicators and evaluation criteria (like those in Table 1) are organized into classes corresponding with these dimensions and the evaluation results are calculated by summing up all these „value loadings“ and mapping the derived numbers onto the orientor dimensions.

The evaluation shows the differences between three types, namely „Generalist“, „Cautious Type“, and „Specialist“ with respect to five orientor dimensions. Whereas the generalist is characterized by high affinity to freedom, and the cautious type to security, the specialist is mainly interested in effectiveness. These are experiments which cannot be applied to real societies up to now. The idea, however, is to take these concepts as a starting point and to develop measures for investigating the stage of development of societies, sub-systems of societies, or to compare different social systems with respect to good performance, for example.

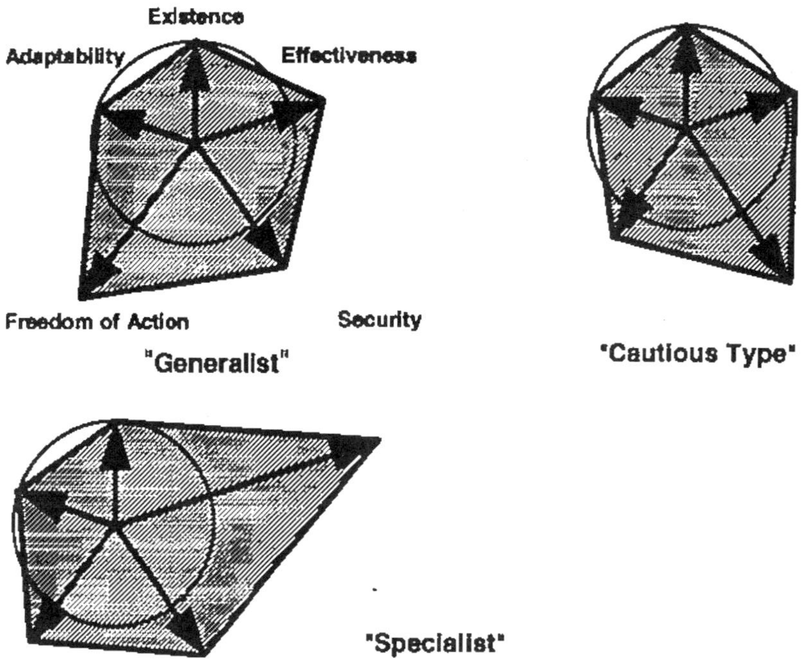

Figure 2 „Orientor Star“ for three „Cultural Types“ (Bossel 1998b)

To sum up: On the one hand, there are promising developments in criteria and concepts for evaluating social systems performance. On the other hand, however, there are also lots of open questions. Some of these questions are:

- Is it feasible to deal with social systems in the very same way as with other types of systems, ecological systems for example?
- How can a more adequate fundament in social theory be achieved in order to connect indicators definition and evaluation of systems performance closer to progress in social theory and guarantee more systematics?
- Then, taken for granted that indicators and criteria for good systems performance can be defined, what consequences are being faced in case of non-attention of these criteria? Is it really possible to draw a clear picture of changes in system behavior?
- What is about the interconnectedness of indicators? Is it possible at all to rely on hierarchical orders of indicators and to arrange them unmistakably into orientor dimensions?

Progress is achievable with respect to these questions, but more integration and cross-cutting hypotheses between social theory, cybernetics and evaluation approaches are necessary.

5. NOTES

1 We avoid to call it theory because the stage of development allows, in our opinion, only a crude pragmatic justification of the concepts which is far away from proper grounding in systematic theory. But taken as a heuristic for systems analysis and developing hypotheses it seems to provide a powerful tool.

2 We decided to exclude a seventh dimension „psychological needs" also part of the orientor systematics in some of the more recent publications.

6. REFERENCES

Bossel H 1998 Earth at a Crossroads - Paths to a sustainable future. Cambridge University Press

Bossel H 1998b Ecological orientors: Emergence of basic orientors in evolutionary self-organization. In: : Müller F, Leupelt M eds. Eco Targets, Goal Functions, and Orientors. Springer

Bossel H 1996 Deriving indicators of sustainable development. Environmental Modeling and Assessment 14, 193-218

Francios C ed. 1997 International Encyclopedia of Systems and Cybernetics. KG Saur Verlag

Fuchs P 1993 Moderne Kommunikation. Suhrkamp

de Groot R S 1992 Functions of Nature - Evaluation of nature in environmental planning, management, and decision making. Wolters-Noordhoff

Hornung B 1998 Orientorentheorie und soziokulturelle Theorie unpublished

Jüdes U 1998 Human orientors: A system approach or transdisciplinary communication of sustainable development by using goal functions. In: Müller F, Leupelt M eds. Eco Targets, Goal Functions, and Orientors. Springer

Krebs F, Bossel H 1997 Emergent value orientation in self-organization of an animat. Ecological Modelling 96, 143-164

Luhmann N 1991/1992 Lectures on „Einführung in die Systemtheorie", audio tapes. Carl Auer Systeme Verlag

Miller J G 1978 Living Systems. McGraw-Hill

Müller F, Fath B 1998 The physical basis of ecological goal functions - Fundamentals, problems and questions. In: Müller F, Leupelt M eds. Eco Targets, Goal Functions, and Orientors. Springer

Müller F, Leupelt M eds. Eco Targets, Goal Functions, and Orientors. Springer

Münch R 1982 Theorie des Handelns - Zur Rekonstruktion der Beiträge von Talcott Parsons, Emile Durkheim und Max Weber. Suhrkamp

Münch R 1992 Die Struktur der Moderne, Suhrkamp

Parra-Luna F 1982 Elementos para una teoria formal del systema social. Editorial Complutense, Madrid

Parra-Luna F unpublished Axiological Systems Theory. Universidad Complutense, Madrid

Parsons T 1966 Societies. Evolutionary and Comparative Perspectives. Prentice-Hall

Chapter 9

Distributed control in social systems

MARKUS SCHWANINGER
Professor of Management, University of St. Gallen

1. INTRODUCTION

The performance of social systems is an object of continuous theoretical debate and of manifold practical conflicts. Many approaches to defining criteria of performance and techniques of measurement have been developed and applied. The fact that the attempts are multifarious indicates that the domain of inquiry at hand is still at an immature stage, with pragmatism and eclecticism dominating the mainstream. Experts have called the subject area an "enigma" (Cameron 1984) and ascertained "a paradox of performance: organizational control is maintained by not knowing exactly what performance is" (Meyer & Gupta 1994: 309). Anyway, also some promising approaches to the conceptualization of organizational performance have emerged (e.g. Venkatraman & Ramanujam 1986, Bamberger & Fiegenbaum 1996). For an overview on the discussion, see Meyer & Gupta 1994 and Lumpkin & Dess 1996.

In the present paper, an effort is undertaken to contribute to the ongoing discussion on the grounds of socio-cybernetics, - more specifically to organizational cybernetics, which has also been called "the science of effective organization" (Beer 1979). Organizational cybernetics is a science which applies principles and models of cybernetics to the design, control and development of organizations of all kinds.

The aim of this paper is to present a Model of Systemic (Self-) Control, and to combine it with the Viable System Model, in order to achieve a framework for distributed (self-) control leading to higher organizational

The Performance of Social Systems
Edited by Francisco Parra-Luna, Kluwer Academic/Plenum Publishers, 2000

fitness or intelligence, and therewith enhancing sustained superior performance.

2. DIMENSIONS OF PERFORMANCE

The limited view which conceived organizational performance exclusively in economic terms has given way to a more differentiated understanding a long time ago. Systemic approaches have emphasized that organizations are systems made up by diverse constituents, - the "stakeholders" whose "stake" in the organization is a function of their specific perspective (Cf. Freeman, Mitroff). Stakeholders incorporate different values, goals and interests. In principle, the performance of an organization can and should be measured continually in relation to these different perspectives. This is necessary because ultimately an organization is only viable as long as it is legitimate. Legitimacy is only given if the legitimate claims of the major stakeholders are fulfilled (Schwaninger 1989). The set of stakeholders varies from organization to organization. For private companies, it usually comprises the customers, staff, sponsors/owners, state, and the public.

This stakeholder perspective leads to a multidimensional conception of organizational performance. A logically sound approach to measuring the performance of organizations is based on the value or benefit they create for the different stakeholders. A distinction in this sense would be:

- Product/service benefits (to customers)
- Economic benefit (to sponsors/owners and customers)
- Social benefit (internally to staff, externally to society, including state and the public)
- Ecological benefit (to nature and society including future generations)

To analyze distinctions of this type is a task undertaken by other contributions to this book.

For this paper an other focus has been chosen: The aspect of control for superior performance, with an emphasis on the self-control of complex social systems. It can be anticipated that this is a multilevel and a multi-recursion issue. With the attribute *multilevel*, a crucial aspect of organizational control is addressed: There are always several logical levels of control, with the higher logical levels exerting a pre-control function with respect to the lower ones. The attribute *multi-recursion* expresses the cybernetic insight that viability is a recursive property of complex systems. Viability resides in (self-)control being a distributed activity (one can even talk of *property*), recursively exerted along the "fronts" on which the complexity confronted unfolds.

3. A MODEL OF SYSTEMIC CONTROL

This section will deal with the multilevel issue of control just mentioned.

The cybernetic concept of control embraces regulating and steering of a system in order to achieve its purpose or goals, or in a more limited sense, to carry out its tasks or activities („what it has to do“ or „what it does“). In all of this, the aspect of intrinsic control (“self-control”) is more important than that of extrinsic control (cf.: Geyer/van der Zouwen 1978 I: 2, Schwaninger 1996, Espejo et al. 1996)[3], an aspect to which I shall revert later (section 4).

Effective (self-) control implies a dynamic equilibrium, at a satisfactory level of performance, between a corporation (or an organisational unit) and the milieu in which it operates. But which variables define adequate performance?

Traditional corporate control models take their bearings largely or exclusively from the goal of „profit“; i.e. they are economistic. It has been demonstrated that such models no longer meet today's requirements. Under the „evolutionary pressure“ of increasing complexity and turbulence the criteria of competent management are rather defined in terms of a comprehensive organizational fitness or intelligence[4].

In this context, a Model of Systemic Control has emerged. It is based on the insight that one and the same system must govern itself with the help of control variables that may contradict each other because they belong to different logical levels: the levels of operational, strategic, and normative management (Schwaninger 1989, Gälweiler 1990).

[3] If Heinz von Foerster (1984) designates social systems as „observing systems“ as opposed to „observed systems“, he therewith addresses such aspects as self-control, self-reference, self-organization and self-transformation, - which are among the main concerns of organizational cybernetics (cf. Schwaninger 1994a).

[4] An intelligent organization shows the following faculties (Schwaninger 1998):

1. to adapt to changing situations,
2. to influence and shape its environmental milieu,
3. if necessary, to find a new "playing field" or to reconfigure itself anew with its environment, and finally,
4. to make a contribution to the viability and development of the larger whole(s) into which it is embedded.

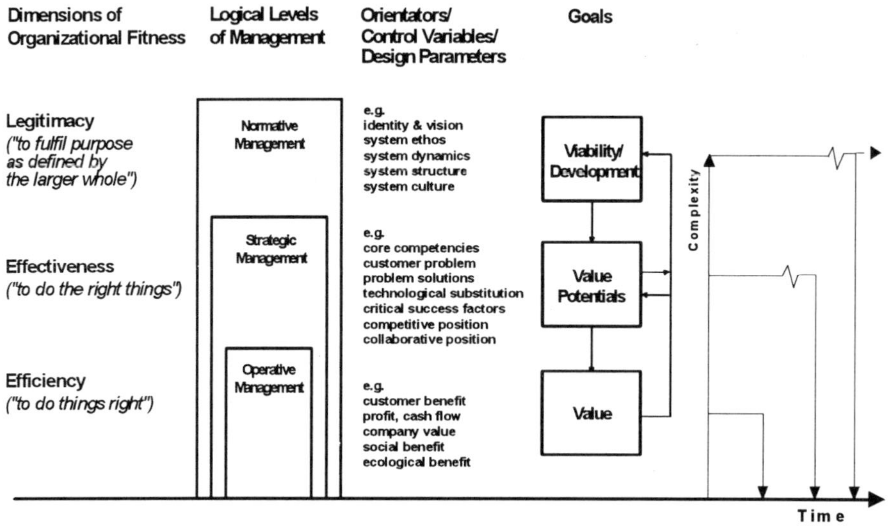

Figure 1: Goals and orientators at different logical levels of management

As shown in Figure 1, there are interrelationships between these levels. Particularly, the control variables of the higher logical levels exert a pre-control influence on those of the lower levels. This has been outlined and specified elsewhere for the case of business organizations in a more detailed way (Schwaninger 1989, 1993).

In section 2 it has been outlined, that the general goal at the operational level is creating value for the stakeholders of the organization (particularly customers, personnel and owners). Specific control variables to achieve this are customer benefit, social and ecological benefits, and shareholder value.

The traditional steering models of firms for example were almost exclusively orientated towards profit and liquidity. But profit behaves in an inherently short-term-mode, and its level is largely predetermined by parameters of another nature (just as liquidity is largely predetermined by profits). The importance of this predetermination or pre-control expresses itself in the fact that – according to comprehensive empirical studies of more than 3000 business units (PIMS database: cf.: Buzzell/Gale, 1987) strategic variables account for 70% of the variance in the ROI (Return on Investment), while only 30% is explained by variables respresenting operative decisions and tactical skills. In a similar vein, a study of a sample of Fortune 1000 firms leads to the conclusion that the impact of organizational factors on profit rates is twice as high as that of economic factors (Hansen/Wernerfelt, 1989).

Good managers have always known that the attainment of operative goals is bound to preconditions that have to be created in advance. Practitioners and writers in many domains of management have become aware of this, more and more. Tentatives have been made, to link the operative and the strategic logic, e.g. in production (Underwood, 1994), logistics (Bowersox et al., 1992), and marketing (Kotler, 1994). Beyond that however, there has emerged a comprehensive theory which provides access to the relevant higher-order criteria.

The value which a company can generate is largely predetermined by the value potentials created beforehand. Value potentials are defined as the set of all applicable business-specific prerequisites (e.g. in the form of resources, competencies or capabilities), that must be fulfilled when value is to be provided (In extension of: Gälweiler, 1990). They represent operational and calculable categories. Their patterns of behaviour can be foreseen and influenced („controlled“). Value potentials must be controlled separately from value, on the basis of independent criteria.

Research on strategic management has clarified the nature of these criteria and shown, for instance, how to apprehend the critical success factors (such as market share, relative market share, quality/-customer benefit, cost/price, speed, flexibility etc.) in a given business system (for details see: Buzzell/Gale 1987). With the help of the PIMS database and a statistical toolbox for quantitative analysis, valuable conclusions about the strategic position of a business and the actions to be taken can be derived. However, such a calculus remains in the domain of *extant value potentials* and is therefore of limited usefulness in turbulent environments.

The management of *new value potentials* includes changing established patterns, taking into account the dynamics of customer problems, problem solutions, technological substitution along the value chain. This involves a sustained long-term effort for innovation, and it often requires a redesign of the business system. Hamel and Prahalad (1994) have emphasized, that too much of strategic efforts rests on established modes, while the essence of genuine innovation is rethinking the system completely, which often leads to the emergence of new modes of doing business, creates new opportunities and may reshape entire industries.

For strategic business units a mature methodology has become accessible. This methodology has made more transparent and controllable the essential variables of the strategic level - as, a long time ago, bookkeeping did for the operational domain. Gälweiler (1990) for the most part, but also Porter (1980, 1985), Schwaninger (1987, 1989), and other authors have elaborated on this methodological concept. The support for strategy at the corporate level has grown into a relatively mature methodology as well (Pümpin, 1991; Hamel/Prahalad 1994; Schoemaker, 1992), although the heuristic devices available must rest more general and abstract. The main idea is building and strengthening core competencies - or capabilities - to ensure new value potentials.

Long term empirical evidence and solid conceptual reflection (cf.: Utterback 1994) cogently suggest that organizations must consistently embrace innovation, even when this appears to undermine traditional strenghts. The respective authors plead for continually renewing core capabilities, while abandoning the logic of past successes: Innovating incrementally is not enough, because in the long run, the regeneration of a corporation's business relies on radical innovation. A "strategic architecture" is needed, "that is less concerned with ensuring a tight fit between goals and resources and .. more .. with creating *stretch goals* that challenge employees to accomplish the seemingly impossible" (Hamel/Prahalad 1994: 23). In the language used here, the requirements of building up new core competencies to ensure new value potentials may contradict those of extant value potentials, but in the long run, it must have priority, if viability is to be warranted.

To resume, in broad outline, profit is not a strategic control variable, and consequently not a strategic aim either. Rather, its appearance or absence is a consequence of good or bad strategies. This divergence from the traditional view, which regarded profit (or other monetary values statically related to it) as the fundamental corporate objective, has also been expressed to a certain extent in more recent attempts to integrate finance theory with strategic considerations. The methodologies developed for assessing shareholder value of companies calculate net present value, which is derived from discounted future free cash flows, a corporation can potentially generate at a certain point in time, in function of possible strategies (cf. Rappaport, 1986; Copeland/Koller/Murrin, 1994). The crux in these assessments is not - as it may seem - the calculus, which relies on more sophisticated accounting techniques, but the proper knowledge of the variables pre-controlling profits, and their interrelationships.

Meanwhile, insights into the referents of normative management have also improved. The research which has led to this is primarily based on systems theory and cybernetics and once more presents independent criteria for the assessment of the viability and development of organisations. Viability, understood as the ability to maintain a separate existence (Beer 1979: 113), i.e. a distinct configuration which makes a system identifiable as such, can be assessed on the grounds of structural considerations which are not bound by the orientators of the strategic and operational levels.

To date, the most advanced theory for assessing the viability of an organisation in functional terms is Stafford Beer's VSM, - to be outlined in section 5. This model is an excellent device for diagnosing and maintaining the degree of viability of an organization, independent of the steering criteria of the lower levels (strategic and operative).

As far as the „soft factors" of organisation are concerned - referred to under the common denominator of „culture" - some models have been elaborated which, for the time being, appear more appropriate for description and diagnosis than for design purposes (e.g. Deal/Kennedy, 1982, Schein 1985). Beyond that, the emerging paradigms of the „learning organisation“ (cf. Argyris/Schön, 1978; Senge, 1992) and of organizational knowledge (cf. Nonaka/Takeuchi 1995) are about to outline a developmental and transformative orientation to structure as well as to culture.

From a systemic point of view, however, we can adopt the position that an organization must and should aim at *viability beyond survival.* Systems thinkers have become more interested in designing evolving structures, in which an organization's identity may completely change, than in sticking to viability in the narrow sense; this, after all, has often led to the self-maintenance or self-production of systems which show a disfunctional behaviour vis-a-vis the larger wholes into which they are embedded. This negative effect on the larger whole is referred to sometimes as „pathological autopoiesis“[5] (cf. Beer 1979). Progressive managers are increasingly adopting a systemic viewpoint in which they enlarge their reference system, eliminating narrow boundaries: This is corroborated by the growing rate of economic, juridical and structural transformations of companies with the aim of creating new viable organizational entities. „Development“ in the sense used by Russell Ackoff is a good term for such viability beyond survival.

At the level of development - defined as a system's growing ability and desire to fulfil its own and others' needs (after Ackoff, 1994: 65) - the quest of an organization is in fact viability beyond survival. At this level, indicators become even more hazy. Yet social system theories provide important insights to diagnose a system's propensity for development, as a function of its ethos (e.g. openness), and the pattern of a system's trajectories or dynamics: criteria such as catalytic reinforcement, instability, consensus, self-governance and learning help to judge whether a change process qualifies as „development“ or not (cf. Etzioni 1968; Jantsch/Waddington 1976).

In sum, the field of indicators at the normative level is multifaceted; social, political, cultural and ecological aspects have to be taken into consideration, giving adequate space to ethical and esthetical concerns for the pursuit of ideals such as beauty, truth, good and plenty (cf.: Ackoff 1981). Multiple constituents and viewpoints ascribe different purposes to a social system, which leads to an emphasis on different criteria of fitness. For a corporation to be viable in the long run, the legitimate claims of these different stakeholders must be met adequately (Cf.: Kotter/Heskett 1992).

The concept of „control“ which applies at the level of normative management is in a certain sense incompatible with the understanding derived from traditional sciences: instability is no longer a feature to be

[5]Pathological autopoiesis, as used here, is the self-maintenance or self-production of a system despite a negative balance of its effects on the larger whole.

eliminated completely. In a sense and to a certain extent it is a necessary and valuable precondition for development (cf. Prigogine, 1989); (self-) control must nevertheless maintain instability within acceptable levels and frequencies. The coming turbulent decades will increasingly demand for „control by development", „control by learning" or „control by transformation"[6].

So much for a survey of the design parameters and control variables. It must be noted that they cannot be compared in every respect, since they belong to three different logical levels; ultimate consistency can be achieved within, but not between, these levels[7]. The variables regulated at one level are the pre-control parameters for the next level down.

Figure 1 illustrates that operational, strategic and normative management are by no means three subsystems detached from each other; rather, each higher level envelops those below it. However, a rigid view of embedment would be too static. If a normative framework expresses a certain identity, this encapsulates a huge set of possible strategies. However, at a certain stage, strategy making may find new ways of relating to the environment, which may reach out beyond the borders of the identity defined by the actual normative framework. In many companies, such attempts are out of the question a priori, - a trait, which sooner or later turns out to be pathological. Also the distinctive features of an identity and its normative implications must be reviewed over time. Most industries are subject to fundamental change; boundaries between industries collapse, so organizational intelligence demands reinventing the company, forgetting („unlearning") outdated recipes of success and building new competencies. Constant creative tension between normative management and strategy making are necessary for a company to evolve. The pertinent connection is not an algorithmic one, but it must express itself in a strenuous process of organizational discourse. In a large empirical study, Collins and Porras have given an empirical account of this „dynamic interplay between core ideology and the drive for progress" (Collins/Porras 1994: 85).

Figure 1 further demonstrates that the relevant time horizon increases from operational to normative management. At the same time, the factual horizon is also extended, as is the complexity which is to be coped with. The dotted arrow indicates that certain principles relevant to normative management (e.g. ethical and aesthetic ones) are largely timeless.

[6] The terms in apostrophes should be read with the prefix "self-", i.e. as "self-control by (self-)development", "self-control by learning", "self-control by (self-)transformation", to indicate the priority of intrinsic control (or eigen-control) over extrinsic control.

[7] In logical terms, the strategic level offers a metalanguage to the operative, the normative to the strategic and the operative levels.

The diagram also shows that the concerns of the higher levels are not detached from those of the lower ones. A company can only survive if it is in possession of value potentials that are actualised; that is, converted into value. An equilibrium between the sacrifices incurred in building up value potentials and reaping their fruits is a further necessary precondition of viability which normative management has to ensure.

At the three levels of management, different *criteria of organisational fitness* - or, to speak more generally, *systemic effectiveness* - apply:

1. 1.at the operational level, the criterion is that of efficiency, mainly in terms such as productivity, profitability and quality;
2. 2.at the strategic level, it is effectiveness in both the competitive and the co-operative sense;
3. 3.at the normative level it is legitimacy, defined as the ability to fulfil the claims of all relevant stakeholders.

The key duty of an integral or systemic management is to meet all three requirements in the long run. In order to achieve such a delicate task a corporation will require - as has been illustrated in detail in Schwaninger (1993) - considerably higher developed mental models than established ones, and more complex control systems than the simple feedback systems traditionally used.

The logical hierarchy of control variables delineated here results in a *multilevel* control structure (cf. Figure 2), which makes clear that one and the same state of affairs cannot be pre-controlled by means of the variable by which it is controlled[8]. If, at the operative level of a firm, control is exercised in the interests of profit - by means of revenues and costs - then profit cannot be pre-controlled by means of these traditional accounting variables. For pre-control other variables are required: Extant and new value potentials, which in turn are pre-controlled by viability and development. This is why I maintain that profit is not a strategic goal. A closure of the hierarchy of control variables is provided insofar as (final) liquidity outcomes at the operational level not only result from higher level pre-controls, but they also pre-control higher level „parameters" to a certain extent. For instance, not only the innovative capabilities, but also the actually available liquid financial resources might play an important role: at least, substantial own funds can add a degree of freedom in the building-up of value potentials.

[8]It has been argued (cf. Jackson 1989: 428) that the use of 'control models' prevents organisational learning, leading to 'increasing dominance of history' (De Zeeuw 1986: 139). A concept of control that embraces development as a goal defies this argument in principle, although it cannot guarantee that there will be learning in every organisation that uses the concept.

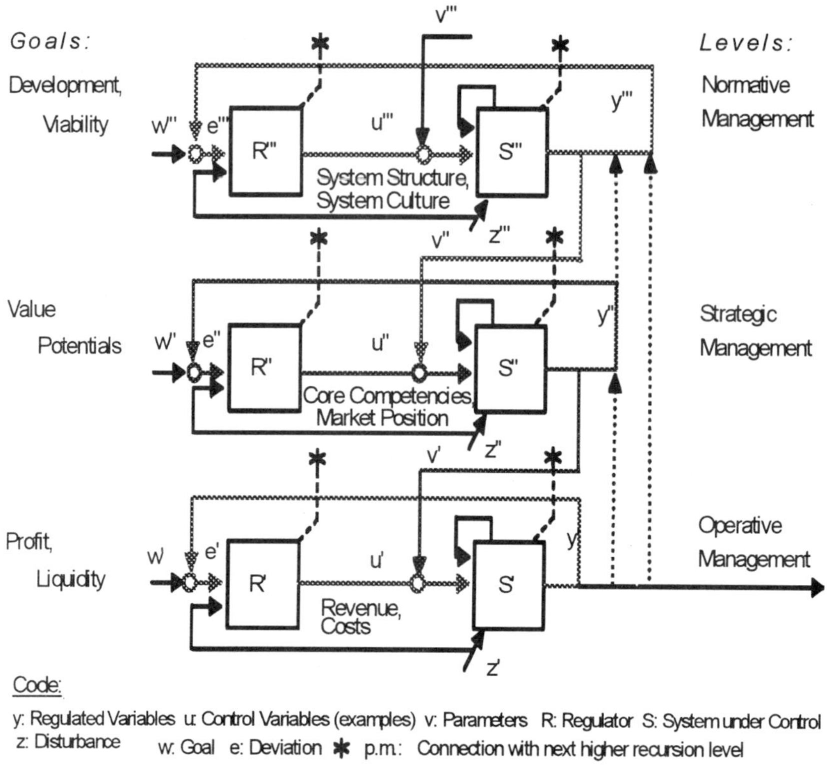

Figure 2: Control as a multilevel cycle with controls and pre-controls

4. INTRINSIC VERSUS EXTRINSIC CONTROL

Cybernetics originated from the world of engineering. The early applications were centered on mechanical systems. With the study and use of cybernetics in the context of social systems some conceptual reframing took place. Social systems are constituted by individuals or groups which have their own goals and adhere to their own values. They are self-conscious and reflexive. Therewith, social systems are capable of self-control, self-reference and self-transformation. This induced a conceptual distinction between first order cybernetics and second order cybernetics. In first order cybernetics, the controller is seen as outside of the system (figure 3). The system is purposive[9], in the sense that goals are established from outside or

[9]Cf. the distinction between purposiveness, as „the effectiveness and efficiency of means or tools", and purposefulness, as „the critical awareness of self-reflective humans with regard to ends or purposes and their normative implications for the affected" (W. Ulrich 1983: 328).

"above". This view is applicable to mechanical systems, but inadequate to capture the purposeful, intentional and also largely voluntaristic character of individuals and groups which constitute modern organizations.

In second order cybernetics the controller is an integral part of the system. The system is purposeful, i.e. it develops ist goals internally and it is made up of subsystems which pursue their own goals. In this conception, goal formation is a process of conversation and reciprocal alignment between a system and the larger whole into which it is embedded. Heinz von Foerster, the eminent cybernetician, defines the difference as follows (von Foerster et al.1974: 1):

- First order cybernetics: "the cybernetics of observed systems"
- Second order cybernetics: "the cybernetics of observing systems" .

Not only the human and social nature of organizations, but also the complexity faced by them makes it necessary to emphasize intrinsic control

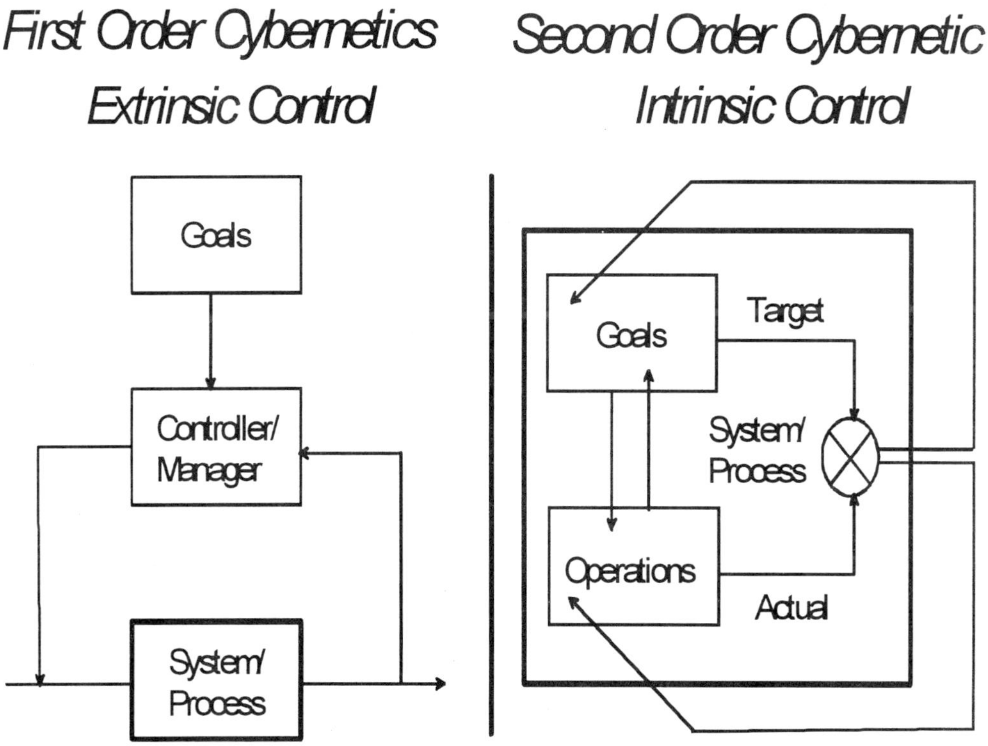

Figure 3: Control according to first and second order cybernetics

(the terms "intrinsic control", "eigen-control" and "self-control" are used synonymously here) as opposed to extrinsic control . If intrinsic control is to predominate, this must lead to a recursive process of intrinsic control, in the case of organizations with several layers of basic units, i.e. units which create the value that is at the core of the organization's mission (e.g. divisions - business units – business segments).

5. A MODEL OF RECURSIVE CONTROL

This section will deal with the multi-recursion issue, which is at the heart of distributed control. The structure-theoretical foundation for distributed control can be found in the Viable System Model (VSM) by Stafford Beer, the founder of management cybernetics. This model answers the question, how an organization must be structured to make its (self-) control possible in a way that its viability and development can be maintained.

A set-theoretic model, in which Beer (1962) defined the organizational prerequisites for the viability of systems, was later operationalized in a topological model, known as the „VSM“ (Beer 1979, 1981, 1985). In this model, a set of functions is distinguished, which ensure the viability of any human or social system (Figure 4). These functions and their interrelationships are specified in a comprehensive theory, the propositions of which can be summarized as follows:

1. 1.) An organization is viable if and only if it disposes of a set of management functions with a specific set of interrelationships, identified and formalized in the model:

- System 1: Regulatory capacity of the basic units, optimization of ongoing activities by which value is created (e.g. the business areas of a private or public enterprise).
- System 2: Attenuation and amplification to damp oscillations and coordinate activities via information and communication (e.g. information systems, service units and coordination teams, standards of behavior)
- System 3: Establishing an overall optimum among basic units, providing for synergies, resource allocation (e.g. an executive corporate management)
- System 3*: Investigation and validation of information flowing between Systems 1-3 and 1-2-3 via auditing/ monitoring activities (e.g. internal auditing, „management by wandering around“)
- System 4: Dealing with long term and the overall outside environment, diagnosis and modeling of the organization in its environment (e.g.

corporate development, strategy, research and much of knowledge management)

- System 5: Balancing present and future as well as internal and external perspectives; moderation of the interaction between Systems 3 and 4; ascertaining the identity of the organization and its role in its environment; embodiment of supreme values, rules and norms, - the ethos of the system (normative management).

Summing up, Systems 1-2-3 (including 3*) make up the operative management, System 4 (in interaction with System 3) strategic management, and System 5 the normative management of the organization.

In this structure, the primary units (basic units with the regulatory capacity supplied by System 1) must dispose of high autonomy in order to be able to adapt to their respective environment or milieu. The combined activities of Systems 1, 2 and 3 (including 3*) provide for cohesion, while System 4 is the fulcrum for long-term adaptation, and System 5 the embodiment of the ethos, - the basic principles and values governing the orientation of the organization as a whole.

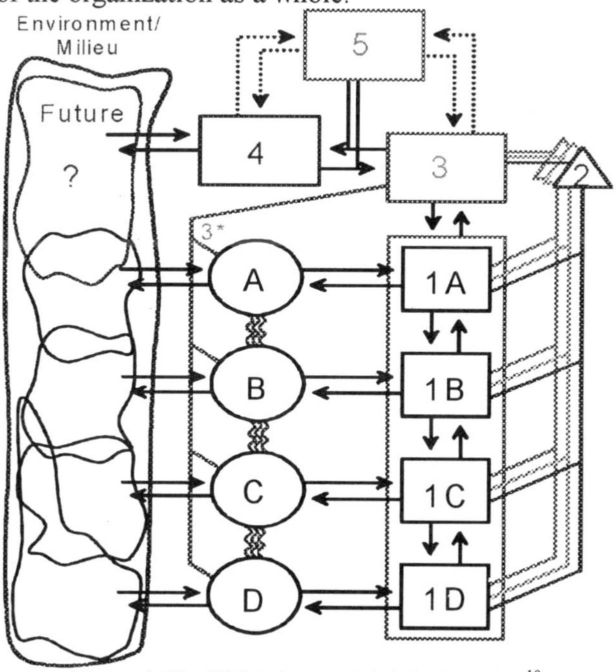

Figure 4: The Viable System Model - Overview[10]

2.) Any deficiencies in this system, such as missing functions, insufficient capacity of the functions or faulty interaction between them impair or jeopardize the viability of the organization.

[10] This is a slightly adapted version. For the full-fledged original, please see: Beer 1985: 136; and: Beer 1979, 1981.

The viability, cohesion and self-organization of an enterprise depend upon these functions being recursively present at all levels of the organization (Figure 5). A recursive structure comprises autonomous units within autonomous units. Moreover, a viable organization is made up of viable units and it is itself embedded in more comprehensive viable units. Each unit, inasmuch as it is producing the organization's task, rather than servicing or supporting this producing, replicates - in structural terms - the totality in which it is embedded: It has all the functions outlined under (1.), to be able to manage, from start to finish, the processes for the purpose of which it exists. If we take such a viable organization as a „system-in-focus", depending on the perspective adopted, it „may have more than one next higher and next lower recursion" (Beer 1985; for a pertinent application, see Leonard 1989).

According to the VSM all three levels of management – operative, strategic, normative – are distributed functions; control, intelligence and ethos being properties of the system as a whole, inherent in all levels of recursion. This model has been transduced into the language of business and also been widely applied, as documented in several books (e.g. Espejo/Harnden 1989, Espejo/Schwaninger 1993, Espejo/Schuhmann/ Schwaninger 1996), and on a CD-ROM edited as a festschrift for Stafford Beer (Espejo/Schwaninger 1998). One of my doctoral candidates has also applied it to the Swiss democracy and discerned important insights (Willemsen 1992).

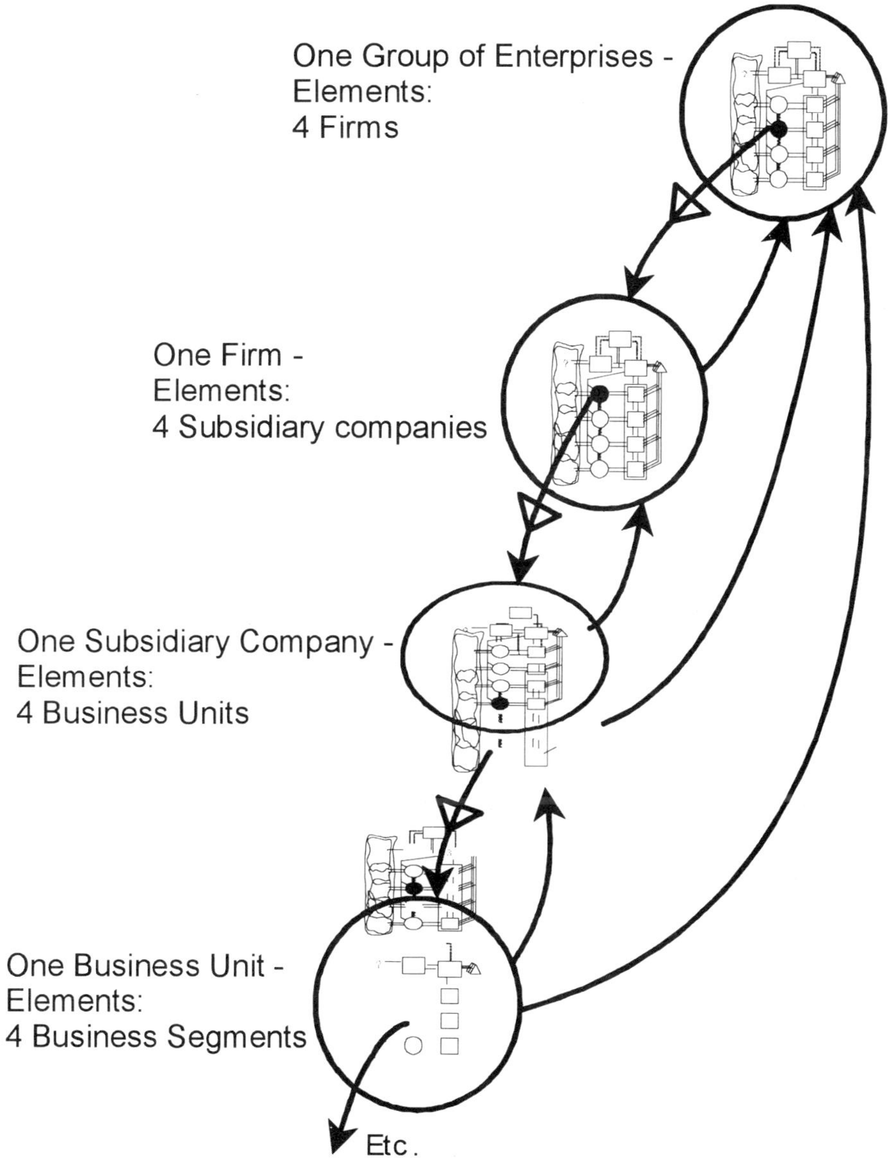

Figure 5: Recursive structure of viable organizations (after Beer 1979)

6. PUTTING THE THREADS TOGETHER

As shown in section 5, the viability of an organization is determined, first of all, by whether the organization has command over certain specific management functions which work together in a manner precisely defined by Beer's theory (cf. Figure 4).

Secondly, this system of management functions can be ascertained recursively. Viability is, in principle, to be structurally anchored down to each work-team or to each and every workplace.

The most important consideration at this point is as follows. There is need not only for operational but also for strategic and normative management in any subsystem of an organization, in so far as it has to meet the criterion of viability.

The cybernetic theory of organization advocated here contradicts the oft-repeated comments that "vision is the concern of the entrepreneur" or "strategy is the duty of the board of directors". Vision is a function of the meta-system: to be precise, it is one of the functions of the normative management of every viable unit. Strategic thinking is necessary even in the smallest units, if such units are conceived as viable wholes.

That is the reason why "meta-systems" have to turn to orientators other than those of the corresponding "systems". In small units, it is often the case that the operational, strategic and normative functions of management as well as the basic activities are discharged by the same persons. In spite of this, it is vital for any unit in focus that it does not restrict itself to the operating functions (always under the assumption that the unit concerned is supposed to be viable). Such a system must be so led that the control variables at all the three levels are taken into consideration and simultaneously kept under control. This holds true even when contradictions arise between these control variables.

It is incumbent upon the management at each recursion level to define for its purposes, the specific orientators in a level-appropriate manner and to realize a correspondingly "integral" leadership. Figure 6 illustrates this aspect in a highly simplified manner. Starting from a more abstract scheme of the control variables than in Figure 1, regulatory variables and orientators are presented as examples which could be of significance for three recursion levels consisting of an enterprise, a subsidiary company and a business unit.

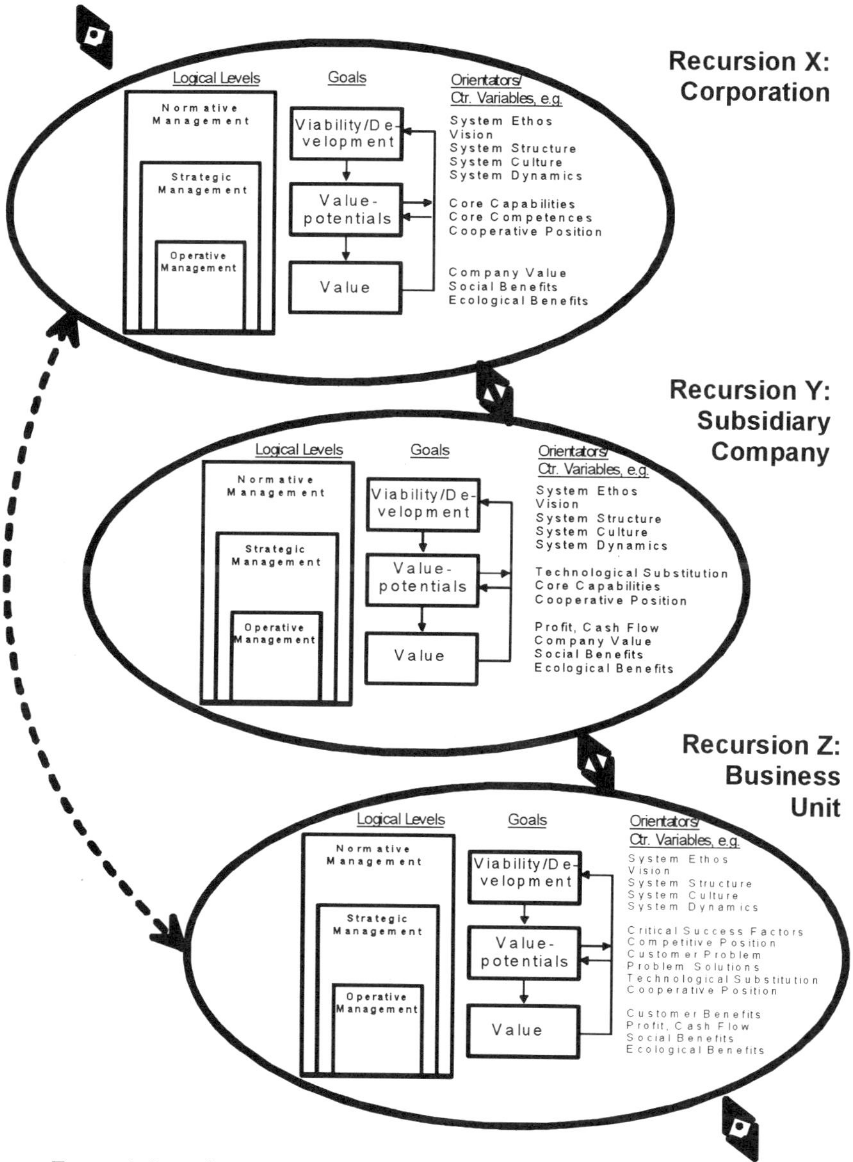

Figure 6: Operative, strategic and normative management are distributed functions

With the help of the broken line an attempt is made to illustrate the point that for a given unit, information about units at recursion levels other than those in the immediate neighbourhood can be of relevance. The dotted line indicates that in principle the logic of the presentation can be transferred to further recursion levels and holds good for other structured, more complex networks.

The same logic applies also to "virtual organizations". In principle, a virtual organization (from "virtus" meaning virtue) is capable of materializing its potential through the most diverse variants. This would mean, for example, that out of the available resources, project teams are formed in many different constellations, tailor-made according to the task that needs to be accomplished. In this connection, the meaning of the word "available" can be expanded almost indefinitely. What is not available "in house" can mostly be acquired through partnerships and cooperation. The goals "viability" and "development" can also be meaningful in connection with projects (compare sections 3 and 4.2). Even projects having a limited life-span need a specific identity during the period of their existence. Looking at the situation from the inside, the question "what is to become of us when the project is completed" can be of burning importance for the members of the project team.

Therefore, the system of regulatory variables for an integral management illustrated in Figure 6 can also be transferred to the project level (Figure 7).

A more detailed account of recursive management and an underlying cybernetic theory of human action in organizations is given in a book on organizational transformation and learning (Espejo/Schwaninger/Schuhmann/Bilello, 1996).

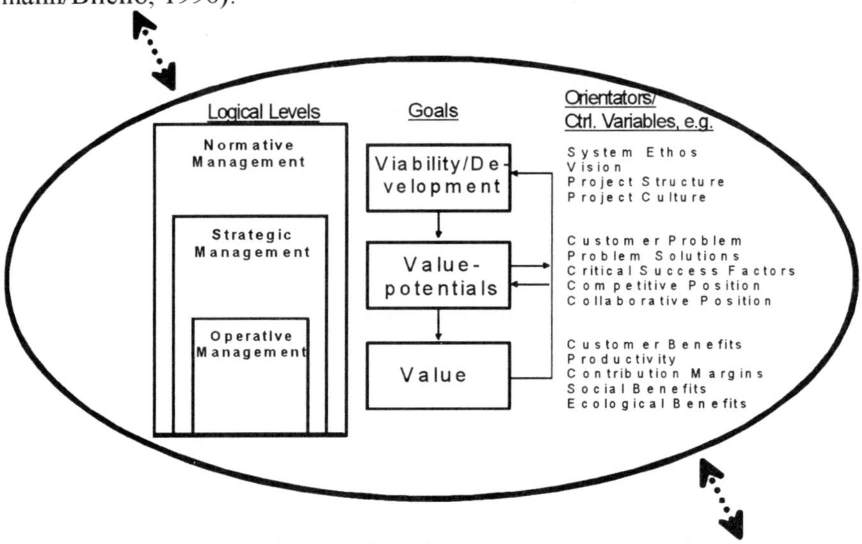

Figure 7: Orientators for an integral management of projects

7. FROM HIERARCHY TO HETERARCHY

It must be added that this concept of layers of basic units, in the face of market complexity, is a multidimensional one. Therefore, in the recent literature on organization, heterarchical networks are being considered as more suitable than "tree-like" hierarchical structures, for handling issues in an environment characterized by uncertainty, complexity and dynamics(Gomez/Zimmermann 1997; Hedlund 1986, 1993; Probst 1992; Schwaninger 1994b; Weber, 1994, and other literature cited therein). Such organizations possess, in principle, a greater capacity for absorbing complexities. To express it in the language of cybernetics, heterarchical organizations are superior to hierarchical systems as far as their "redundancy of potential command" (McCulloch 1988) is concerned, i.e. in their excess capacity to control themselves. However, networks are not necessarily heterarchies, although all heterarchies are structured in the fashion of networks. For this reason, it is important in what follows to make precise the concept of heterarchy in contrast to that of hierarchy. Both of these concepts are derived from the Greek "archein" which means 'to reign'; hierarchy (derived from the word "hieros", meaning holy, belonging to the gods, priestly) stands for the command of the few, "initiated" ones, and heterarchy (derived from the word "heteros", meaning different, many) refers to the command vested in different and distributed instances.

The structure and the culture of an organization are very closely interconnected. The ideal-types of two models of organization are presented below (Figure 8).

Each of them embraces both of the aspects mentioned, namely the cultural and the structural:

- - "HAT": the hierarchical-authoritarian type,
- - "HPT": the heterarchical-participative type.

Hierarchical structures are associated with autocratic patterns of behaviour including monological commands emanating from one single centre. As opposed, a heterarchical organization can have many centres of command and it is characterized by a culture involving a high degree of participation and dialogue. In such a form of organization, participation and "empowerment" are so deeply entrenched that any unit whatsoever within the organization is in a position to take command of the entire organization, with reference to a specific aspect, while at the same time in a subordinate role with reference to other aspects (cf. Hedlund 1986, 1993).

The HAT is characterized by a high degree of control (extrinsic control) and very little autonomy in the workplace. The emphasis is on long-lasting structures and a high division of labour. The HPT, in contrast, endows units at all levels with considerable autonomy, which encourages high levels of

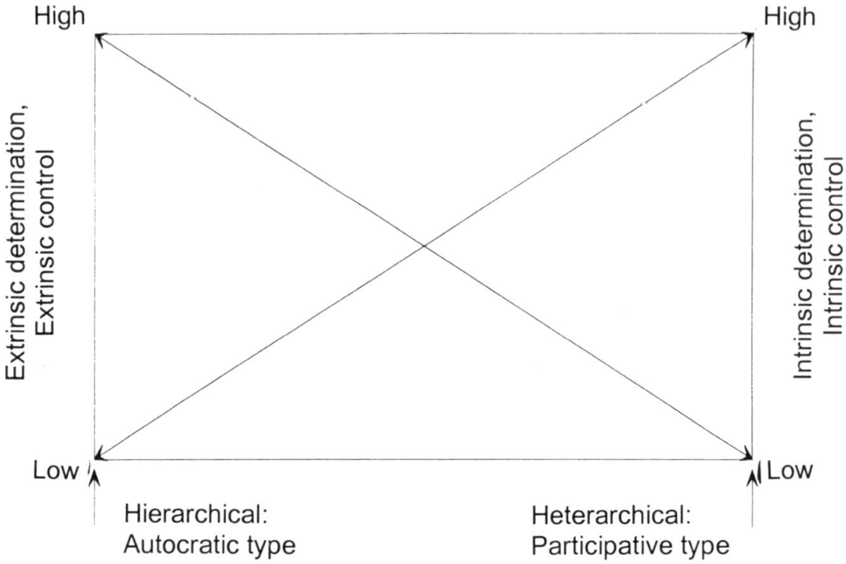

Figure 8: Two ideal-types of organization

self-determination and intrinsic control (Figure 8). Here, the emphasis is laid on temporary structures, on multiplication and recombination. While hierarchies are strong at utilizing available resources, the power of heterarchies lies in their flexibility and innovativeness.

At this point it needs to be added that in reality, organizations do not conform rigorously to one or the other of these ideal-types. Hierarchical systems will always exhibit some elements of a dialogue, even if they operate in a rather placid environment. On the other hand, network-type heterarchical systems will never be completely free of some characteristics of hierarchies, since certain units within the organization will have to assume the global responsibility in some specific area in their relation to the superordinate systems. Earlier publications by the author have elaborated on this aspect (Schwaninger 1994b, 1996, 1999).

At this point, it is important to emphasize that the framework presented here is fully compatible with the need for heterarchical control. In the first place the control structure outlined is essentially one of a network type (see figures 2 and 3).

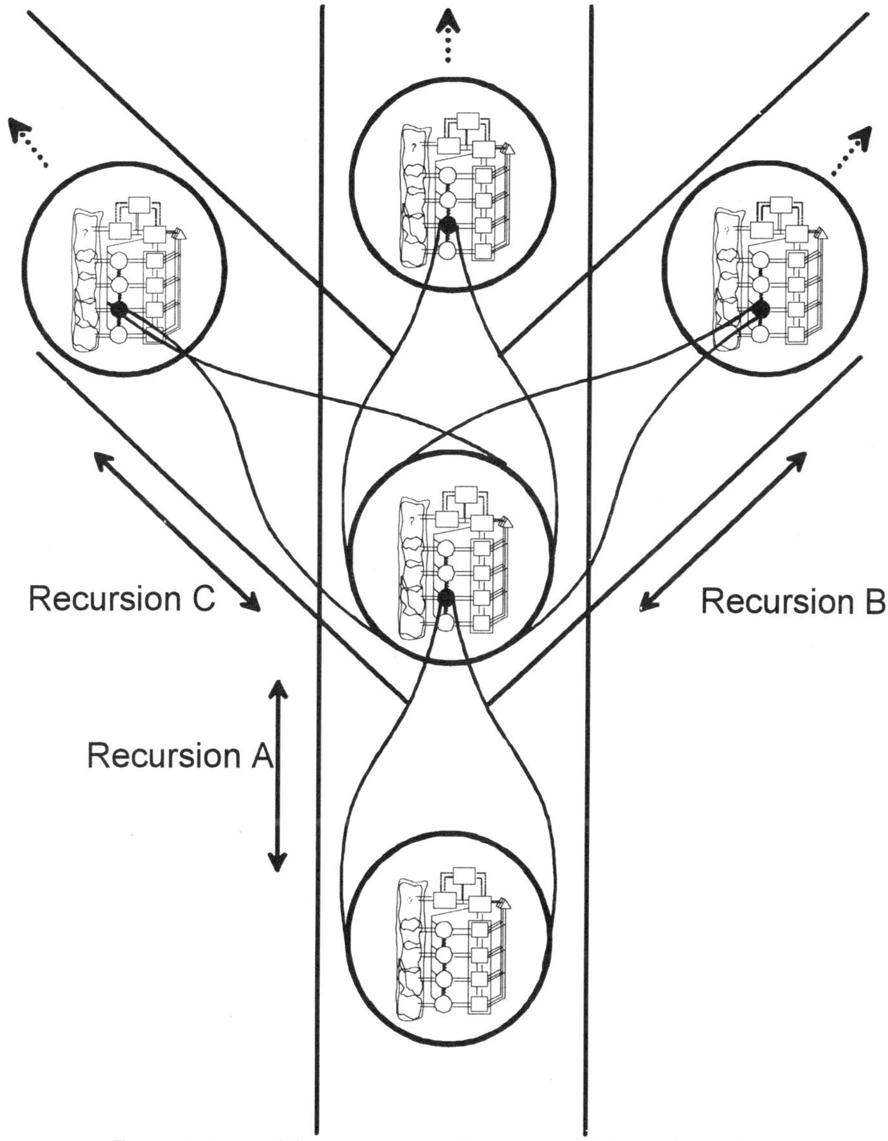

Figure 9: One and the same organization as part of different recursions

Secondly, the concept of recursive control is multidimensional, and therewith heterarchical. One and the same organization or organizational unit can function simultaneously both as a sub-system and a super-system within the framework of different recursive organizational configurations (see Figure 9). Not only can it "function" in this way, but it can also be conceived or can perceive itself in this fashion. A simple example illustrating this point is an enterprise which is a part of a concern, and is itself composed of several divisions (recursion A in Figure 9). In order to cope with ecological challenge, the firm in focus joins other enterprises to form a suitable association (recursion B). Additionally, this enterprise is a member of a consortium for research and development (recursion C). It is possible to conceive of other recursions. A recursion D, for example, can be introduced in the form of companies in which the enterprise has a share, to keep the enterprise close to innovative developments in its areas of interests.

Thirdly, networks should – at least as far as their durable manifestations are concerned – be conceived as viable systems, and viable systems are also networks of networks, where these networks can (although they need not) be of a heterarchical nature. Pertinent examples were given elsewhere (Schwaninger 1994b, 1996, 1999),

Finally, as already shown earlier (Schwaninger 1994b), recursions do not necessarily run "from top to bottom". In a system of concerted mutual controls, e.g. in the context of alliances and holdings, it is possible to have a configuration as illustrated in Figure 10. This scheme, which shows a subset of a more comprehensive network, illustrates how companies can be "locked" into each other across the recursion levels V to Z. The company at the recursion level W is a subsidiary of the company at the recursion level V, and so on. The company at the recursion level Y is a joint venture of companies at the recursion levels X1 and X2.

A paradox crops up in this scheme in the sense that company 1 is at once "great-grandmother" and "daughter" of company 5. An analogous situation characterizes companies 2 to 5. This situation is in conformity with the paradox of the recursion levels relating to each other as follows:

$$V > W > X1 > Y > Z < V$$

It is obvious that the solution of the paradox lies in the fact that here we have a case of a circular recursion. (From a certain point of view, this can also be called a self-referential recursion).

Figure 10: Recursive and heterarchical relationships of control are compatible

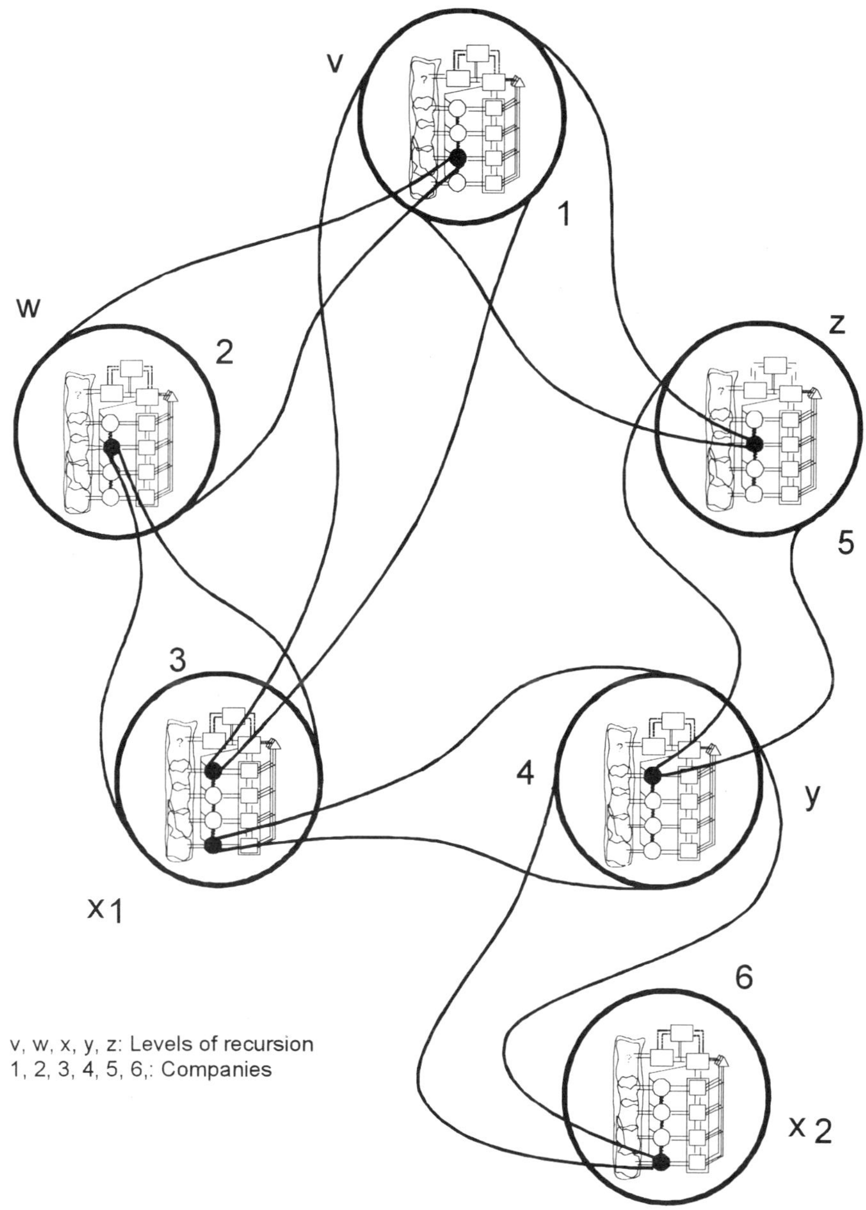

8. CONCLUSIONS

As demonstrated, the viability of an organization rests on multilevel control in recursive structures. "Viability" as understood here means more than "autopoiesis" (self-production). This term stands for more than survival at any price: Intelligent organizations are committed to the goal of "development". In borderline cases, they eliminate themselves when, with their given identity, they are not capable of fulfilling meaningful functions in the service of the more comprehensive whole, (i.e. in the service of the systems at the next higher recursion level in which they are embedded, to put the matter in the language of cybernetics). In the last consequence, intelligent organizations strive for viability beyond survival.

As empirical studies suggest, viability is also grounded in certain recursive ("self-similar") socio-cultural properties. Kotter and Heskett subsume them under an ideal-type which they call "adaptive culture"[11] (Kotter/ Heskett, 1992; see also: Espejo/Schuhmann/Schwaninger/Bilello, 1996, Chapter 8; Schwaninger, 1989).

Organizational viability is recursive, and organizational intelligence is recursive, as well. If organizational viability is a recursive property, it requires recursive organizational intelligence as a necessary prerequisite. The key to organizational intelligence is a distributed control in both senses: It must be a dominantly self-controlled process of recursively viable units, which in each case maintain the control variables of all three logical levels – operative, strategic, and normative – under control, simultaneously. As shown, these propositions are compatible with the need for heterarchical control in complex settings.

Finally, it must be said, that the models outlined and integrated in this paper have been applied in all kinds of organizations. This has been documented in an extensive literature There is a large body of writings which document applications of the VSM, e.g. Beer (passim), Espejo/Harnden 1989, Espejo/Schwaninger 1993, Brocklesby & Cummings 1996. A smaller set of sources on integrated applications of both models is also available (Schwaninger 1988, 1989, 1993). Many of these sources contain empirical evidence about the effectiveness of the models and framework discussed in this paper.

[11] In a nutshell this is a culture that emphasizes close attention to all the key constituencies of the firm (customers, stockholders, employees). It also values people and processes that create change when needed to serve the legitimate interests of key constituencies. Finally, adaptive cultures are based on a leadership from managers at all levels, who convey these concerns convincingly to their staff.

9. REFERENCES:

Ackoff, Russell L. 1981. Creating the Corporate Future, New York etc.: Wiley.

Ackoff, Russell L. 1994. The Democratic Corporation, New York/Oxford: Oxford University Press.

Argyris, Chris/Schön, Donald A. 1978. Organizational Learning: A Theory of Action Perspective, Reading, Mass.: Addison-Wesley.

Bamberger, Peter & Fiegenbaum, Avi 1996. The Role of Strategic Reference Points in Explaining the Nature and Consequences of Human Resource Strategy, in: Academy of Management Review, Vol. 21, No. 4 (926-958).

Beer, Stafford 1962. Towards the Cybernetic Factory, reprinted in: Harnden, Roger/Leonard, Allenna, eds., 1994: How Many Grapes Went into the Wine. Stafford Beer on the Art and Science of Holistic Management, Chichester: Wiley.

Beer, Stafford 1979. The Heart of Enterprise. Chichester: Wiley.

Beer, Stafford 1981. Brain of the Firm, 2nd edition, Chichester: Wiley.

Beer, Stafford 1985. Diagnosing the System for Organizations, Chichester: Wiley.

Beer, Stafford 1994. Beyond Dispute. The Invention of Team Syntegrity. Chichester: Wiley.

Bowersox, Donald J., et al. 1982. Logistical Excellence, Burlington, MA: Digital Press.

Brocklesby, John & Cummings, St. 1996. Designing a Viable Organization Structure, in: Long Range Planning, Vol. 29, No. 1, (49-57).

Buzzell, Robert D., Gale, Bradley T. 1987. The PIMS Principles, New York, London: Free Press, Collier.

Cameron, Kim S. 1984. The Effectiveness of Ineffectiveness, in: Staw, Barry M. & Cummings, L.L., eds.: Research in Organizational Behavior, Vol. 6, Greenwich, Connecticut/London England: JAI Press (235-285).

Collins, James C./Porras, Jerry I. 1994. Built to Last. Successful Habits of Visionary Companies, Harper Business.

Copeland, Thomas E./Koller, Tim/Murrin, Jack 1994. Valuation: Measuring and Managing the Value of Companies, 2nd ed., New York etc.: Wiley.

Deal, Terrence E., Kennedy, A.A. 1982. Corporate Cultures. The Rites and Rituals of Corporate Life, Reading, Mass.: Addison-Wesley.

Espejo, Raúl/Harnden, Roger, eds. 1989. The Viable System Model. Interpretations and Applications of Stafford Beer's VSM, Chichester: Wiley.

Espejo, Raúl/Schuhmann, Werner/Schwaninger, Markus & Bilello, Ubaldo 1996, Organizational Transformation and Learning. A Cybernetic Approach to Management, Chichester: Wiley.

Espejo, Raúl/Schwaninger, Markus, eds. 1993. Organizational Fitness - Corporate Effectiveness through Management Cybernetics, Frankfurt/New York: Campus.

Espejo, Raúl/Schwaninger, Markus, eds. 1998. To Be and Not to Be, that is the System. A Tribute to Stafford Beer, CD-ROM, Wiesbaden: Carl-Auer-Systeme-Verlag.

Etzioni, Amitai 1968. The Active Society. A Theory of Societal and Political Processes, London, New York: Collier, MacMillan.

Foerster, Heinz von 1984. Observing Systems, 2nd ed., Seaside, Cal.: Intersystems Publications.

Foerster, Heinz von, et al., eds. 1974. Cybernetics of Cybernetics, Urbana, Illinois: The Biological Computer Laboratory, University of Illinois.

Freeman, Edward R. 1983. Strategic Management: A Stakeholder Approach, in: Lamb, Robert (ed.): Advances in Strategic Management, Greenwich: JAI Press, (31-60).

Gälweiler, Aloys 1990. Strategische Unternehmensführung, 2nd edition, Frankfurt/New York: Campus.

Geyer, R. Felix/van der Zouwen, Johannes, eds. 1978. Sociocybernetics, 2 Vols., Leiden/Boston/London: Martinus Nijhoff Social Sciences Division.

Gomez, P. & Zimmermann, T. 1997. Unternehmensorganisation. Profile - Dynamik - Methodik, St. Galler Management-Konzept (3rd ed.), Frankfurt, New York: Campus. (1st ed. 1992).

Hamel, Gary/Prahalad, C.K. 1994. Competing for the Future, Boston, Mass.: Harvard Business School Press.

Hansen, G.S., Wernerfelt, B. 1989. Determination of Firm Performance: The Relative Importance of Economic and Organizational Factors, in: Strategic Management Journal, 10 (399-411).

Hedlund, G. 1986. The hypermodern MNC - A heterarchy? Human Resources Management. 25(1) 9-35.

Hedlund, G. 1993. Assumptions of hierarchy and heterarchy, with applications to the management in the multinational corporation. In S. Ghoshal & E.Westney (Eds.), Organization theory and the multinational corporation: 211-236. London: MacMillan.

Jackson, Michael C. 1989. Evaluating the Managerial Significance of the VSM, in: Espejo, Raúl/ Harnden, Roger, eds.: The Viable System Model. Interpretations and Applications of Stafford Beer's VSM, Chichester: Wiley (407-439).

Jantsch, Erich/Waddington, C.H., eds. 1976. Evolution and Consciousness: Human Systems in Transition, Reading, Mass.: Addison-Wesley.

Kotler, Philip 1994. Marketing Management: Analysis, Planning, Implementation, and Control, 8th edition, Englewood Cliffs: Prentice-Hall.

Kotter, John/Heskett, James 1992. Corporate Culture and Performance, New York: Free Press.

Leonard, Allenna 1989. Application of the VSM to Commercial Broadcasting in the United States, in: Espejo, Raúl/Harnden, Roger, eds.: The Viable System Model, Chichester: Wiley (175-209).

Leonard, Allenna 1995. Team Syntegrity: Planning for Action in the Next Century, Working Paper, Toronto: The Complementary Set.

Lumpkin, G.T./Dess, Gregory G. Clarifying the Entrepreneurial Orientation Construct and Linking it to Performance, in: Academy of Management Review, Vol. 21, 1996, No. 1 (135-172).

McCulloch, Warren. 1988. Embodiments of mind (reprint). Cambridge, Mass./London: M.I.T. Press.

Meyer, Marshall W. & Gupta, Vipin 1994. The Performance Paradox, in: Staw, Barry M. & Cummings, L.L., eds. Research in Organizational Behavior, Vol. 16, Greenwich, Connecticut/London England: JAI Press (309-369).

Mitroff, Ian I. 1983. Stakeholders of the Organizational Mind, San Francisco etc.: Jossey Bass.

Nonaka, Ikujiro/Takeuchi, Hirotaka 1995. The Knowledge Creating Company, Oxford: Oxford University Press.

Porter Michael E. 1980. Competitive Strategy, New York: Free Press.

Porter, Michael E. 1985. Competitive Advantage, New York: Free Press.

Prigogine, Ilya 1989. The Philosophy of Instability, in: Futures (396-400)

Rappaport, Alfred 1986. Creating Shareholder Value: The New Standard for Business Performance, New York: Free Press.

Probst, G.J.B. 1992. Organisation. Landsberg a. Lech: Moderne Industrie.

Pümpin, Cuno 1991. Corporate Dynamism, Aldershot: Gower.

Rappoport, Alfred 1996. Creating Shareholder Value: The New Standard for Business Performance, New York: Free Press.

Schein, Edgar H. 1985. Organizational Culture and Leadership, San Francisco/London: Jossey-Bass.

Schoemaker, Paul J. H. 1992. How to Link Strategic Vision to Core Capabilities, in: Sloan Management Review, Fall (67-81).

Schwaninger, Markus 1987. A Practical Approach to Strategy Development, in: Long Range Planning, Vol. 20, No. 5, (74-85).

Schwaninger, Markus 1988. Anwendung der integralen Unternehmungsentwicklung, Bern und Stuttgart: Haupt.

Schwaninger, Markus 1989. Integrale Unternehmungsplanung, Frankfurt/New York: Campus.

Schwaninger, Markus 1993. A Concept of Organizational Fitness, in: R. Espejo/M. Schwaninger, eds. Organizational Fitness - Corporate Effectiveness through Management Cybernetics, Frankfurt/New York: Campus, (39-66).

Schwaninger, Markus 1994. Die intelligente Organisation als lebensfähige Heterarchie, Diskussionsbeiträge des Instituts für Betriebswirtschaft an der Hochschule St. Gallen, Nr. 14.

Schwaninger, Markus 1994a. Managementsysteme, Frankfurt/New York: Campus.

Schwaninger, Markus 1996. Structures for Intelligent Organizations, Diskussionsbeiträge des Instituts für Betriebswirtschaft an der Universität St. Gallen, Nr. 20.

Schwaninger, Markus 1998. Managing Complexity - The Path Towards Intelligent Organizations, Contribution to: Complex Systems Revisited, Conseil Suisse de la Science, Hearing at the University of Neuchâtel, Switzerland, September 4 and 5 (Publication forthcoming: Th. Bernold, editor).

Schwaninger, Markus 1999 (in print). Intelligente Organisationen: Strukturen für organisationale Intelligenz und Kreativität, In: Papmehl, André/Siewers, Rainer, Hrsg.: Wissen im Wandel. Die lernende Organisation im 21. Jahrhundert, Wien: Ueberreuter,

Senge, Peter M.1992. The Fifth Discipline. The Art and Practice of the Learning Organization, London: Century Business.

Ulrich, Werner 1983. Critical Heuristics of social Planning. A New Approach to Practical Philosophy, Bern/Stuttgart: Haupt.

Underwood, Lynn 1994. Intelligent Manufacturing, Wokingham etc.: Addison-Wesley.

Utterback, James M. 1994. Mastering the Dynamics of Innovation, Boston, MA: Harvard Business School Press.

Venkatraman, N./Ramanujam, Vasudevan 1998. Measurement of Business Performance in Strategy Research: A Comparison of Approaches, in: Academy of Management Review, Vol. 11, No. 4 (801-814).

Weber, B. 1994. Unternehmungsnetzwerke aus systemtheoretischer Sicht. Zum Verhältnis von Autonomie und Abhängigkeit in Interorganisationsbeziehungen. Diskussionsbeiträge, Nr. 9. Hochschule St. Gallen: Institut für Betriebswirtschaft.

Willemsen, Maarten Helmut 1992. Die Schweizerische Eidgenossenschaft als lebensfähiges System. Kybernetische Analyse des schweizerischen politischen Systems, Dissertation Nr. 1314, Hochschule St. Gallen.

Zeeuw, Gerard de 1986. Social Change and the Design of Enquiry, in: Geyer, Felix/van der Zouwen, Johannes

Chapter 10

Space of state and performance's optimisation

Lessons of Russians' failures

VLADIMIR DEGTIAR
Social Informatic Department of Moscow State University

1. INTRODUCTION

When "reorganisation" began in Russia a factory from the middle region of Russia invited Japanese experts to help decide where to begin. These experts looked over a shop of a factory. They saw a neglect of the equipment, a cluttering the shop and a lot of refuse. They advised the employees to begin by cleaning this shop and putting the equipment in order. An analogy to medicine it is like to a doctor advising to a bed-patient to begin treatment with dances. It displays the incompetence of the experts. A real doctor would endeavour to make a diagnosis. If he found out that a source of illness is not in an infrastructure, but in the people, which by the existing mechanism of social control were so alienated from manufacture and so hated authority manipulating them that they could not force themselves to carry out obvious actions. He would advise them to change their system of social control, so that it would help the following step to change their state. If he were really an expert, he would also write out a prescription for them showing how to make this work.

The experts in the fields of economy and management have got used to considering that if there is a purpose and a plan for its realisation, all that is necessary for its achievement is a presence of enough material resources. For example, if we want to optimise the performance of organisation, it is enough to have the optimal plan and the necessary resources. It would be fair if the set of the executors could be limited to ourselves. However in the economy we deal with rather large organisations. This demands a necessity to considerably expand the set of executors.

The Performance of Social Systems
Edited by Francisco Parra-Luna, Kluwer Academic/Plenum Publishers, 2000

The purpose of this paper is to show that *the basic problem* is not contained in the development of the optimal plans and the receiving of the investments for their realisation. It is included in the creation of a system of social control capable to diagnostics of the states of organisations and their transition into a state necessary for the performance of these plans. Either we shall change the system of social control with the purpose to realise the social transformations necessary for a society, or the complete failure with huge waste of means or lives lie in wait for us. The failures of the Russian reforms will be considered as an example.

2. SPACE OF STATES AND DIAGRAM OF STATES

The set of states of "bioobject" is formed at all levels of its organisation and all temporary scales of its life. A state should be described by some area of change of values of a vector in space of coordinates (attributes) describing the state [4]. These vectors describe *a space of states*. At a cell level states of object are connected to state of its genetic network, to the expression of a subset of genes. At a level of a nervous network they are connected with the development of a set of centres, at a level of organisations – with states of individuals, their systems of values and mechanisms of the social control.

Let S be a set of states of a bioobject, and Y_s be a set of strategies accessible to it in a state $s \in S$. The adjustment of the state of the bioobject on performance of a command "a" involving a choice of $y(a) \in Y_s$ is that it is necessary to transfer it from the current state s_0 to one the of states of a set K, having

$$Y_s = \bigcap_{k \in K} Y_{ks} \subseteq Y_s, \qquad (1)$$

i.e. containing s, $y(a) \in Y_s$ on the diagram of transitive states of a bioobject.

The logical diagram of the states of an object is a graph containing units – states and arches – transitions between states. The description of a state consists of three sections: 1) names of a state, 2) lists of attributes, 3) lists of operations started by event causing transition to other state [1]. The variable of a state is described by the list of attributes connected to the given state or to its substates. The values of attributes are defined for each state of the diagram of states. They influence the flows of the managing influences. The transitions between states are caused by events. To execute a command "a" at a reception of the message m(a) and to realise strategy y(a) by the mentioned object there should be a certain event before. Together with a name and attributes of the event appropriate operations Y_s initiated at the occurrence of the event are reflected on the diagram. The conditions are specified also at which the object in a state s realises y(a) at the presence of event. Thus, the occurrence of the event is necessary, but is not a sufficient condition of reaction $y(a) \in Y_s$.

To transfer an object from an initial state s_0 to a desirable state s_n it is necessary to transfer it through a set of transitive states $s_1, ..., s_{n-1}$. They are reflected in the diagram of states. For example, because of a shortage of resources or an absence of necessary strategies it is not possible at once to transform the initial ordering of needs or values of individual into a desirable state. It is necessary to transfer it through a number of orderings by performing the way certain under the diagram of transitive states. For this purpose in each state s_i it is necessary to realise certain energy operation, or influence h_i: $s_i \rightarrow s_{i+1}$. In natural scientific areas the first problem is the rather exact description of a set of transitive states and influences capable to proceed to the necessary transition. The second problem consists in the search and concentration of resources for realisation of the desirable transition.

If our purpose is that the object should have executed a command "a" (for example, optimal plan as a complex of strategies y(a)), it is necessary: 1) to define an initial state s_0 of an object and to find out whether the object in this state has a necessary complex of strategies y, 2) if no, then to reveal a set of states $K \subseteq S$ in which the object has a necessary complex of strategies $y(a) \in Y_s$, 3) to reveal events which should take place for the transition of the object from a state s_0 in each of the states of set K, and also conditions at which it will in a state s realise y(a) at presence of the event, 4) on the basis of an estimation of an opportunity of initiation of these events, the mentioned conditions of the realisation y(a) and the expenses for initiation in set K to find a state **s**, in which it is possible to transfer the object from a state s_0 with the least expenses, but by complete observance of declared ethics, 5) to transfer the object from an initial state s_0 in the state **s** through a command "a".

As conditions that should be checked after the emergence of an event, the conditions of individual rationality of the recipient should appear. It is necessary that the choice y(a) in the given moment satisfies these conditions or the sender is distinguished by the recipient as the sender of commands which with high probability will satisfy these conditions during the future, foreseeable at a temporary horizon of the recipient. If we want to plan precise performance of a command, it is necessary for us to collect and to estimate the appropriate information.

3. STATE OF AN INDIVIDUAL AND ITS ADJUSTMENT

The set of states of an individual depends on both a scale of acceptance of the decisions, and a set of the needs that have prevailed a competition on this scale, and a role connected to this set. For example, the role of dominant differs from the role of dominated. It is linked to a set of strategies existing on this scale and in this role, or operations within the framework of a terminology of the diagram of states. It can be described by attributes connected both with the cognitive sphere, and with the emotional sphere of an individual. The set of states on more long-term scales of training and planning is connected with the learning of knowledge, skills **(competence)** and gaining of resources. On a scale of the current decisions a state of an object, i.e. the set of strategies $\mathbf{Y}_s$ accessible to him is determined not only by his motivation, but also by the available internal and external **resources**. The significant waste of these resources during the conflicts without their restoration conducts to a state of **exhaustion**.

The state of an individual on the long-term scale of decision - making is determined by his system of values and also by his knowledge and skills. Let criteria (motivation, value) $w_1, \dots, w_n \in \{W\}$ compete for a choice of a strategy from non-intersected subsets $\mathbf{Y}_1, \dots, \mathbf{Y}_n$. Then the win of a competition by criterion w_s on a long-term scale means a choice from $\mathbf{Y}_s$, i.e.

$$y_s \in \operatorname{argmax}\{\max_{y\in Y_1} w_1(y), \dots, \max_{y\in Y_n} w_n(y)\}.$$

In this case state s of the object determines the set of strategy $\mathbf{Y}_s$ and criterion w_s. If there is a set K of criteria $w_{1s}, \dots, w_{ks}$ in s co-operated during the mentioned competition for a choice of strategy from intersected subsets $\mathbf{Y}_{1s}, \dots, \mathbf{Y}_{ks} \subseteq \mathbf{Y}_s$, then a win of a competition by a set $w_{1s}, \dots, w_{ks}$ determines a state s of object by a set of strategies Y_s of (1) and criteria $\{w_{1s}, \dots, w_{ks}\} = w_s$. As the state s of bioobject is characterised by a set of potentially performed strategies Y_s and motivations w_s determining a choice, it is determined by the results of co-operation – competition of his needs. The reordering of needs changes his state on a long-term scale.

Result of a competition $w_1(y, h(y))\dots, w_n(y, h(y))$ usually depends on influence h(y) of the environment on various temporary scales. For rather fast reordering of needs it is necessary to create the rather exact mechanism h of powerful influences. Usually it is based on the demonstration of force and will. The change of expected values of satisfaction of needs more slowly influences the result of reordering and the state of **readiness** of object to execute a command. As these expected values depend on local changes of environment in a point where there was a bioobject, they can be estimated by bioobjects from different points differently and to cause different resulted states. Therefore the same command can be executed by one individual and is not executed by another.

Each object has a set of states. As it usually contains a set of subobjects in all levels of an enclosure and the state of embracing objects depends on a state of even part of those included, then in such a case we have an opportunity to operate by a choice of object at a level of his subobjects, it is meaningful to describe him by a set of substates at all levels of an enclosure. For example, some subobjects can be in a state of readiness, others – unreadiness.

It was already spoken that the performance of a command in each state is conditional. For example, the state of bioobject is characterised by a variable of probability of rather good and fast indemnification of all expenses of the recipient or by a level of expectations of future indemnification. If their values surpass a threshold for the moments of time within the limits of his temporary horizon, the condition of individual rationality allows the recipient to pass to a state of readiness by this condition. It allows him to accept the decision on the performance of the required operation at the satisfaction of other conditions. The satisfaction of all conditions for all substates will transfer him into a state of *complete readiness*. However, when the recipient gets into a desperate situation when his actions during a significant time did not result in indemnification of expenses, his temporary horizon is narrowed almost to zero, and he passes into a state of unreadiness in which the condition of individual rationality does not allow him to invest the work in the performance of any commands of any sender.

This results that the concept of a state appears a key one at consideration of problems of management. Only after transition into a necessary state the choice of optimal or of quasioptimal strategy for criteria (values) linked with this state is possible. It means that the designing of declared criterion system should occur only in the framework of the state linked to it in the space of states, set of strategies, which has the object in this state and possibility of transition in the target state. If one designs it abstractly on a basis of "common values" without taking account of the state, it will appear in contradiction with the behavioural strategy linked to a state and will not be optimised.

The nucleus of declared ethics as a system of declared values should be defined on set of states and to set preferences for generation of new states and for transitions between existing ones. The system of values, or value orientation should differ from criteria system not simply by wider area of its definition, but also by more general or more abstract objects of this area. It provides an opportunity to be limited by more rough estimations admitting a considerably smaller number of gradations than criteria systems.

The object can not have a state necessary for performance of a command, if it was not generated on a scale of development or training. It is enough to recollect a huge quantity of the decisions of bodies of authority which have remained outstanding, since they assumed implicitly a presence of other systems of values of individuals or of other states of the system of social control, at which it is capable of support of the accepted decisions. Any script, operation, law, etc. should be carefully co-ordinated to the diagram of states of a considered class of objects, and before the edition of a command it is necessary to be convinced that the object is in a required state. Otherwise it is necessary to transfer him to a target state.

In a basis of stability of a state the following phenomenon lays. If some need has won a competition against other needs and has had an opportunity to create feedback in an external environment, it will use it with an aim of a choice such behavioural strategy by an individual which will ensure it a subsequent win of competition. If there will be a choice before an individual, at which another need may be dominant one the given need uses strategies controlled by it to not admit it. If there is no need at which an individual can in the given situation win a competition against the mentioned need and ensure an alternative choice, the prevailing need can support domination until it reaches a state of saturation.

A method of stimulation of an individual (organisation) for transition to a target state should be based on the substantial growth of a level of claims of realised cognitive needs of individuals at less high levels of claims of material needs. If we shall find the powerful cognitive stimuli ensuring even one win of cognitive needs at a certain part of individuals and feedback in external environment will be created, the significant part of these individuals can keep domination of this need.

The state of an individual can be described by attributes of list 1. The **target** state of an individual adequate to a level of his claims can be defined as a desirable set of values of variables of his "file" (for example, character, or kind of **competence**). This state is connected with the set of strategies needed for performing the task. The state of **readiness** can be estimated by values of the special attributes connected by opportunities of realisation of planned transformation. The exact diagnostics of a state is necessary to carry out optimal choice of a state in which it is possible to transfer the system within the framework of an available set of criteria, resources and physical restrictions imposed by the given state to construct the diagram of transitions.

Let us sum up. To transfer an individual from a given into a target state it is necessary to apply to him certain sequence of influences. This sequence for different individuals can differ very strongly both by the nomenclature, and by value of influences. Therefore the diagram of states is necessary to build not only for the organisation, but also for every individual. Perhaps the *ethically allowable scenario* of influences capable to transfer concrete individual to the necessary state does not exist, or the scenario for a given individual capable to execute this task in general does not exist. It means that a given individual has not a necessary set of transitive states, and they still need to be generated. The set of transitive states of an individual depends on his genetic abilities, his age, the mechanism of social control and resources available for this mechanism. Therefore it can appear empty. Having the description of the set of states of an individual, it is possible to solve the problem on the expediency of his attraction to the decision of a considered problem.

Some states are **critical** ones in which an individual can not accept ethically or professionally competent decisions. On short-term scales they can be states of strong frustration and stress, on long-term scales **pathological** states arise. The critical and pathological states are characteristic not only for individuals, but also for any objects. The living objects differ from technical ones that they have a set of **irreversible** states among which there are pathological ones. An individual requires protection against transition to irreversible pathological states.

In a set of states of nervous system it is possible to distinguish subsets of **conscious** and **unconscious** states. The conscious states and correspondent conscious needs are connected to a set of specific nucleuses of the brain cortex. The rest nucleuses are connected to unconscious states and correspondent unconscious needs. The wider the conscious set of needs of an individual and the more he has to harmonise his satisfaction, the more specific complex strategies he should use. As conscious needs are complementary to unconscious ones in a common complex of needs, they are closely connected with conscious ones. The neglect by unconscious needs at planning the conscious behaviour conducts to **hypertrophied** satisfaction of some needs and the occurrence of pathological unconscious states with additional violation of harmony. It leads also to overestimation of influence of knowledge accumulated in declarative memory on forming of short-term behavior, to underestimation of necessity of forming of environment on long-term scales on the base of knowledge and training in this environment. Let us remember witty, but deep words by O. Wild that true gentleman need not knowledge. It means that knowledge influences only on long-term behavior, but does not influence on short-scale behavior.

4. STATE OF ORGANISATION

What we told above is fair for organisations embracing an individual. In spite of the fact that the states of embracing objects are not so irreversible, as states of separate individuals, the comprehension of some states and necessary sequence of transitions to a desirable state represents a heavy dynamic problem. The irreversibility is determined by the results of the selection on an evolutionary scale, e. g., by genocide of carriers of the certain genotypes. The degree of irreversibility of a state of population, if there is a significant number of individuals which are appeared in a state with an undesirable set of expressed genes on a scale of development, can also be great. But it can be overcome on a historical scale at the expense of the creation of new mechanisms of social control. The creation of the diagram of states for the complex scenarios requiring large power and emotional expenses can help both to bypass undesirable subsets of states, and to concentrate efforts on an exit from the existing critical state on temporary scales, which it is possible to leave from these state. The space of states of organisation can be described by attributes of above-mentioned list 1.

The embracing organisation can change the set of the states as a result of the transition to substate of the included, but the dominant object that is not yet created as a substate of its state. This change can grow out of an transition of the included dominant object both to a state of degradation, and to a state of development. It is connected to the initiation of strategies of development or introduction. In a state of development the transitive states can correspond to the separate stages on a way to a target state.

The strategies of transition between states of an object can be divided into three classes: strategies of degradation of the structure of an object, strategies of support of the structure of an object and strategies of improvement of the structure of an object with the purpose of growth of fitness of organisations, including him. Usually strategies of transition of the first class are carried out without external and internal inflow of energy. However in crisis situations they are realised by inflow of energy destroying structure. The strategies of the second class demand the inflow of energy and its direction according to the influence of the existing system of social control. The strategies of the third class also demand the inflow of energy and its direction. But considerably large expenses of energy are necessary than in the case of strategies of second class since except energy for support of the system of the social control we need vast energy for the development and introduction of a new system of social control.

The strategies of degradation of structure of an object work constantly, and as soon as the system of the social control spends energy, the object quickly begins degrade to a new steady state. As an example it is possible to consider the process of ethical degradation of organisation. It represents the process of gradual alienation of its majority from the observance of declared ethics. In the beginning the dominant is alienated from the performance of a rule of declared ethics interfering to him. The dominated individuals observing the growth of his local fitness and impunity copy his behaviour.

As a result the greater part of organisation is alienated from performance of this rule. Then there comes a turn of the next interfering rule, etc., while the system of social control will not show the support of some stayed rules. During the expansion of the process of alienation of an object from the ethics it appears frequently that the kept rules are not enough for the performance of a condition of the individual rationality for object, and he leaves organisation. In result ethical degradation of an organisation conducts to its structural degradation. Ethical degradation of organisation beginning from its "head" proceeds to a level kept by its support because in the case of neglect of all rules, a threat to its survival connected to its complete destruction arises.

The events are the phenomena that can switch states in the diagram of transitions. On the diagram of transitions they are designated by names accompanied with Boolean expression. The event occurs when the expression accepts value "true". The default of a condition can result in the event being lost: the transition will not take place. Some events change a state of system by change of parameters (adjustments) of the system, which is included in the mentioned Boolean expression installing an allowable range for their change. Example of such parameter is degradation.

The state of an organisation can also be changed at the change of a sign of an inequality in conditions of an individual and collective rationality that conducts its formation or disintegration. Some events transfer system to a new state others do not transfer. The events of the formation or disintegration of organisation change a set its emergent strategies, i.e. switch a state. The state of social system is connected to the dominant system (structure) of needs. Therefore the transition from an initial state to a target one is connected to the transformation of the initial structure of needs to a target one through an optimal intermediate state.

So, the transition between the states is carried out at occurrence of events. The first natural event is **degradation**. At excess of the certain threshold this event initiates a state of **frustration**. In the presence of the specially created mechanism of diagnostics the special part of the diagram of states submitting to the conscious control is emerged. If the mechanism of diagnostics has ability to distinguish states of frustration of the separate members of organisation and aggregate them in the value of its frustration, then at excess of a threshold it initiates the clear comprehension of same event at a level of an organisation. This event can at the presence of the special mechanism of social control cause an operation of transition to a state to a start of initial **researches**.

The operations caused by transitions between states of an object such as social system can represent the large projects connected to research, development and introduction. A main objective of these projects is the creation of an infrastructure for its support. On the basis of the diagnostics of an end of the first stage the transition to a state of development and accumulation of knowledge by a population can be caused. Further, the transition to a state of acceptance of the decisions in a direction of the overcoming of the crisis and readiness for its overcoming can be caused, etc.

The initial and target states of an organisation can be determined as sets of states of key subsystems. As the system of social control is key one, the major state is a state of its mechanism of aggregation of the individual decisions into collective one on co-ordinate "authoritarism – nonauthoritarism". Division of various architectures of the mechanism aggregation on the most distinguishing basic characteristics can be a basis for the marking of its possible states.

Because 1) each state differs from any other state of the list 1 by the characteristics, 2) in each state the population has a specific set of strategies linked to a state distinct from the set of strategies in other states, 3) in the set of the mentioned characteristics and strategies the population can have systems of preferences, the question about a possible exact diagnosis of the states receives major importance for it. Exact diagnosis can allow to establish current (initial), desirable (final) and transitive states of a population.

List 1: Attributes of the state of organization (individual)

- Name of organisation
- Type (state, private)
- State
  - name of state
  - type of state (initial, target, transitive)
    - subtype of state
      - researches
      - development
      - diagnostics
      - introduction
  - temporal scale (current, intermediate term, long-term decisions)
  - characteristic of a state
    - critical level of maintenance by resources
    - pathology
    - harmony
    - remoteness from a level of survival
    - type of degradation
      - degradation of resources
      - degradation of spirit
      - degradation of knowledge
      - degradation of health
      - degradation of culture

        - ethical degradation
            - reversibility
            - sensibleness
    - substates
        - readiness for transition
        - type of readiness
            - research
            - development
            - introduction
            - support of the given state
        - presence of resources
        - competence
        - safety
        - psychological readiness (comprehension and spirit)
        - collective readiness (portion of the members of collective in a state of readiness)
        - organisational readiness
        - an opportunity of performance within the framework of existing system of the social control
            - conformity to declared ethics including the contractual obligations
- co-ordination of declared and realised ethics

The decision on a transition of an object to a new state must be accepted within the framework of the system of social control, if as a result of diagnostics of an existing state it admits unsatisfactory from the point of view of a set of clearly formulating criteria of declared ethics of the system. In the case of non-developing system as key criterion can serve degradation of the current state in comparison with what existed earlier. The worst current state on the diagram is "degradation (resources, energy, knowledge, spirit, ethics)" with a small portion of ethical and professional part of the population, the irreversibility of degradation of ethics of a significant part of the population, unconscious of degradation (list 1).

The planning of a state to which it is necessary to proceed can be carried out on the basis of comparison of the current values of the attributes of the object getting diagnosis with the appropriate values of a target state and fixing a class of strategies of transition to a desirable state as a result of comparison. The planning should be based both on declared ethics, on estimation of a temporary and spatial scale of change of existing behavioural strategy, accessible a set of operations of each state, and with the presence of resources. In the case of internal inability of behavioural strategy to one-step transformation to the desirable state or absence of necessary for this set of operations, and also in the case of shortage of resources the intermediate states are planned. First should be the transition of a population to a higher level of knowledge and comprehension of an essence of degradation, then it is necessary in a cycle to increase the spirit, ethics, operations, resources and portion of ethical and professional part of the population by planning the next states for this purpose.

For simplification of diagnostics it is necessary to describe the map of the set of the basic diagnoses of object of degradation on the set of strategies of transition to a desirable state. Further it is necessary to develop the tests establishing the diagnosis within the framework of collective feedback of an object. The most simple description of the diagnosis from a set of the diagnoses of degradation can be given on the basis of a set of attributes allowing to specify a level and the name of the degraded object, the name of the sphere of degradation, type of the degradation, and also its intensity. The name and possible values of these attributes are:

1. Level of object of degradation: from a top to a bottom level (individual)
2. Name of sphere of degradation
- Resources
- Spiritual sphere
- Behavioural sphere (realised ethics)
3. Type of degradation
- Reversibility (reversible, irreversible)
- Sensibleness (conscious, unconscious)
4. Intensity of degradation (weak, moderate, strong)

On established by a way of aggregation in the channel of a collective feedback value of the diagnosis the mentioned map in the case of realised degradation at once offers strategies directed on its overcoming. An example of the social diagnosis and the restrictions on the strategies of transition to a desirable state from the given one in a case of consciousness can be written down as follows: “degradation of behavioural sphere of organisation K, realised, but strong and irreversible” → “the conscious development of strategies directed mainly on young generation of the organisation K”. Further it is necessary to describe the rather deep decomposition of an offered complex of strategies on the programs of learning and training for various groups in view of a level of professional and ethical competence and age features, and also methods of diagnostics of a state of degradation of separate individuals for the division among them on the mentioned groups.

In case of unrealised degradation initial strategies should be directed on its comprehension. In case of spiritual degradation initial strategies should be directed on the spiritual rise of the organisation which the rise of knowledge should precede. It is natural that it is possible more precisely to specify a class of recommended strategies if to take advantage of a wider set of attributes of the characteristic of a state from list 1. The names of the operations and initial stage of their decomposition in the given examples speak that in case of social systems behind many operations there can be powerful projects, that transition of social system to a new state is far from transition of the technical device. Therefore the diagram of states can send the designer to the diagrams of uses, interactions, etc. [6]. Similarly it is possible to describe a set of the diagnoses of readiness and the diagnostics of readiness.

5. STATE OF THE SYSTEM OF SOCIAL CONTROL

If an exact diagnostics of states of a social system is necessary, there is a question whether a set of the experts participating in diagnostics has to be limited to a narrow group of competent experts or should be rather wide. The answer is dictated by 1) vast distinctions of spatially – temporary scales of a population and any expert, 2) the mentioned necessity of harmony (balance) of stay of any object in a role of dominated and nondominated for harmonising of the state of population.

Let's begin from first. Even the most remarkable expert is compelled to internalise spatially - temporary scales and priorities of needs of the local organisation ensuring him a certain standard of living, but is not capable on the internalisation of a set of needs of all organisations created by a population, and the population as a whole in any way may not stimulating to him. Even when the expert is formally stimulated by global organisation, an actually narrow group of the system of social control accepts decisions on his stimulation, and he is compelled to represent its interests. At diagnostics and planning of states it is necessary to remember that many states of organisation as social object are pareto – unoptimal ones. Therefore a sharp struggle, which results can considerably break the requirement of harmony, for the diagnosis of the given and target state, and for transition to target state will go.

For diagnostics of a state of a population the set of the experts reflecting a state, preference and knowledge all of its objects is necessary. *The declared ethics should demand a recognition of every individual to be competent even on one spatially - temporary scale though one problem areas though one level of detailed elaboration though one organisations including him. It should demand the creation of the mechanism of revealing of his level of competence in the field of his claims so that he can join in the mechanisms of decision - making in the field of the competence even by participation in the appropriate circuits of collective feedback.* The adequate expansion of a set of experts included in the mechanism of social control except of intellectual function will execute also the function of harmony stay of an object in roles dominated and nondominated, participating in management accessible at his level of competence spatially - temporary scales of the problem area accessible to him of an object accessible to him. If we thus shall manage to create and to introduce in the mechanism of the social control co-ordinated with declared ethics the normative base of the system of encouragement – punishment of the experts and nonmanipulated mechanism of its use, we can co-ordinate areas of claims of the expert on a rather wide set of high and low needs to his abilities and to correct his behaviour on observing the declared ethics.

Therefore previously before diagnostics of a state of a population and its objects it is necessary to create nonmanipulated mechanisms of aggregation of the individual decisions in a collective one, underlying systems of diagnostics not only of population, but also its organisations, including diagnostics of the mechanism of social control based on declared ethics. These mechanisms should be a part of the system of social control. On their basis it is necessary:

- To develop architecture of decomposition of an organisation on objects of all levels of an enclosure and the architecture of the mechanism of social control appropriate to this decomposition determining interaction between separate subobjects at the acceptance of the decisions,
- To create mechanisms of diagnostics of a level of competence of individuals to ensure the formation of personnel structure of the mechanism of social control, including the distribution of individuals on the channels of collective feedback,
- To develop structure of normative base of system of encouragement - punishments and the mechanism of its support within the framework of the mechanism of social control.

As by the development and the introduction of the mechanism of social control there will be significant conflicts for certain, it is necessary to develop and to introduce it with an aim of step-by-step development within the framework of an evolutionary cycle, using the previous version of this mechanism for the development of a subsequent one so long as it will not be accepted by the significant majority. For example, the initial development of the mentioned three tasks can be carried out before diagnostics of a state of the system of social control. Normative base should be developed after the mentioned diagnostics.

The mechanism of social control should represent multilevel highly structured object consisting of two subsystems: the people and the infrastructure. Elements of the first subsystem are the organisations of all levels down to individuals, elements of the second one are technical and software facilities (information systems, system of support of the decision-making, diagnostics etc.), which design does not depend on a state of the mentioned object and its subobjects. As this infrastructure is necessary for diagnostics of the mechanism of social control and population, just from its creation the transformation of the social system should begin.

The mechanism of social control should constantly be improved independently of the current state of a population. Diagnostics of its state can be carried out within the framework of the procedure of monitoring of the members of an organisation and its clients during periodic interrogations on a scale of development. As individuals included in the mechanism of social control made up only by small part of a population, for the change of a state of the first subsystem of the mechanism of social control in contrast to a change of a state of population it is not necessary to change a state these individuals. Using the mechanism of diagnostics of the system of social control created independently of the current state of a population, it is possible to exchange individuals which are in an inadequate state to the social control for individuals co-ordinated with this state that will facilitate transition and will reduce time of transition. However opportunities for personnel changes are always limited. Therefore the projects of operations of transition of other individuals of the system of social control to a necessary state are necessary also.

Initial states of creation of the mechanism of social control are the states of research, development and diagnostics (list 1). These operations should be carried out in a cycle up to the achievement of the moment at which diagnostics will accept the decision on the beginning of preparation of introduction. The first problem that is necessary for deciding at the moment of a beginning of development is the problem of decomposition of this operation on key elements, since this decomposition should determine the distribution of resources. From the previous it follows that the elements of this decomposition should be:

- Creation of nonmanipulated mechanisms of aggregation of the individual decisions in collective one,
- Creation of mechanisms of diagnostics of the level of competence of individuals and the formation of personnel structure of the mechanism of social control,
- Development of the architecture of splitting up of an object into subobjects of all levels of an enclosure and architecture of the mechanism of social control determining the interaction between separate subobjects at the decision-making.

6. IMAGE OF ORGANISATION AND STATE OF REALISATION OF THE OPTIMAL PLANS

The criteria of an estimation of any object are defined by the observer, his localisation, interests and role. If this observer is a set of the external users on which the survival of organisation depends on, then organisations adapt their declared image for it. This image allocates to this set a role of the dominant client. The image of an organisation on a scale of the current decisions is represented as the serving mechanism, i.e. an object which is in a role of dominated (fig. 1). The market organisations in a pursuit of the client proclaim that for them there is no greater satisfaction than the best way to satisfy all needs of the clients. Even the bodies of authority assert that they serve.

There is some truth in all these statements. The organisations are designed as dominated, serving the client on a scale of the current decisions mechanisms, to a certain degree carrying out these functions to ensure a flow of resources necessary for their survival. The lie a of proclaiming image of an organisation consists in its incompleteness and accordingly in nonadequateness to a nature of its behaviour. A source of nonadequateness is that the maintenance necessary for a survival on a scale of the current decisions of a flow of resources after excess of some threshold ceases to be the unique need of individuals. The set of basic needs is initiated which basically can not be satisfied by transferable usefulness, for example, by salary being functions of a created flow of resources. Among such non-recognised needs the key role is played the need of domination and of autonomy.

Fig. 1□ represents the sole dominant forcing other members of an organisation to execution of a role of dominated concern of the client. Non-recognition of their needs in domination and in autonomy, non-discussion of these needs supersedes them from consciousness, but due to their evolutionary stability [2, 5] does not eliminate them. Therefore individuals are compelled either to suppress their satisfaction or in some degree to satisfy them unconsciously. But to satisfy them illegally because they are not obviously reflected in declared ethics of organisations and are not ensuring with the legal mechanisms supporting a certain level of satisfaction. The illegal satisfaction is a use of a specific complex of strategies not co-ordinated with declared ethics. It occurs in conditions of growth of cognitive dissonans between declared and realised ethics of

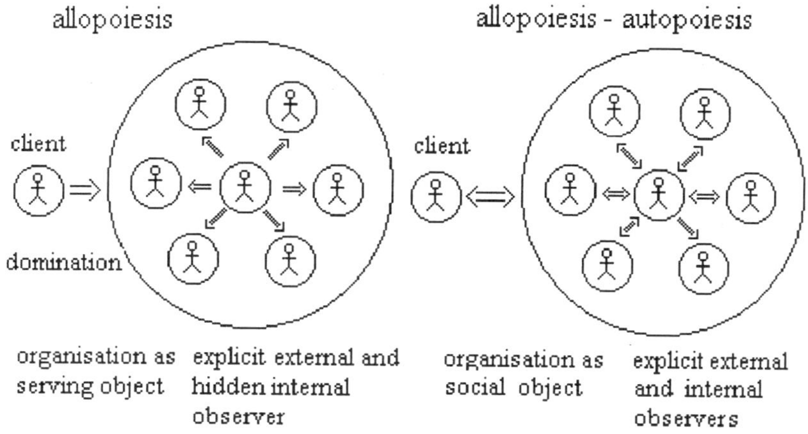

Fig. 1

individuals and conducts to the formation and the development of the deformed structure of the nervous centres – carriers of this complex. The growth of a dissonans constantly increases his frustration, deforms his mentality and behaviour. This deformation of structures of the nervous centres with age becomes irreversible. The appropriate complex adaptive systems are not capable any more to readaptation.

The suppression of satisfaction of needs in domination and in autonomy due to their evolutionary stability results in the following. Like hypertrophy satisfaction of the need for food after long hunger, the win by the need of domination of a competition at other needs after accumulation of frustration because of the attempts to follow the contradicting of a nature of behaviour the requirements of organisation initiates a burst of hypertrophy satisfaction of this need on a scale of the current decisions, which is possible only at the expense of the partners and clients. Such phenomenon was investigated by A. Uhtomsky, which developed the notion of "dominantis" as a system of nervous centres connecting with the winning motivation and temporary determining character of respond of an organism on internal and external influences at simultaneous suppression of other motivations. This respond results in the development of crises of the relations of an individual both with the partners inside organisation, and with the clients. Due to the constant exchange by strongly unsatisfied personnel between organisations these deformations of behaviour are distributed on all population destroying as ethics of behaviour of market and unmarket objects, and reconstructing the structure of their nervous system.

So, the set of roles of organisation is not limited to a role of the dominated object. Any bioobject can not exist only in one role, in this case in a role of dominated even if it just provides him with resources. The resources are necessary to him not only for the sake of survival, but also for the satisfaction of a wide set of needs. He should satisfy the need of correction elements of the structure (for example, nervous centres) with the purpose of the creation of the optimal interface between his own genetic structure and external environment, the need of following to the genetic program shown as genetic switching on a scale of development, initiating dynamic dependence of motivation structure upon age. For this purpose he should exist also in a role of nondominated which can be submitted as an association of two roles: roles of autonomous and of dominant.

For the restoration of elements of its own structure there is no necessity for autonomous to co-operate with external environment in general. In this role the bioobject serves the internal needs instead of the needs of the client (autopoises [7]), and from the point of view of advertising (declared ethics) does not carry out the obligations. Being in this role the organisations change a priority of the purposes directing the efforts not for better quality and greater assortment of services, but on the greater income. In a role of dominant the bioobject influences external environment in order to adapt for itself instead of to adapt itself. In result the image of his behaviour observable from the outside sharply changes. Even when he forces dominated individuals (fig. 1□) to adapt for external environment of the clients, he pursues his own purposes (for example, profit), i.e. has nothing common with image declared by him.

For the harmonising of behaviour it is necessary for bioobject to observe balance between allopoeisis of a role of dominated and autopoiesis of a role of nondominated. Evolutionarily stable mechanisms are based on the switching of roles as a result of the change of result of a competition of the needs connected to these roles [5]. If declared ethics and the mechanism of social control support extremely long stay of individual in a role of dominated, as soon as the long unsatisfied needs of domination and autonomy win a competition against need of getting a flow of resources, the restoration of balance occurs by a sharp change of behaviour of individual perceived by partners as "revolt by the ship". The organisations did not create mechanisms allowing every member of the organisation to harmonise the satisfaction of his needs so that the harmony on the set of mutual relation of the members of organisation were achieved also. If the bottom levels of hierarchy of organisations are compelled the most part of life to stay in a role dominated, the top levels appear in a role of dominant. For it those and others pay often by "revolts", internal pathological shifts and external deformation of behaviour.

The consideration of organisation from the point of view of long-term scales changes its image since on these scales the need of a set of members of an organisation emerges to define its future. Now image of the observer divide in two. By consideration of organisation as social object (fig. 1b) the root client part of the observer moves inside of an object (autopoiesis [7]), and the unroot client part of the observer is kept outside of it (allopoiesis). The more steadfast sight on the internal structure of organisation speaks that the root client part is distributed inside organisation. It means that the main requirement to a mechanism of social control should be to ensure harmonising the functioning of mechanisms of autopoiesis supporting some level of satisfaction of needs of domination and autonomy, and allopoiesis supporting adaptation to the requirements of the external clients.

The normative image of organisation is reflected in the consciousness of the internal observer as a *social object*, or environment in which the members of an organisation will spend a vast part of their life. The estimation by the internal observer of an organisation as social object certainly includes his estimation of the organisation as an object that serves the client since he understands that the client satisfies the need of organisation for getting a flow of resources. But he also understands that the significant part of his own environment is created by the organisation, and if it wants him to use an optimal strategy for co-operation with its external client it should consider him as the internal client with which it is necessary for it to co-operate optimally providing the certain level of satisfaction of his needs. Therefore a complex estimation of an organisation should include not only the component connected with service of the external clients, but also the component connected with the satisfaction by an organisation of needs of the internal client.

The satisfaction of needs of domination and autonomy in non-hypertrophied variant demands the granting of an opportunity of a choice of area of domination and realisation of a flow of the decisions to the individual, which intensity is determined by a level of his claims, co-ordinated within the framework of the information conflicts to the level of claims of the partners. The need of a choice of area of domination can externally be shown as the need of stable development of the abilities of an individual. The unconscious need of domination and autonomy is shown in the requirement of the satisfaction of such conscious needs as professional and intellectual growth in the chosen area, maintenance of opportunities of his self-actualisation and improvement of an economic situation, maintenance of participation in management. The high complex estimation should testify that the organisation meets the image of 1) complex adaptive system both in relation to the members, and in relation to the clients, 2) unadaptive system in relation to declared ethics of society. The reduction of estimation is connected to the frequency and value of stresses received inside organisation and value of the expected accumulation of frustrations.

The advantage of change of a role of organisation as serving object on a role of social object is that the last will increase intellectual power of the collective and will execute also the function of harmonisation of stay of every individual in roles of dominated and nondominated, participating in management accessible at his level of competence spatially - temporary scales of problem area, of accessible object to him, by the inclusion of him in processes of acceptance of the decisions. It stabilises his psychological sphere and health.

For harmonisation of satisfaction of needs in resources and needs inherent to organisation as social object members of an organisation have necessary to pay dearly in the work, and everyone will be compelled to define *a point of harmonisation* accessible to him and the appropriate level of a payment for the transition to this point and the existence in this point accessible to him. All members of organisation will be compelled to co-ordinate among themselves points of harmonisation to ensure *a common point of harmonisation* not causing powerful conflicts. The expenses necessary for such co-ordination will depend on the type of social system, which can both stimulate the growth of professional and ethical competence necessary for participation in decision-making and to counteract it.

The attributes of estimation of organisations given in list 2 represent the important addition in relation to attributes of estimation of quality of organisation as social object of list 1. On conformity of their values with the norms one can judge whether there is a necessity for the transition of organisation to a new state. A small number of gradations can be used at estimation of an existing (initial) and target (final) states of object.

It is visible from list 2 that the estimation of organisation as social object includes its estimation as the serving economic object since its survival is connected to the performance by it of these functions. The designing of organisations as social objects means that the balance between actually social and economic functions will be dynamic, that the point of balance between the mentioned functions should be displaced in this or that direction depending on what scale (long-term or short-term) the organisation will be compelled on to make a decision by a situation in which it was put by external environment. The organisation designed and realised, as a social object will be compelled to create till now not existing mechanisms based on collective participation in diagnostics of states. It will be compelled to include these mechanisms in the system of social control that will facilitate the opportunity to create the mentioned displacement of a point of balance without the powerful social conflicts that arise with their absence. With the same purpose organisations existing on rather long-term scales (e.g., a state) will be compelled to create and to include in the system of social control mechanisms to its self-perfection.

In list 2 the nomenclature of the aggregated parameters is given on which individual intuitively estimates the organisation. It is possible to develop ball estimations for it and then the strategies of aggregation of estimations of individuals of organisation for the getting of their collective estimations by the members and for opportunity of their discussion. Use of computer technologies can ensure an actualisation of these estimations for organisations of all levels that include individual. Thus it enables their members and the system of social control to know about their collective estimations and collective requirements to a direction of reorganisation of system of social control.

List 2: Estimation of organisation as social object

- estimation of development
- estimation of the harmonisation of the satisfaction of needs
- estimation of an opportunity of professional and intellectual growth
- estimation of an opportunity of self-actualisation
- estimation of an opportunity of satisfaction of other needs ensuring harmonious development and growth of fitness
- estimation of parameters of economic development on long-term scales
- estimation of the growth of the quality of production, assortment of services, value of the cost price, value of the income and estimation of growth of own economic parameters connected to them
- estimation of a degree of safety, or survival
- estimation of an organisation as serving object on short-term scales
- estimation of quality of production, assortment of services, value of the cost price, value of the income
- estimation of an organisation as a social object on long-term scales
- estimation of the frequency and values of stresses received inside an organisation, of sum frustration, connected with the wreck of the majority of hopes during work in organisation, estimation of the probability of irreversible pathological changes
- estimation of the mechanism of social control on long-term scales
- estimation of quality of system of decision-making
- estimation of system of distribution of encouragement
- estimation of system of interactions
- estimation of realised ethics
- estimation of sensibleness of their procedures

7. DIAGNOSTICS OF THE STATES AND OPTIMISATION OF PERFORMANCE

The structure of the normative aggregated estimation of organisation should be developed so that when the organisation appears on a threshold of survival because of low efficiency as an economic object on a short-term scale, the greatest share of its estimation as the social object accounted for its estimation as an economic object. Thus the basic resources will begin to be allocated on the improvement of its economic estimation. The necessary normative structure can be achieved by a choice of coefficients of aggregation of estimations making variable depending on a state in which there was an organisation. It is natural that for various states the coefficients of aggregation should be chosen apriori on a normative basis. When an organisation will proceed to a state at which the more important will become the satisfaction of high needs of a long-term scale the structure of the aggregated estimation should change.

If the organisation appears on a threshold of a survival on a long-term scale because of its low efficiency as an economic object it is necessary to find out the reason of such a fall of an economic efficiency. It can appear that the reason of economic fall is the social inefficiency. Thus the coefficients of aggregation of estimations should be those that the greatest share of its estimation as a social object should account for a social component of estimation. In this case the resources will be allocated so that for the improvement of an economic estimation only minimal resources necessary for a survival would be used and the other resources would be used for the improvement of social efficiency.

So, the state of organisation as an economic object should be described not only by the value of excess of stability concerning a threshold of survival, but also the scale on which decrease of an economic efficiency was observed. The normative structure of aggregation of collective estimations w_j of separate variables in the aggregated estimation of organisation

$$w^s = \sum_{j \in J} \alpha_{sj} w_j$$

is connected to each state $s \in S$. It means that the value of the same variable is considered in different states differently. For some s $\alpha_{sj} < \varepsilon$ that makes attribute j insignificant. Such an approach can allow on the one hand to have a rather large set of attributes for an estimation of organisation and on the other hand to have small set of "active" attributes connected to an estimation of each concrete state of organisation. It is natural that at this approach it is necessary to have the mechanism of diagnostics of states that should be developed on the basis of use of computer information systems and methods of aggregation of a collective estimation of a state. It is necessary to notice that the states of embracing objects depend on the states even parts of included ones. Therefore it is necessary to describe a state of embracing object by set of substates of objects of all levels of an enclosure also and on this basis to create the aggregated mechanism of diagnostics.

Every individual, each organisation depending on the professional orientation and external environment have distinctions in the structure of their needs and values. As the individual is the carrier not only his own values, but also the values of the organisations including him, unconsciously weighing all values he has to weigh values of all objects embracing him. In different states he makes it differently.

Let's consider a case when the most important attribute is j named by productivity of organisation. It requires the realisation of the optimal plan of growth of an estimation $g_j = Y_j / X_j$ of the ratio of it output Y_j to input X_j [8]. In economics the output Y_j refers to the production function and represents regression model on a set of expenses X_j. In the rather general form the production function is represented as the following model: $Y_j = qX_j^{\alpha}M_j^{\gamma}e^{vt}$. In this model M_j is value of the production assets, X_j is an expense of living work, q, a, g, n = const > 0, t is time. The output Y_j of system at any updating of the model of the production function is positive monotonous function of an input X_j, i.e. of a set of expenses, and $X_j \to 0 \Rightarrow Y_j \to 0 \Rightarrow g_j = qX_j^{\alpha-1} M_j^{\gamma}e^{vt} \to 0$ at $\alpha > 1$. We see that the ratio of output Y_j to input X_j of system is directly determined by expenses of living work.

If an organisation proceeded from a state s – 1, in which it was at a previous moment, to the state s, then expenses X^s_j, which object is ready to spend, depend on its subjective estimation $w^s_j(g^s_j, g^{s-1}_j, r)$ what it will receive from organisations in the state s with a level of trust r embracing it and demanding its contributions in exchange for its own expenses. Differently, they depend on the performance of a condition of individual rationality: $w^s_j(g^s_j, g^{s-1}_j, r) > X^s_j$. For example, as soon as the Russian banks transferred to a state s, at which they have ceased to return the contributions to the population ($w^s_j = 0$), its readiness to bear expenses $X^s_j > 0$ has become zero.

Designating the number of embracing object in state s through n and subjective estimation w^s_j of what included object receives from an organisation n embracing it through w^s_{jn}, it is possible to write down an estimation of the individual in the state s as

$$w^s_j = \sum_{n\in N} w^s_{jn}.$$

Similarly it is possible to define a collective estimation of organisation by a set of organisations included in it for which the optimal plan is made. Before to engage in realisation of the optimal plan it is necessary to transfer the organisation to a state s, at which the estimation w^s_j will appear maximal on a set of states S. If organisation remains in the state, at which the estimation w^s_j in individual and collective estimations will be small, it will be impossible to achieve the realisation of the optimal ratio of an output Y_j to an input X_j of a system because the selected share X^s_j of average individual and collective resources on its realisation will be small.

In USSR in the eightieth years the powerful specialised organisations having the competent staff designed complex criteria of optimisation of a national economy "reflecting" its production functions and developed mathematical methods and computer programs of their optimisation. However, they did not take into account that for the realisation of the decisions received on their basis (optimal plans) it was necessary that the social system was in a state co-ordinated with the offered criteria and was ready to go on expenses connected with the use of strategies necessary for the realisation of the optimal plans.

In the considered period ethical and connected with it structural degradation of social system occurred. Ethical degradation represents process of gradual alienation of the majority of object from the declared ethics. In the beginning from the performance of a rule of declared ethics interfering to him is alienated dominant which demonstrates neglect by it. Dominated individuals observing growth of his local fitness and impunity copy his behaviour. As a result of an avalanche copying the most part of an object is alienated from the performance of this rule. Then there comes turn the next interfering dominant rules, etc., while the system of social control will not show the support to some stayed rules.

Long degradation of ethics internalises in degradation of realising it the system of the centres, i.e. physical degradation. If it occurs in mature age behavioural degradation becomes irreversible**. The more age, the more energy is necessary for an individual on formation of new structure of the nervous centres. Eventually, this ability is gradually lost. If the percent of mature individuals in a population is great, ethical degradation takes the irreversible form at the expense of its reproduction during copying by young.**

First of all, alienation from needs of optimisation of fitness of organisation as a whole is concerned rules, which observance demands considerable expenses. Just such is the requirement of the maximisation of a productivity of an organisation. The alienation also has resulted in a sharp decrease of a voluntary allocated share X_j of average individual and collective resources on realisation of the plans. It was possible to observe the neglect by the global optimal plans, both from the side of bodies of authority, and from the side of hierarchy of organisations and separate individuals. There was a sharp localisation of interests, i.e. frank egoism increased that undermined bases of co-operation. The society was not ready for the realisation of these plans not because it did not have physical resources, not because it needed the investments, but because of the latent sabotage both by the management of organisations, and by their employees.

The developers were not ready also. Many developers including the author of this paper [3] knew about it, but had not chances to be heard by the mechanism of social control keeping till now "deafness". Others were more and more alienated from global needs of a society, aspired only to the satisfaction of the needs connected to the achievement of personal success in narrow subject domains, for example, optimal planning and mathematical statistics. It is possible to count hundreds of books issued in these areas under the guise the decision of this problem at those times. And not a single book devoted to the problem of change of states of social systems. It is possible to count many tens of computer centres and thousand programmers involved in processes of optimisation of the plans, but not in processes of transition of system in a state, at which the optimisation could be carried out.

To avoid similar troubles it is necessary to create mechanisms of diagnostics of states of social system and substates of all organisations included in it, of check of a co-ordination of offered criteria system with the revealed state of system of the social control of the country and the systems of social control of organisations, i.e. with readiness of the staff for the realisation of these plans. In case of revealing inconsistency it is necessary to postpone offered works before transition to a necessary state. The mechanism of diagnostics of states on base of all attributes should be based on rather the representative and competent examination. It should guarantee safety of the interrogated experts and rather high reliability of diagnostics. It can be developed on the basis of use of the specialised information systems serving for use in the channel of collective feedback.

The mistakes in the diagnosis of illness of Russia have continued with coming to authority "democrats" and their advisers from the international financial organisations. Again the patient was announced the infrastructure, i.e. national economy, rather than ill society and ill the system of social control, which is disease-breeder. Accordingly, as means of treatment was announced investment and sufficient budget. As parasites destroying the ill organism the corrupted economists, financiers, managers started to multiply, but activity which could result even in a small step in a direction of change of the state of the system was suppressed. The mechanism of social control continued to corrupt, its officials continued to be involved in plundering the credits and criminality. It was found out that the international financial organisations do not have at all mechanisms of diagnostics of states of social system, that their understanding of a problem and the advice "to have budget without deficit" does not differ by "brilliance of idea" from advice of the mentioned Japanese experts. It means that their interaction can be effective only with healthy organisations. It is possible to give a lot of examples of the incompetence of democrats, which have not understood what means the term "a state of a system", for what it is necessary to create purposeful transitions between states. They wanted to achieve purpose by the loud declarations instead of by development and introduction of strategies that would transfer a system to a necessary state.

8. REFERENCES

1. Booch, Grady. Object-Oriented Analysis and Design with Applications. 2nd ed. Redwood City, CA.: Benjamin/Cummings, 1994.
2. Dawkins, R. The Selfish Gene, Oxford, Oxford University Press, 1976 and 2nd Ed. 1989
3. Degtiar V. U. Crises, Mechanisms and Strategies to Overcome Them; The Example of Russia. Cybernetics and Systems '94. Ed. by R. Trappl, Singapore: World Scientific Publishing, 2, 1303-1309, 1994.
4. Harel D. and Gery E., "Executable Object Modelling with Statecharts," Proc. 18th Int. Conf. Soft. Eng., Berlin, IEEE Press, March, 1996, pp. 246-257.
5. Maynard Smith J. Evolution and game theory. Cambridge: Cambridge Univ., 1982.
6. Martin J. and Odell J., Object-oriented Methods, A Foundation, ISBN: 0-13-630856-2, Prentice Hall, 1995
7. Maturana, H. R. and Varela, F. Autopoiesis and Cognition: The Realization of the Living, Dordrecht, Reidel, 1980
8. Parra-Luna F. Hacia la medida del Desarrolo Sostenible: una tentativa metodologica. Sustainable development: towards measuring the performance. Eds. Elohimjl, Parra-Luna F., E. A. Stuhler. Ruiner Hump Verlag. Munchen und Mering, 1997.

Chapter 11

Conditions for testing performance

IRIS BÁLSAMO
Instituto de Derecho Público, Ciencia Política y Sociología. Academia Nacional de Ciencias de Buenos Aires

1. SYSTEMICS

1.1 Philosophical and epistemological considerations

Systemic approach is recognized as the most apropriated for scientific research by the realist philosophy of science. It is useful for interdisciplinary research and is called to be the protagonist of the unity of science (Bunge, 1996).

Regarded as superior to its alternatives - holistic and individualistic approaches - the systemic approach hands wholes and studies individual components at the same time. Its analytic power is particularly useful for social sciences in which these approaches are still employed. Its superiority is due to that the truth is reached in the latter (Bunge, 1996: 281).

Its scope is universal while holistic and individualistic approaches are not. The systemic approach is tacitly or explicitly employed in the scientific study of systems of all kinds: conceptual, concret and symbolic, natural and social - from atoms to societies. It admittes many levels of analysis as many levels of reality (Bunge, 1996: 281).

Last but not least, systemic approach lets tackle the remainder quota of irrationalism of the modern scientific tradition. It is the assumption of a knowledge-independent reality which generates the epistemological problem of infinite regression (Aristotle, Met.; Hartmann, 1949; Popper, 1935). The assumption becomes unnecessary when the terms of the scientific knowledge

The Performance of Social Systems
Edited by Francisco Parra-Luna, Kluwer Academic/Plenum Publishers, 2000

(reality inclusive) are operatively defined and intersubjectively (Popper, 1935) or socially (Maturana y Varela, 1984) validated.

The consequences for the nature of values are remarkable. They do not exist *per se* as Platonic ideas, they are relative to any referential framework (Bunge, 1996: 220).

1.2 An optimum model

In a general way, systems are things or objects, and systems science search to describe and to explain outputs of systems appealing to inputs and complex processes of systems.

An optimum systemic model is dynamic. Not only operational definitions of *composition*, *structure* and *environment* are necessary but also a theory of *social change* has to be implied (Bunge, 1996: 270-277).

A useful systemic model makes possible to distinguish if change is refered to components, properties, structure or identity of system. It implies a clear distinction between the system and its domain of change of state by one side, and the agent or operator by other side. The agent or input could be a motive inside the system. So the distinction is operative. What or who originates the change is the agent and what or who suffers the change is the system. While the results are the outputs.

A systemic model which keeps the difference between inside and outside makes possible to distinguish between personally and socially refered values. They are values refered to the system and the environment.

The passive role of things or objects inherited from scholastic is regarded as active determinant of change by the modern systems science. What the system does?. It specifies by its own structural pattern of behavior if a stimulus of the environment is a threat, a challenge, an opportunity, or something irrelevant. It may be observed to any scale from microparticles to complex systems as societies. The opening of new markets is a well-known example. They are the firms, companies, associations, societies which determine how a new market stimulates them.

It introduces a basic difference in the way of understanding the performance of a system. They are not stimuli or inputs whose determine results or outputs. They are the systems which specify how inputs affect the outputs. It is not new. The Harvard's law - different members of the same species answer in a different way to the same stimulus - is the individualistic antecedent. Then the causal argumentation, outputs are function of inputs becomes affected. It is the system which specifies the outputs under certain conditions (inputs). So, the law of cause and effect may be amplied and refined by systemic modelization, without underestimating the conditions proposed by Albert Einstein and Max Planck (Planck, 1933) for solving the modern causal problem.

Finally, an optimum modelization may distinguish between formula of codification and registry; and formula of calculation and computation.

1.3 Definitions

By realists, a system is a complex object whose parts or components are related to other components of the same object (e.g. atoms, organisms, families, schools, firms, informal networks, societies, theories, classifications, computers) (Bunge, 1996: 270).

A concrete system at any time is represented by its composition, environment, and structure. It is called CES model and written in this way: m(s) = { C(s), E(s), S(s) }. Composition *C* is the collection of the parts of system *s* at the given time (e.g., members of a community). Environment *E* is the collection of things not in *s* that are connected to parts of *s* (e.g., the physical surroundings of a community plus the human outsiders with whom members of the community have relationships). Structure is the collection of relations among members of *s* plus the relations among these and those of *E*. The former is called the 'endostructure' and the latter is the 'exostructure' of the system (e.g., the kinship relations inside the community and the trade relations between the community and the outsiders, respectively). The part of the system whose components are linked directly to environmental items may be called the system boundary (e.g., the representatives, salesmen, and PR officers of a business firm constitute the latter's boundary) (Bunge, 1996: 270).

A matrix of components and an evolution operator as an agent of change are basic elements for a mathematical representation of changes of system (Bunge, 1996: 274-277) and basic concepts to understand the dynamic of systems as eigenproblems.

Also by realist position, values are relational properties of system, not intrinsec ones (Bunge, 1996: 220). They may be measured with reference to scales for classifying systems as more or less good for something.

Similarly, by constructivists the first step is to make distinctions. This everyday activity by which we separate unities from a background is the base of scientific knowledge. Unities are simple or complex entities distinguished as simple, and then discompose in their components. A *system* is a complex unity defined by *structure* and *organization* (Maturana, 1984: 17).

The *structure* is formed by the components - with their properties - plus the relations among them. The *organization* - whose definition evokes the

Aristotelian *organon* - is defined by those structure implied in the identity of class of a complex unity (Maturana, 1984: 17-19).

As elemental example, a table is a system whose structure is formed by one or many legs and a board. They are the components whose properties to verify may be their matter, color and size. The specific disposition between leg/s and board by which we have a table and not other system is the structure implied in the class identity of system, or organization.

Due to the changes that any system could suffer (including static systems like the table) it has been said in strict sense that any system is dynamic (von Foerster, in von Foerster and Maturana,1988).

Then, any system as structurally determined complex unity specify four domains:

i. *Domain of change of state or structural change*: those differences that a system may suffer without loss of identity. Structural changes are differences in the components and/or in the relations among them. While changes of state are differences in the properties of the components.

ii. *Domain of destructive changes*: those differences in the organization of a system which produce loss of identity.

iii. *Domain of productive interactions*: those actions between the system and the environment which produce changes of state or structural changes.

iv. *Domain of destructive interactions*: those actions between the system and the environment which produce changes with loss of identity (Maturana and Varela, 1984: 65).

In the example of the table, the break of its board by a tool becomes a destructive structural change by a destructive interaction if consequently objects can not rest on the board. If this is the case, the table loses such identity due to a difference in the organization. Usually, it is said that the table loses its value of table. Inversely, if the board could be repaired as manner as objects could rest over it, the structural change could become reversible. Then, the value of table would be retrieved.

The example illustrates that the four *domains specified by the system* make possible to pursue what changes and what is kept constant or invariable, with reference to the properties, components, structure or organization of a system, distinguishing unreversible and reversible changes with theorical and/or practical proposes. Really interesting is to pursue at what level of reality or observation changes are reversible or unreversible.

But really interesting is the dynamic of systems whose changes are produced by themselves or internal stimuli. By their own operations these systems generate under certain environmental conditions, or external stimuli, their evolution, mutation or destruction. The history of any kind of organizations, societies, firms and enterprises is full of these cases. The central concept is the operational closure of system by which is determined what can and not be made, and what could and not be made by a system.

Some misunderstandings are associated to the concept of operational closure. Every system is opened to matter, energy, information, new components, or could constitute new relations, but is constrained at individual (e.g., the prescription of roles and functions in a company) and global level (e.g., the Pauli principle in atomic physics) (Bunge, 1996: 391-392; Maturana, 1984: 47). None system can or could make any thing (e.g. Ford could not produce Volvo cars at least both companies are joined; the European Community could not extend its limits beyond the geographical, economical, political, cultural, legal Europeanization). It means that organization is not only a logic tool but also a physical concept by which something is or not in quantum sense (Maturana, 1984: 18-19). Their consequences are measurement and computation of systems.

By the cybernetics defined by three conceptual axes - paradox, causation and computation (von Foerster, in von Foerster and Maturana, 1988) - productive and destructive interactions are refered to an $operator_1$ (the observed system) which recursively operates over an $operator_2$ (the observed system) which also recursively operates over the $operator_1$, in a circular causation. A kind of recursion of recursion which solves the problem of infinite regression. Briefly, the recursion of recursion closes a cycle of argumentation. It indicates the epistemological sense of the second order cybernetic by which it is not necessary for validation of knowledge to multiply the ontological commitments proposing more cybernetic orders - as clear example of Occam Razor.

The key of second order cybernetics by contrast to first order cybernetics is to translate the focusing from control of independent reality to coevolution of observed and observer systems. Performance test illustrates it when it is operatively taken into account that the inputs of the system to test are the outputs of the tester; and conversely, the ouputs of the tested system are inputs for the tester.

Second order cybernetics computates domains of productive interactions and changes (*supra*) as stable eigenvalues, and domains of destructive interactions and changes as unstable eigenvalues (von Foerster, in von Foerster and Maturana, 1988).

2. PERFORMANCE

2.1 Test conditions

Basically, Performance Test seeks to answer the questions: 'What something does' and 'What it would do'. Also, 'How something or somebody works' and 'How it/she/he/they could work'. The question on possible or probable stages, situations or sceneries implies the variation of stimuli (inputs) with research and/or development purposes, and could be subjected to experimental treatment quantifying outputs.

When patterns of behavior - given answers (outputs) to certain stimuli (inputs) in a regular way - have been identified, they act as control conditions. While the variation of inputs is the experimental condition. The systemic modelization lets attribute the changes operated in the system by the inputs to specific relations (structure); parts or individuals (components); emergent properties of system or component´s properties; and/or relations implied (structure) in the class identity of system and/or components. These changes - or their observational consequences when they are formulated as hyphoteses - are the outputs. Hypotheses of level of reality may be formulated with reference to individuals (components), social relations (structure), economic, legal, politic, educational, scientific systems (subsystems), and/or society (system). These are structural couples by constructivist terms: structural couples of order II - organisms with nervous system-, and order III - cultural and social relations between individuals (Maturana y Varela, 1984). Likewise, a quadripartite model of human society called BEPC and composed by four subsystems - biological or kinship; economic; political and cultural - is proposed in realistic terms (Bunge, 1996: 271).

Whether realist or constructivist, social system research and development may be focalized on cultural, human and social communications (Luhmann, 1984; Teubner, 1989a, 1989b). In this case, messages may be regarded as components, and social systems become the context in which messages adquire specific meaning or are re-specified. Measurements of systems of communication also can be made (Bálsamo, 1986; 1987).

When certain outputs are sought, it is necessary to know if inputs act at level of components, properties of components, relations among components, structure, emergent properties, and/or organizational structure (*organon*). Technology of materials, industry of food and drugs, production of energy, demographic movements of emigration and immigration, access to new information technology by enterprises, government and families, are examples.

The value judgement about the goodness (or not) of certain inputs for a given system in certain circumstances implies a comparison between inputs and outputs with reference to a value scale whose empirical consistence about it measures must be demostrated. For example, the judgements about the goodness of decentralization of political power for increasing local autonomies and democratic participation.

Due to outputs of social systems have plenty of values for the members of the society (e.g., reduction of inflation, reduction of unemployment, control of natality, reduction of violence, reduction of contamination) it is regarded that the outputs of social systems are values.

The improvement of certain properties of systems also implies an evaluation situation in which present and future - or past and present - inputs and outputs are compared. For examples, the learning of new skills by students, innovations in customer services and institutional procedures in public sectors as Health, Education and Justice.

Failures in the acquisition of certain results are on account of constraints at different levels (e.g., prescription of roles in a company, bureaucratic hierarchy, systems of ideas and beliefs of individuals, social groups and societies).

All that requires a basic universal tool which can be specified to any area of knowledge for scientific and technological purposes. Such a tool would have to comprehend systems in their actual changes, and prospectively in their possible and probable changes by different stimuli. Such a tool would have to make possible comparisons not only inside a discipline but also in transdisciplinary and interdisciplinary way. Such a tool would have to make possible codification, registry and measurement of inputs and outputs pointing out causal relations, and also it would have to admit formula of calculation and computation. Finally, such a tool would have to be compatible with different epistemological positions if keep the condition of unity of science by systemic approach.

2.2 Causal Legal Tool

The proposed tool satisfies all those conditions. Its formula of codification and registry is compatible whith quantitative treatments of imputs and outputs, as well as properties, components, relations, structure and identity. It is formulated as strict law, satisfying the scientific desideratum of searching laws, and the technological ends of prescriptions for doing things. Last but not least, this strict law has been proposed (Bálsamo, 1997) for solving the modern causal problem formulated by Albert Einstein and Max Planck (Planck, 1933).

In the first half of the century Albert Einstein recognized the insufficiency of the modern law of cause and effect and asked for a more strict law of causality (Einstein, in Planck, 1933).

Since then, the main strict causal law formulated by Donald Davidson (1970,1980) is holistic and with the restriction (Anomalism) of the mental. Instead, the causal law proposed is systemic. It means holistic plus individualistic, and does not mean the restriction of the mental. It is illustrated by the application of the proposed causal law to the study of the cognitive modality (Bálsamo, 1987), and basically implies that the mental level (by realists) or mental phenomenon (by constructivists) may be explained as emergent property of organisms with nervous system.

The proposed causal law is formulated by the four senses of scientific law (Bunge, 1958):

1.objective: the law with reference to objects;

2. nomological: the law of laws or second order law;

3. nomopragmatic: the law with technological purposes;

4. metanomological: the epistemological sense of law for theoretical purposes.

Causal law also comprehends the most systematic and complete causal codification of the history of knowledge. It is the classical four Aristotelian causes - efficient cause (ec); formal cause (foc); material cause (mc); and final cause (fic) (Aristotle, Met.).

Each one of them codifies a specific information in the construction of knowledge or discovery of the truth: formal cause (foc) answers the question what is?, it refers to the thing; efficient cause (ec) is the agent, it refers to the stimuli, perturbations, interactions, inputs; material cause (mc) codifies the space in which something exists, its constituent, it refers to properties of components, while final cause (fic) answers the question for what?, it is defined by the purposes, objectives, functions, results and outputs.

So then the four senses of *Causal Law* are the follows (Bálsamo, 1997: 188-192):

Causal Law_1: any system (foc, mc) specifies which interactions (ec) can destroy it or produce changes of state or structure without loss of identity (fic).

The law embraces infinitude of stimuli, perturbations, opportunities, interactions, able to produce changes of state or structure without loss of identity.

Causal Law_2: evolution = conservation + variation. Conservation is defined by those structures that can not change without loss of identity. Variation is defined by those structure that can change without loss of identity.

Causal Law_3: for the evolution of a system do not interact destructively with it.

Destructive interactions produce disintegration and loss of identity.

Causal Law_4: Coordination of action of coordination of action. The four Aristotelian causes (ec; foc; mc; fic) are organized in two ordinate pairs. It produces twelve sequential combinations. The relation R between the two ordinate pairs is inclusion or belonging.

Causal $Law_4 = \{ (c1; c2) \, R \, (c3; c4) \}$

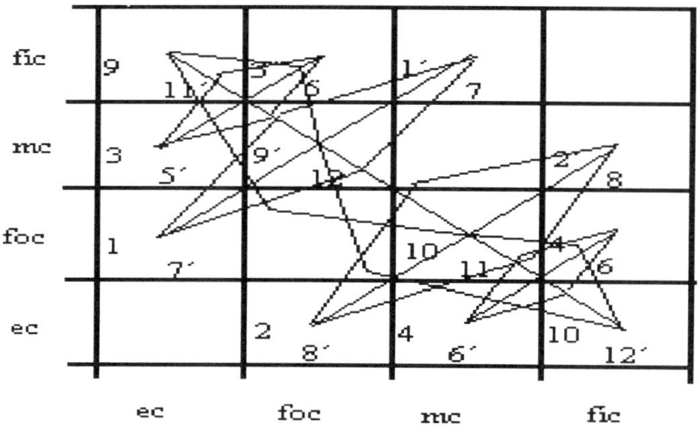

Fig. Combinatories of Causal Law$_4$ (Bálsamo, 1997: 192)

The experimental character of the Causal Law refers to the different meanings that the four causes can and could assume and the frequency of empirical observations to them associated by statistic, intersubjective and social validation. Falsability is also admitted.

According the classical example of the fire, the question would be: how often the fire (foc; mc) is extinguished (fic) by water (ec); by sand (ec), or by fuel (ec)?. Other question could be, under which circunstances (ec) people (foc; mc) asks for more justice, freedom or solidarity (fic)?. Values like these may be measured and compared as inputs objectively scarce (ec) and outputs subjectively expected (fic). In a process, when these valuable objectives (fic) are finally reached as an increase in the quantity of justice, freedom or solidarity, they become means (ec) for reaching new goals. Whether efficient or final causes, values must be refered to societies whose members (foc; mc) specify with their perceptions and actions the state of them. A complete and true knowledge only arises searching the four causes.

The given knowledge of only one cause - as the efficient one (e.g., lack of justice) - makes possible to search, to research, and to experimentate the other three causes in different circumstances and with different purposes.

As tool of knowledge, Causal Law can be applied to any type of concrete systems - social, physical, biological -; or abstract systems as theories, approaches, cognitive modalities; patterns of behavior of any type - biological, psychological, social, cultural; also devices, machines, products of any type - technological, cultural, scientific - with purposes of performance test. Likewise, it can be applied to collections of properties, aggregates of elements, et cetera, under condition that they are subjected to the systemic formulation - as the application of causal law to the typical case of indeterminist physics, the Uncertainty Principle, illustrates it (Bálsamo, 1997: 193-194).

In Social Sciences the causal law may be applied to design, improvement, evaluation and evolution of theories and politics of different type - societal, scientific, legal (Bálsamo, 1997: 197-201; 243-261). A scale of evolution theoretically founded (Luhmann, 1984; Teubner, 1988b) may be used to evaluate social systems as Economy, Politics, Law, Science, Education. When social systems become functionally differentiated among them - a modern success -, other functional differences may take place inside them. When the functions of selection, variation, and retention or re-stabilization are structurally differentiated in a social system, it is regarded as an evolutionary social system. For example, in the legal system, normative structures take over variation, institutional structures - specially procedures - take over selection, and doctrinal structures take over retention (Teubner, 1988b: 228). When these functional differences have not been introduced yet, it is regarded as a pre-evolutionary social system. When these differences were reached and then lost, it is regarded as a post-evolutionary social system - a postmodern success. It makes possible to explain disfunctional effects or outputs which have plenty of values - as corruption of social systems - and also to design and to implement strategies to search its reduction as social re-engineering (Bálsamo, 1997: 189).

Causal law is useful in case of orthogonal intervention. They are interventions that operate perpendicularly to the plane of relations which define a system as member of certain class (Maturana, 1984: 127). They are prescribed to revert courses of action of systems. Applied to a community with economic depression, problems of emigration, and many antagonic subgroups, the orthogonal intervention made possible: i. to interact at level of familiar networks as manner as the members of the community perceive themselves as such, and apply their own decision criteria (by affective and physic or geographic proximity) to design the sequence of interviews; and ii. to identify the main social network of emigration, number of members, relationships, frequence of coming back for participating in the 'Patronal Celebrations', a folk and religious annual meeting; iii. to point out the *via regia* of cultural access to the community as manner as the mayority of its members participate joint to the research team in the establishment of an Archeological and Natural Park in the area (Bálsamo, 1986). In cases like this, causal law lets identify patterns of cultural and social behavior necessary to design and to implement such a intervention. At the same time, it lets increase the values of mutual acceptance, participation and cooperation among the members of the community and between the community and the research team. These values are measured as outputs of the human social system and compared in their states before and after the orthogonal intervention.

Causal Law may also be applied to the measurement of the social impacts by the introduction of a new information technology. Such the case answering the question, how internet affects the social relations?. A simple protocol for internet users asking them over the time devoted to internet, the frequency of use of its options, and the activities modified and substituted by internet, would let to know by statistics what social structural changes (outputs) are produced in certain social relations - work, trade, recreation, study, learning - and/or social systems - economy, education, science, politics, law - by internet (input). The results would draw a previous stage 1 without internet, and a stage 2 with its introduction (including its progression) and would identify niches of accessibility and resistence to internet in the different social relations and systems. The mathematical and computational formulation as eigenproblem would let answer the discussed question, if internet changes or not the society?. Operatively, society is the whole of social relations and social systems taken into account, while its changes are represented by unstable eigenvalues when its constituent structure or subsystems mute in a destructive way (variation without conservation, see causal law_2), and by stable eigenvalues when they mute in a productive or evolutive way (variation with conservation by causal law_2).

3. EVALUATION

Systemism = holism + individualism, by philosophical definitions (Bunge, 1996: 241- 281).

The proposed causal law, in its reference to systems, falls under this definition. According to epistemological considerations, causal law combines systemic and cybernetic definitions by constructivist position, and legal formulations by realist position. Methodologically, causal law may be applied by realist or constuctivist positions. The triangulation between realist and constructivist positions by causal law, potentiate the usefulness of both of them, instead to produce unfruitful competence.

So realistic and/or constructivist, Performance Test requieres access to the dynamics of systems as manner as to decide if interventions produce changes of system at level of properties, components, and/or structures. It not only implies a theory of systemic change but also a clear distinction between domain of changes in the system and domain of interactions with the system; and a causal legal tool to design and to implement interventions which combines the scientific requirements of law and the technological requirements of rule, as well as a solution to the modern causal problem. All that, without refusing, but assimilating the controversial study of values. Finally, such a causal legal tool may be applied to tested and observed systems, and/or tester or observer systems; and admittes the mathematical and computational formulations of eigenvalues.

4. CONCLUSION

A systemic approach called to play the protagonic role in the Unity of Science not only has to optimize its modelization as manner as epistemological differences become fruitful, but also to develop analitically powerful tools able to afford inclusively the study of values in Social Sciences. The conditions for Testing Performance illustrate that.

5. REFERENCES

Aristotle, Metaphysics, London, Heinemann; Cambridge, Massachusetts, Harvard University Press, 1947, Loeb Classical Library, 2 vols.

Bálsamo, Iris B., 1986: "Ensayo de comunicación transcultural generado en el acceso a un sistema social humano", Consejo de Investigaciones Científicas y Técnicas CICyT, Universidad Nacional de Córdoba, Argentina.

1987: "Test of Cognitive Modality in the descriptive analysis of a video record", Postgraduate course "Ethology", Laboratory of Analysis and Registry of Anthropological Data - LARDA -, Natural Sciences Faculty and Museum, National University of La Plata.

1994a: "La noción de evolución en Weber", Cuadernos Weberianos I, II, III, IV, Compilador Dr. Juan Carlos Agulla, Estudios de la Academia Nacional Ciencias de Buenos Aires, Instituto de Derecho Público, Ciencia Política y Sociología, Tomo I, 51-64.

1994b: "El sentido de la función humanista á la Luhmann", curso doctoral "Teoría Sociológica III", Facultad de Postgrado, Universidad de Belgrano, Buenos Aires.

1994c: "Los tres Estadios de Modelización Sistémica", curso doctoral "Sistemas Políticos Comparados", Facultad de Postgrado, Universidad de Belgrano, Buenos Aires.

1995: "Causalidad. Su relevancia para la literatura", Revista TRAMAS, Año 1, N° 3, Alción Editora, 104-107.

1997: "La Evolución de la Causalidad en la Ciencia Moderna", tesis doctoral en Sociología, Facultad de Postgrado, Universidad de Belgrano, Buenos Aires.

1998a: "La Evolución de la Causalidad", Evolucionismo y Racionalismo, Eustoquio Molina, Alberto Carreras, Jesús Puertas (Eds.), Institución "Fernando El Católico", Universidad de Zaragoza, 31-37.

1998b: "Causal Law. Classical Innovation", Joint Sessions Sociocybernetic Methodology, Research Committees Logic and Methodology and Sociocybernetics, 14th World Congress of Sociology, Montreal.

Bunge, Mario, (1957, 1958, 1959), 1993: La ciencia, su método y su filosofía, Ediciones Siglo Veinte, Buenos Aires.

(1969), 1989: La Investigación Científica. Su Estrategia y su Filosofía, Editorial Ariel, Barcelona.

1973: Method, Model and Matter, Dordrecht, Reidel.

1958: Metascientific Queries, Springfield: Charles Thomas.
1959: Causality. The place of the causal principle in modern science, Cambridge, Massachusetts, Harvard University Press.
1996: Finding Philosophy in Social Sciences, Yale University.
1998: Social Sciences under Debate. A Philosophical Perspective, University of Toronto Press, Canada.
Davidson, Donald, (1966-7), 1993: "Causal Relations", in Causation edited by Ernest Sosa and Michael Tooley, Oxford University Press.
1970: "Mental Events", Lawrence Foster, J.W.Swanson (eds.): Experience and Theory Amherst: The University of Massachusetts Press.
1980: Essays on Actions and Events, Oxford: Clarendon Press.
Hartmann, Nicolai, 1949: Grundzüge einer Metaphysik der Erkenntnis, 4. Aufl., de Gruyter.
Landsberg, P.T., Ludwing, G., Thom, R., Schatzman, E., Margalef, R. & Prigogine, I., 1986: Proceso Al Azar, Tusquets Editores, Barcelona.
Luhmann, Niklas, 1984: Soziale Systeme. Grundisse einer Allgemeinen Theorie, Suhrkamp Verlag.
1996: "Causality in the South", Soziale Systeme Zeitschrift für soziologische Theorie, Vol.1, H.1: 7-28.
Maruyama, Magoroh, 1963: "The Second Order Cybernetics: Deviation-Amplifying Mutual Causal Processes", American Scientist 51: pp. 164-179; pp.250-256.
1980: "Mindscapes and Science Theories", Current Anthropology, 21 (5):589
Maturana, Humberto, 1970a: "Neurophysiology of cognition" in Cognition a Multiple View, Paul Garvin (Ed.): 3-23, New York: Spartan Books.
1970b: Biology of Cognition, BCL Report N°9.0, Biological Computer Laboratory, Department of Electrical Engineering, University of Illinois, Urbana, 95 pp.
1973: De máquinas y seres vivos, Editorial Universitaria, Santiago, Chile
1978a: "Biology of Language: The Epistemology of Reality", in Psychology and Biology of language and Thought; G. A. Miller & E. Lenneberg A. P. (Eds.):
27-63, New York: Academic Press.
1978b: "Cognition", Wahrnehmung und Kommunikation / hrsg. von Gerhard Roth - Frankfurt am Main. Bern. Las Vegas. Lang.
1982: "Reflexiones: aprendizaje o deriva ontogénica" en *Archivos de Biología y Medicina Experimentales*, vol.15:261-271, Chile.
1983a: "What is it to see?", en *Archivos de Biología y Medicina Experimentales,* vol.16:255-269, Chile.
1983b: "Fenomenología del conocer", *Revista de Tecnología Educativa*, Vol.8, Nos.3/4: 228-252.
1984: "Diálogos con el Dr. Humberto Maturana", *Cuaderno de Terapia Familiar*, Año 2, No.2, Centro de Estudio de Terapia Familiar (C.E.TE.FA.), Rosario.
1986: "Epistemología", *Conferencia*, Facultad de Medicina, Universidad Nacional de Rosario.
1988: "Ontología del Conversar", *Revista Terapia Psicológica*, Año VII, N°10:15-23, Santiago, Chile.
(1988), 1990: *Emociones y Lenguaje en Educación y Política*, Hachette/CED (Centro de Estudios del Desarrollo), Chile.
(1975,1981-1990), 1992: *El Sentido De Lo Humano*, Hachette, Chile.
1994b, "Conocimiento y Realidad", *Seminario*, IV Congreso Internacional sobre Constructivismo en Psicoterapia, Buenos Aires.
Maturana, Humberto and Varela, Francisco, 1972: "Discussion: Mechanism and Biological Explanation", *Philosophy of Science*, Vol.39, No.3, September.

1984: *El árbol del conocimiento*, Organización de Estados Americanos, OEA/Editorial Universitaria Santiago, Chile

Maturana, H., Varela, F., & Uribe, R., 1974: "Autopoiesis the organization of living systems; its characterization and a model", *Biosystems*, 5, 185.

Maturana R., Humberto y Verden-Zöller, Gerda, 1994: *Amor y Juego. Fundamentos Olvidados de lo Humano. Desde el Patriarcado a la Democracia*. Editorial Instituto de Terapia Cognitiva, Santiago, Chile.

Nagel, Ernest, 1961: *The Structure of Science*, Harcourt, Brace & World, New York.

Pareto, Vilfredo, 1966: *Sociological Writings*. Selected and Introduced by S.E. Feiner. Totowa, N.J.: Rowman and Littlefield.

Parsons, Talcott, 1951: *The Social System*, New York: Free Press; London: Collier-Macmillan.

Planck, Max, 1933: *Where is Science going?*, with a Preface by Albert Einstein, London: Allen & Unwin.

Popper, Karl, (1945), 1962: *The Open Society and Its Enemies*, 2 vols. Routledge & Kegan Paul, London.

(1935), 1959: *The Logic of Scientific Discovery*, London: Hutchinson.

1972: *Objective Knowledge*, Oxford: Clarendon Press.

Russell, Bertrand, 1961: *Basic Writings, 1903-1959*, Simon and Schuster.

Teubner, Gunther, 1987: "Juridification. Concepts, Aspects, Limits, Solutions", in G.Teubner (ed.), *Juridification of Social Spheres*, Berlin: de Gruyter, 3.

1988a: "Hypercycle in Law and Organization: The Relationship between Self-Observation, Self-Constitution and Autopoiesis", *European Yearbook in the Sociology of Law* 43.

1988b: "Evolution of Autopoietic Law", in G.Teubner (ed.), *Autopoietic Law: A new approach to Law and Society*, Berlin: de Gruyter.

1989a: "And God Laughed ...: Indeterminacy, Self-Reference and Paradox in Law", in C.Joerges and D.Trubek (eds.), *Critical Legal Thought. An American-German debate*. Baden-Baden: Nomos

1989b: "How the Law Thinks: Toward a Constructivist Epistemology of Law", *Law & Society Review*. Volume 23, Number 5.

1996: "The King's Many Bodies. The self-deconstruction of Law's Hierarchy", *Soziale Systeme Zeitschrift für soziologische Theorie*, Vol.2, H.2:229-256.

von Foerster, Heinz and Maturana, Humberto, 1988: "Cybernetics Biology and Communication", *Seminary*, Sociedad Argentina de Terapia Familiar - SATF - y Asociación Sistémica de Buenos Aires - ASIBA - , Buenos Aires.

Weber, Max, (1904-5), 1920-21: "Die protestantische Ethik ubd der Geist des Kapitalismus". In Weber, *Religionssoziologie*, 1:1-206. Tübingen: Mohr.

Wilden, Anthony, 1972: *System, and Structure - Essays in Communication and Exchange*, Tavistock Publications, London.

Part III

Operational Approaches to the Concept of Social Performance

Chapter 12

Environmental balance and the performance of social systems

ANTONIO SÁNCHEZ SÚCAR
Aeronautical Engineer, Head of Department of Ecology and Environment, Iberia, S.A., Madrid

1. FOREWORD

This paper intends to present approaches made to integrate environmental aspects with economic decisions at macro and microeconomic levels and create a debate on how to proceed for improving their performance: should private economic agents from the productive sectors ranging from large multinational corporations to small family companies foster this process or should it be promoted by politicians or administrative organisations? We will take a look at some examples of theoretical developments intended to be applied to National Accounting Systems, and will pay attention to the current situation of environmental involvement and see how it is accounted for at corporate level, particularly in the Spanish case. With several years of experience in the air transport sector, the author also examines the sustainability of aviation, how airlines tackle this problem today and considers the future outlook. Finally, through sociocybernetic theory, the environmental performance of a business organisation is approached using a transport company as an example.

The Performance of Social Systems
Edited by Francisco Parra-Luna, Kluwer Academic/Plenum Publishers, 2000

2. THE INTEGRATION OF ECONOMIC, SOCIAL AND ENVIRONMENTAL SYSTEMS WITH SUSTAINABLE DEVELOPMENT.THE ROLE OF TECHNOLOGY AND THE NEED FOR GLOBAL DECISIONS

In recent years an increasing number of contributions have been made towards integrating ecology and the economy . Jiménez Herrero (1) describes the changing attitudes of scientists who in the past have been reluctant to admit a clear convergence of ecology and the economy, until a new paradigm of sustainable development has become prevalent in the belief that common aspects of the two disciplines were more important than their different origin. What then is sustainable development?

World Bank economist Herman Daly (1990) has provided a often cited definition of sustainable development, whose material and energy throughputs would have to meet three conditions:

-Rates of use of non-renewable resources may not exceed their rates of regeneration.

-Rates of use of non renewable resources may not exceed the rate at which renewable substitutes are developed.

-Rates of polluting emissions cannot exceed the assimilatory capacity of the environment.

It is apparent that the word "rate" appears in the three paragraphs of the definition. Rate is a variable factor or "a relation between two magnitudes that usually change in nature". So, a given non sustainable process can be reversed by persons using the proper technology. In the following paragraphs we will refer to the concept of Best Available Technology, as " the technology that provides maximum output with minimum environmental burden". There is still no sustainable technology but this is the best available for now.

Sustainability may adopt the following approaches: a) limit economic growth according to the environment's capacity to absorb undesired by-products, or b) develop technology to prevent, reduce or reuse emissions, wastes, and so on, improving efficiency of non renewable energy processes, or stimulating those based on renewable energies. Therefore, technology (Fig.1) constitutes a strong basis at the point where economy and ecology must relate, refuting the Malthusian idea of incompatibility between economic development and sustainability.

It is often forgotten that economic activity varies according to different levels of economic growth. Natural resources intensively used in industrial activities lead to services- based economies with lesser consumption per production unit. Furthermore, the public's "willingness to pay" additional costs to prevent environmental deterioration is becoming more common as incomes rise in developed societies, where consumers and ecological organisations pressure administrations for environmental policies. These pressures tend to be more intense at local or regional level, where the environment is conceived as a community asset instead of its more ambiguous or uncertain position in global nature.

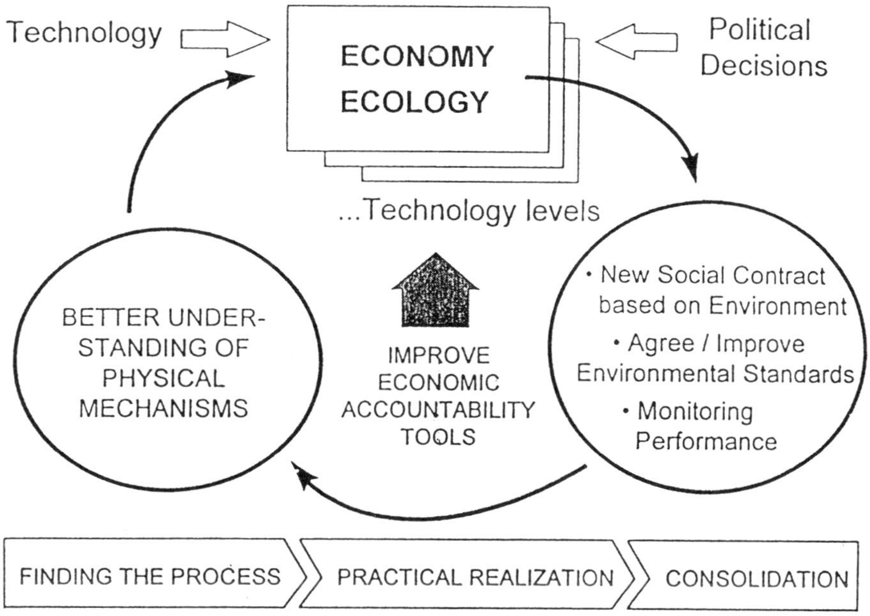

Figure 1 The role of technology in the relationship between economy and ecology

3. PROVIDING ENVIRONMENTAL ACCOUNTABILITY AT MACROECONOMIC LEVEL. NEW METHODOLOGY AND PRACTICAL PROBLEMS

It has been mentioned previously that processes to account for environmental changes can be promoted by the administrative departments that regulates the Systems of National Accounts or at corporate level in profit/ losses form or by Statement of Income modifications. In the following section, several attempts to account for environmental considerations that may affect former GDP accountability will be presented.

Under the United Nations Rio Conference Program 21, a commitment to use new Systems of National Accounting was agreed. The basic principle on which the UN Statistical Office has been working (2) has been that National accounting should be modified to include the fact that the environment is not a gift and that its utilisation should be measured, accounted for, and listed as a national asset in GDP. This can be achieved by monetary or physical accounts (the "satellite approach"). In the European Union, a forum for Accounting Advisors has been formed to make recommendations that include environment issues in National Accounts from 1999. Carsten Stahmer (3) has proposed several measures to improve GDP which include the removal of "regrettable" parts such as the by-products of economic activities like the depletion of natural resources and land degradation. Economic production inputs must contain the costs of using produced assets and natural assets, too. Based on this consideration, the ecodomestic product can be calculated as follows:

Gross Domestic Product(GDP)-Depreciation of produced fixed assets caused by economic activities = Net Domestic Product (NDP)

-Depreciation of non-produced Natural Assets caused by Economic Activities= Eco-Domestic Product (EDP)

Such depreciation of non produced natural assets, must only include assets used in economic activities and market valuation, and/or contingent values. In this approach, withdrawals or even additions to GDP can be found if protective efforts lead to a better environmental quality.

By using "Sustainable Indexes" we can compare the historical evolution of sustainability, or welfare parameters. An example could be the ISEW, the Index of Sustainable Economic Welfare, published in the U.S. by H. Daly from a diverse set of activities (4) where 21 parameters have been tracked since 1950, including costs of transportation, car accidents, education, health and water, air and noise pollution. The author accepts the limitations of such an approach, due to the difficulty of establishing standardised criteria, quality measurements or the many physical processes that govern environmental changes.

Combination of GDP values and sustainability factors. Parra Luna (5) has set out the "Nature Reproduction Index" (NRI), thus:

$$\mathbf{NRI=1/[1-(NT\ (t+1) - NT(t))/NT(t))]}$$

where NT(t+1) is the nature level at a given time and NT(t) at the preceding one. Having values from 0 to NRI>1 would indicate an increase of natural elements (water and air pollution, forestation, energy used, etc) and could be used in conjunction with GDP to obtain the Sustainable Economic Index (SEI) as follows:

$$\mathbf{SEI=GDP*NRI}$$

We believe that attempts to modify SNA or GDP with environmental variables is, now at the stage of setting appropriate methodology, but still important problems of practical nature remain to be resolved. The ultimate goal being a reasonable monetarisation of such effects (since most accounting variables are monetarised as GDP), the problem at this point is to understand the physical effects, in particular those of a global nature that are supposedly responsible for major economical impacts. The problem referred to is the one reproduced at corporate level as described later. So, what is the social cost of releasing into the atmosphere a ton of CO_2 or depleting the Ozone Layer by a mm.? Widely used mathematical models are compared with relatively few "in situ" measurements. At this stage of the investigation, NO_x produces or depletes tropospheric ozone (aggravating or alleviating the greenhouse effect) according to altitude and concentration. An EPA expert recently confirmed to us that "true value" externalities are not available, although a US organisation (The New York State Public Service Commission) has sponsored a study to estimate quantitative, monetarised externalities on dollars per ton of several common pollutants. Unless this ratios can be demonstrated and generalised, only part of the externalities can be internalised if not over-costed with dissuasive taxes, or nearly arbitrary money-gathering penalties.

Deep research is still required to know the effects, as is the case of aviation. The United Nations Intergovernmental Panel on Climate Change (IPPC 1990), considered the most authoritative summary on the state of knowledge in this field, in 1990 published estimates that maintain that if no mitigating actions were taken, global mean temperatures could rise from 1990 onwards by between 0,2 and 0,5 ° c per decade, that is a deviation of 150%. Mathematical models are still not sophisticated enough to predict the future and have to be taken carefully. During a recent European Commission of Companies of General Economic Interest (CEEP) meeting, actual inventories showed differences when compared with World Economic Council (WEC) model values. Two Spanish scientists C. Duarte and S. Agosti (6) have concluded that predictions of the CO_2 sink effect of oceans must be revised to have zero net effect due to underestimates of water surface fauna.

Other sources of problems are related to monetary values . It is generally accepted that environmental impacts can be calculated with a reasonable degree of accuracy when market prices mechanism is used. However , in the long term or with large geographical scales, results have to be taken carefully. The case of deforestation caused by timber factories can be evaluated at the market prices for timber as a raw material, but other aspects such as CO_2 sink effect, soil degradation, and so on, must also be considered.

4. THE SOCIAL CONTRACT AND THE SUSTAINABILITY AT CORPORATE LEVEL

During the 1960s, enterprises and academics realised that, in addition to other obligations to shareholders to produce profits, companies should have commitments to other members of society, including their own employees, and they should determine, measure and report corporate social impacts that would be reflected in the Company Social Balance.

However, the company's Social Balance, alone, has not been as successful as expected decades ago and its scope of interest has been absorbed by the broader Health, Risk Prevention and Environmental framework. Several reasons could explain this trend: a) a surge in neo-liberalism theories, the welfare crisis and subsequent company social budget and wage cuts. b) The new role of Unions in companies following market deregulation, with outsourcing strategies and rising unemployment and c) The social balance was erroneously used as a public relations tool and not sufficiently enough to make fundamental changes in corporate culture. Increasing environmental concern, the risk of corporate and executive legal prosecution due to non observance of regulations and the possibility of costs reduction have led to a new approach in social responsibility particularly when addressed to environmental issues.

How do companies perform environmentally nowadays? We realise that, due to the rapid evolution of this issue, what is asserted today may not be valid tomorrow. This is the result of a study using several sources. One is the survey undertaken in 1992 by the Institute of Management Accountants, including more than 100 regional , national and multinational, public and private North American companies. It was compiled by Marc J. Epstein, from Stanford University (7) and showed that US administration efforts to improve accountings at "national green GDP " levels were been correlated into corporate "green accountability". The first of these reports was published by the Department of Commerce in 1994 and recognised the poor development of these issues. The second source is an UNEP survey dated 1995 on environmental reporting excellence of more than 100 pioneer companies (8), and finally, we have studied and compared the 1996/97 Environmental Reports published by 15 recognised companies (9). We have also been able to see not only the differences in the origin and sizes of companies but also the evolution of the environmental strategies and reporting systems:

Environmental Technics and Community Participation Companies are improving the scope of their environmental performance with the involvement of internal and external agents. The aim is to achieve greater participation of all company stakeholders to reach a widely accepted

Corporate Environmental Strategy Implementation. A good example of stakeholders participation is provided by Madrid's Barajas Airport, where a new runway has been constructed. Airport Authorities, state administration and mayors of 17 towns located around the airport are participating in several Committees to co-ordinate noise and other environmental programmes for which Pta17,000 million. are allocated.

Informational and Reporting Systems The UNEP survey ranked companies reporting excellence in five categories with the following results:

-Many of the report makers (39%) are still at stages 1 or 2.

-A quarter (25%) of those surveyed are at stage 3, with a .further 11% making the transition to Stage 4.

-Only 5% have reached stage 4.

-Stage 5, which will be based on the extensive use of quantitative methods (such as life cycle assessments and mass balances) and on strong links with industry-wide and national sustainable development reporting remains largely unexplored territory.

A comparison of the UNEP survey with the efficiency of the 15 company Reports for 1996/97 shows a noticeable change in the period, with the following results:

-All Reports are published annually, and include text ,figures, corporate and Site reports, annual environmental targets and programmes, and details of energy and water consumption. They also include links of environmental social and staff aspects as well as safety and health matters.

-One Report includes information both on paper and in CD-ROM format.

-Five include costs information on eco-programmes and internal investments, and three mention programmes related to Life Cycle Assessment. In one case, information on taxes and fines paid is given.

Therefore, all these Reports can be assigned to cat. 4 or 5 according to UNEP ranking, which shows a clear improvement in environmental management at company level.

Lack of Standards Generally accepted standards are insufficient to conduct audits and reports which makes it difficult to compare company performance. In the U.S., the Environmental Protection Agency (EPA), the Financial Accounting Standards Board (FASB) and the Securities and Exchange Commission (SEC), are major organisations with published audit standards, and we have listed 7 more organisations with the same objectives. It would be desirable that the International Standard Organisation (ISO) 14000 norm, published in October 1996 , could help in this direction.

Cost Accountability Although ESI requires that environmental costs, risks and liabilities be determined and integrated in the general accounting system, Epstein stated that "one of the most disturbing findings of the field research project was that most companies do not have an adequate system for identifying and measuring environmental costs. In most instances they do not track or accumulate these costs separately and thus do not know the total corporate environmental costs or their causes". This finding in the U.S. companies' survey has been confirmed through the Environmental Reports of companies surveyed. Only 3 of the 15 mentioned the costs of some environmental programmes and 1 disclosed recovery savings. The problem of environmental costs accountability is far from being resolved and this can be attributed to the following:

Cost imputation of the environmental expenditures. When a capital expenditure is selected, benefits can be counted in other areas such as better product figures (longer life, reliability, performance, etc.), company image or productivity, so that costs must be attributed to all improved areas. The following approach could be suggested:

Total Costs=f(Total Benefits) or $\sum C_i = f(B_1, B_2, B_3, \ldots B_n)$, being

B_1=Benefits due to product improvement:: sales for market share increase.

B_2=Savings on new equipment maintenance

B_3=Environmental savings due to: less cost of waste management, ecotaxes and other externalities, green image, energy efficiency, etc.

It could be assumed that::

$$B_n/\sum(B_i) = C_n/\sum(C_i)$$

Time and space scale. Environmental impacts may take a long time to be perceived. Many externalities were caused years ago, where sustainability was not a worry and must now be confronted with current expenditures. The question is how these costs can be ascribed to recover lost environmental performance. Furthermore, costs such as those for local remediation programmes (for soil or forests degraded in the proximity of a production plant) can be evaluated at market prices and costs allocated accordingly, but this is not the case for global effects.

Monetarisation and physical units monitoring Due to difficulties found in monetarising environmental impacts described earlier, the method used by companies to monitor environmental impacts relies on the control of physical units. Historical series can be plotted and compared with programmes targets. Some companies disclose environmental programme costs but very few the attributable benefits. When they occur, benefits do not include the "hidden" effect of the externalities. Energy efficiency is one of the most common targets of environmental programmes. It can be stated that, as physical units monitoring is almost the only way now of measuring environmental performance , this will have to coexist with monetarised accounts in the future.

5. CORPORATIONS AND THE ENVIRONMENT. THE SPANISH CASE.

In 1996, a Ministry of the Environment was created for the first time to upgrade environmental strategies and strengthen the capabilities of the Spanish administration. Since then, consumer demands and competition with more environmentally advanced companies in other European countries has meant that companies in Spain have made significant progress in this field.

The consultancy firm KPMG begun a survey of the environmental situation in Spain in 1995 that was updated in 1997 (10), and included more than 200 of the most representative companies. A more comprehensive study was undertaken by Fundación Entorno (11), compiling information from 4384 production centres during 1997, in addition to qualitative interviews with managers and business association leaders. Conclusions can be summarised as follows:

Implementation of Environmental Management Systems. The rate at which companies are being registered for environmental performance certificate is high . In 1996 only 21 centres reported plans for registration . The number of registered centres was 18 by mid 1997 with 46 applications and 29 and 51 respectively by October of last year . There is still some way to go before European averages are reached.

Reasons for environmental involvement As in other European countries, legislation seems to be by far the main reason for environmental concern shown by companies in Spain. Interviews with managers show that the environment is not a priority, unless other economic benefits can be derived.

Company size There is a clear tendency to implement Environmental Management Systems in large companies rather than in small ones. This is because the environmental function is considered as an overhead, instead of direct productive labour. Environmental functions are less common in companies with less than 20 employees.

Environmental Expenditures As far as capital and operating expenditures are concerned, the 1996 KPMG report showed that only 2 companies, out of the 200 approached, reported environmental costs in their financial statements, 3 mentioned some capital expenditures used to correct air emissions and water wastes, and 2 listed environmental liabilities. Managers cited lack of criteria for breaking down environmental costs, as the main cause for their omission in the financial statements. However, the White Paper, surprisingly , states that 86% of the total number of productive centres surveyed reported having some kind of capital expenditure in that year. It can be concluded again that companies increasingly make environmental expenditures but that they do not account for them. The capital expenditure destination shows in Fig. 12 that the lowest investment was used to the least regulated areas: noise and soil remediation.

6. ENVIRONMENTAL EFFECTS OF AVIATION

The aviation main environmental impacts can be basically attributed to atmospheric emissions due to combustion of fuel and noise produced by the engines. The emissions have global as well as local effect at the airports, and engine noise mainly affects the airports and their vicinities.

Emissions.- Aviation's share to aggregate global emissions is low, but it is expected to grow in the next decades faster than the economy as a whole. There is a relative consensus that increase on aviation CO_2 emissions will triple in absolute terms in 2015 versus 1990 , according to VROM 1995 (12) but they are more pessimistic than prospects made by IATA (International Air Transport Association).

Regardless the energy efficiency achieved so far, several European Union proposals to apply restrictions and market instruments to aviation (13) are being discussed. While agents involved are not in principle opposed to the internalisation of externalities created by air transport, a demand of transparency and credibility is becoming a must .

NoiseThis is the other major environmental impact of aviation, and perhaps the most controversial. While the benefits given by an airport can be extended to large geographical areas, the environmental impacts are borne by residential areas close to them. The most relevant problem of an airport is then its economic and social cost/benefit geographical distribution.

Operations at most airports, specially in crowded Europe, are in any case, restricted by the authorities. Attempts to deal with this problem are based on the following instruments:

Restriction or banning of old noisy aircraft operation (called chapter 2 aircraft) to be replaced by newer (Chapter 3) aircraft. This must be done before 1.1.2000 in the U.S. and 1.4.2002 in the E.U. Internalisation of better technology cost is made by aircraft manufacturers and the air transport industry and the environmental impact in terms of footprint noise exposure is reduced by 50%.

By application of taxes and charges to aircraft operations. An Airport Council International (ACI) survey (14) of all member European airports, 104 of which answered, showed that market instruments are frequent to discourage the use of noisy aircraft. However, a majority of them use it as "general finance" (22 answers) which is against the "polluter pays principle" jeopardising the transparency. Others use it for special issues like insulation programmes, community work or other environmental protection measures (5 answers). The vast majority of airports didn't provide any answer on how collected money is used. A database available on Boeing's Internet Web Page of Boeing Company, including 590 world-wide airports (294 U.S. and 297 non U.S.) indicates that 55% of foreign airports (mostly in Europe) and 17 % in the U.S. have or are planning compatibility programmes, that would indicate a different degree of public pressure in crowded Europe. Madrid Barajas Airport and other environmental advanced airports, like Manchester Airport (15) have adopted insulation plans for houses exposed to certain noise level thresholds.

Land use Planning and Zoning To minimise the airport effects and prevent distortions caused by land speculation around them, long term land planning must be carried out. 55% of the respondents to this element of the ACI survey had some role in the planning process but only 24% (11) stated that they have the power of veto on plans which do not conform to their land use plan for the immediate vicinity.

Summarising, it can be assumed that while aviation is still one of the most energy consuming collective transport mode, it has no substitute in distances above 500 kilometres. Enhanced technology has produced in the past decades much more environmental friendly aircraft and will continue to do so in the future. Since companies cost efficiency have permitted a widespread social access to this transport mode, we believe that strategies to reduce air transport mobility in segments with no existing alternatives will constitute unaccepted social burdens. There is a living debate on how companies should deal with externalities, and how they will have to be borne by the sector, as in others, in a more transparent way.

7. IS AN APPROACH TO THE ENVIRONMENTAL PERFORMANCE OF COMPANIES STARTING FROM SOCIOCYBERNETIC THEORY POSSIBLE ?

In recent years, companies understood from the capitalist viewpoint as social organisations that seek profits through processes of transformation, have acquired greater environmental awareness in response to regulatory and social demands. Is an approach to the problem in sociocybernetic theory possible?. The company is a system (social organisation) that, by using a series of resources from the start, transforms them (productive process) to produce controllable finished results, according to a finalist criteria (teleologic). According to Parra Luna (16), as the capitalist company is a system of production values, it can be assumed not only the existence of those values it pursues in an intrinsic form in accordance with capitalist logic such as profitability, but also others such as security, prestige, nature conservation, self-fulfilment and power (17), among others that would be connected with a broader societal conception.

From our point of view, economic and societal values are not homogeneous nor can they be juxtapositioned, rather the former are of higher ranking for the capitalist company than the later, which are also desirable. Thus, one can speak of intrinsic or underlying values (making profits), and of others that are not pursued in a direct way, but whose degree of attainment affect the former. So, starting off with some initial resources, there are two transformations, $\mathbf{T_1}$, or a productive transformation that connects the value output (underlying values) and $\mathbf{T_2}$, of societal value, whose production is necessary-although insufficient-for the survival of the company.

A greater market share would reflect in greater power in a certain segment of that market, but power is only sought for greater sales and profits. Self-fulfilment is sought to the extent that a good working atmosphere satisfies employees, reduces conflict and boosts productivity. Environmental improvement, as has been seen previously, would be connected with care for nature to the extent that it would decrease fines and environmental charges and create a better image, and therefore improve sales by increasing the preferences of consumers sensitive to how this brand affects the environment, and so forth. Finally, it is possible to define a third relation of transformation $\mathbf{T_3}$, that connects both groups of output values.

The following problem is how to measure and control these outputs, in particular the "care for nature" that is to what extent has the system been able to satisfy the needs or values according to the relative importance of each. Therefore, the aim is to a) achieve the **axiological integration** of environmental needs in the company ´s standard of values and give them due consideration within this standard and b) to establish a methodology to evaluate compliance of this needs, that is an **efficiency calculation**.

Within what we have called societal values, nature care could be one of the values most suited to treatment from a cybernetic point of view. To evaluate environmental management in companies, two world-wide documents have been published: the International Standarization Organisation's ISO 14000 norm and the European Ecomanagement and Ecoauditing Regulation 1863/93, in which emphasis is placed on obtaining environmental registrations as a procedure for continual improvement (companies are being viewed as feedback systems with historic memory).

Calling the productive values vector $\overrightarrow{\mathbf{YE}}$, and societal values vector $\overrightarrow{\mathbf{YS}}$, for a vector **X** entry of the system, the three controlling transformations can be expressed as follows (Fig.2)

$\overrightarrow{YE}=T_1(\vec{X})$; $\overrightarrow{YS}=T_2(\vec{X})$; $\overrightarrow{YE}=T_3(\overrightarrow{YS})=T_3 \, . \, T_2(\vec{X})$ and then

$$T_1 = T_3 \, . \, T_2$$

where $\vec{X}=(X_1, X_2, ...X_i)$; $\overrightarrow{YE}=(YE_1, YE_2 ....YE_r)$ and $\overrightarrow{YS}=(YS_1, YS_2 .....YS_q)$ are vectors of **i, r,** and **q** dimensions.

But the company system can be seen as a set of **n** subsystems, which is why the previous vectors are applicable to each one of them , in such a way that the system can be written in matrix form:

$$[YE]_{r,n} = [T_1]_{r,i} \cdot [X]_{i,n}$$
$$[YS]_{q,n} = [T_2]_{q,i} \cdot [X]_{i,n}$$
$$[YE]_{r,n} = [T_3]_{r,q} \cdot [YS]_{q,n} \quad \text{and}$$
$$[T_1]_{r,i} = [T_3]_{r,q} \cdot [T_2]_{q,i}$$

For a subsystem **n=x**, the **YS** vector column can be expressed as

$$\overrightarrow{YS}_x = (YS_{1x} ...YS_{ax} .....YS_{qx})$$

where the component YS_{ax} represents the environmental performance of this subsystem, while for the whole system , this performance would be the vector row

$$\overrightarrow{YS}_a = (YS_{a1}, YS_{a2}, ...YS_{an})$$

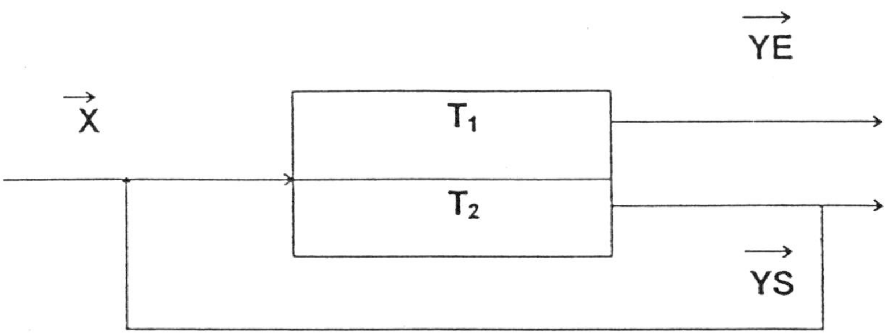

Figure 2 Transformations in a capitalist economy

As has been stated previously, any environmental performance must be measured in relative terms, not in absolute values, for which reason efficiency parameters have to be defined to determine the efficiency of an output value. It can be done by expressing its relationship with other societal values (other components of the **YS** vector) or with the totality of the societal vector which would be represented by the modulus of **YS.** Efficiency can also be measured in relation to the productive values that represent the securing of economic values, **YE**, or to the entry to the system, vector **X.** In relation to the **YE** exit, the efficiency of the **x** subsystem in environmental matters, will be given by

$$R = YS_{ax} / |\overrightarrow{YE}_x|$$

We have considered in (Fig.3) an example of subsystems and values that can be seen in environmental performance in an airline company and which is also a way of measuring efficiency according to comparative criteria with economic or productive values. To do this, subsystems affected by the "Care for Nature" in this type of companies need to be defined. Generally accepted criteria determine the following subsystems:

Subsystem 1): Flight Operations

$\mathbf{YE_1 = (\overrightarrow{RPK, RTK})}$ **;** $\mathbf{YS_{a1} = E+N+F}$ **;** $\mathbf{R=E+N+F/RPK+RTK}$

where RPK=Revenue Passenger Km. RTK=Revenue Ton. Km.

EM=Aircraft emissions ($CO_2 + H_2O + SO_2 + NO_x$) in Tonnes.

N=A/C noise level (in Decibels)

F=Fuel used

Subsystem 2): Cabin Operations

$\mathbf{YE_2 = (\overrightarrow{PPS, SOQ})}$ **;** $\mathbf{YS_{a2} = W\text{-}RW+Wa+Wsan}$

R=W-RW+Wa+Wsan/PPS+SOQ

where

PPS= Passenger Perceived Satisfaction ; SOQ=Service Observed Quality;

W=Generated Wastes; RW=Recycled Wastes; Wa=Waste Water

Wsan=Sanitary Water.

Subsystem 3) Catering Operations

$\mathbf{YE_3 = (\overrightarrow{M, SI})}$ **;** $\mathbf{YS_{a3} = E+W\text{-}RW+Wu+Wa+EN}$

R=E+W-RW+Wu+Wa+EN/ M+SI where

Wu=Water used;EN=Nonrenewable energy used; E=Thermal Energy Emiss.

M=N° of meals ; SI=N° of Service Items

Subsystem 4) Industrial (maintenance) Operations

$\mathbf{YE_4 = (\overrightarrow{AR, ER})}$ **;** $\mathbf{YS_{a4} = E+W\text{-}RW+Wu+Wa+EN}$

R=E+W-RW+Wu+Wa+EN/AR+ER, where

AR=N° of aircraft revisions ; ER=n° of engine revisions

Subsystem 5) Airport Ground Operations

$\mathbf{YE_5 = (\overrightarrow{AC, PAX, CB})}$ **;** $\mathbf{YS_{a5} = E+n+B+EN}$ **;** **R=E+n+B+EN/AC+PAX+CB**

where n=noise produced by ground vehicles (decibels) and B=unserviceable car batteries; AC=N° of aircraft assisted; PAX=N° of Passengers and CB=N° of tons of cargo and baggage.

Subsystem 6) Corporate

$\mathbf{YE_6 = (\overrightarrow{GPP, TC, I, IN, AL, SP})}$ **;** $\mathbf{YS_{a6} = E+EN+BS+Wu+W\text{-}RW}$

R= E+EN+BS+Wu+W-RW/ GPP+ TC+ I+ IN+ AL+SP, where

GPP=n° of General Policies and Procedures; TC= N° of training courses; I= Company Image ; IN=N° of publications and reports ;

AL=Employees Attitude Level ; SP=N° of sponsorship activities;

BS=Building surface used (square meters)

$\vec{X}, \vec{Y}$ and YS_{ax} IN AN AIR TRANSPORT COMPANY

(Y_{ax} = Environmental components of system X)

SYSTEM	X	YE	YS_{ax}	R
1 Flight operations	Aircraft Crews Traffic control	Revenue passenger km Revenue tonnes km	Atmospheric emissions Noise Fuel used	E+N+F/RPK+RTK
2 Cabin operations	Meals, beverages Service items Cabin crew	Pass, perceived satis-faction Service observed quality	Waste, recycled waste Waste water Sanitary water	W-RW+WA+WSAN/ PPS+SOQ
3 Catering operations	Food Unprocessed service items Staff	Meals Service items	Atmospheric emissions Waste, recycled waste Water used/waste H_2O Non–renewable energy used	E+W-RW+Wu+EN/ (M+S)
4 Industrial operations	Unservic. aircraft/eng. Spareparts Chemicals Staff	Serviceable aircraft & engines	Atmospheric emissions Waste, recycled waste Water used/waste H_2O Non–renewable energy used	E+W-RW+Wa+Wu+EN/ SA+SE
5 Ground operations	Ground cars Mobile stairs Ground power units Cargo and baggage trucks Staff	Assisted aircraft Passengers transported to planes Cargo and baggage transport	Atmospheric emissions Noise Unserv. car batteries Non-renewable energy used	E+n+B+EN/AC+Pax+CB
6 Corporate	Staff Capital stock Buildings	General policies and procedures Training courses Company image, information, attitude, sponsorships	Atmoshperic emissions Non-renewable energy used Building surface managed Water used Waste-recycled water	W-RW+EN+BS+WU+E+BS/ AL+TC+I+SP+IN+GPP

Figure 3. Vectors X and YE and YS_{ax} in an air transport company

8. CONCLUSION

One might think that it would be difficult to reach definite results in such a multifarious field as sustainable development. The amount of effort being dedicated by scientists, businessmen and ordinary citizens, and the attention being paid by the media to environmental issues is a clear indication of its importance today.

If we refer to the different stages that any developing interrelated discipline must face-finding the process, reaching a practical outcome and aiming for undisputed consolidation- it could be said that significant progress have been made to establish grounds for a methodology to integrate consolidated sciences such as physics, chemistry, sociology and economics, but practical problems remain to be addressed in the future.

Accountability in the use of the environment is a paramount issue. A better understanding of the cost of externalities will lead to the introduction of a new decision-making process for economic agents, both government organisations and business enterprises, to decide own matters with environmental implications. This process is in its infancy but to acquire credibility, a deeper understanding of the complex physical interrelations existing at the biosphere, will be needed.

Another encouraging indication is the amount of environmental information that is being generated and demanded now, both for personal or research purposes, compared with what was available a few years ago. In this paper we have seen that information systems and more comprehensive reporting standards are becoming more frequent and which , in time, will generate new challenges, as part of a continuing and promising way ahead.

Finally, the possibilities that sociocybernetic theory offers for an environmental analysis and evaluation of companies should be stressed. This aspect is also just starting but will expand alongside a growing social sensitivity towards the environment.

Madrid, January, 1999

9. REFERENCES

(1) Jiménez Herrero, Luis. "Desarrollo Sostenible y Economía Ecológica". Ed. Síntesis. Madrid 1996. Pág. 180-197.

(2) "UN SNA Handbook on Integrated Environmental and Economic Accounting" UN. New York . 1990.

(3) Carsten Stahmer "Integrated Environmental and Economic Accounting". UNEP Environment Programme (Feb 1996). Paper 17. Chapter III.

(4) Daly, H.E. and Coob J.B. "For the common good". Beacon Press. Boston. 1989.

(5) Parra Luna F. "Hacia una medida del desarrollo sostenible: una tentativa metodológica".WACRA. Madrid. September 1997.

(6) C. Duarte and S. Agosti: "The oceans and CO2 sink effect". "Science" Magazine.Vol. 27. Pag.10. July 1998.

(7) Epstein Marc J. "Measuring Corporate Environmental Performance". Irwing Professional Publishing. IMA. Montvale, NJ. 1996.

(8) UNEP. Industry and Environment. Technical Report nº 24. 1994.

(9) Environmental Reports 96/97 of the companies: Air France, American Airlines, British Airways, Endesa, IBERIA Airlines, IBERDROLA, IBM, Dupont, KLM, Lufthansa, SAS, SONY, Swissair and Unión Fenosa.

(10) "Empresa y Medio Ambiente". KPMG. Servicio de Estudios. 1996

(11)"Libro Blanco de la Gestion Medioambiental en la Industria Española". Fundación Entorno. Mundiprensa. 1998.

(12)VROM. Government Policy of the Netherlands on air pollution and aviation. 1995.

(13) Communication from the European Commission on Transport and CO2" COM (1998) 204 Final. Brussels March 31, 1998.

(14)"Environmental Handbook ". Airports Council International (ACI). European Region. 1995

(15) Callum Thomas. "Noise Related to Airport Operations. Community Impacts", article from "Environmental management at Airports" pag.8-34. Thomas Telford Publishing. London.1996.

(16) Parra Luna." El balance integrado de la Gestión Estratégica". Ediciones Deusto.1993

(17) Parra Luna. "Axiological Systems Theory. The role of values in the Evaluation of Social Systems".

Chapter 13

The performance of education systems

RAF VANDERSTRAETEN
Faculty of Social Sciences, Utrecht University

In the past decades, national governments have strived for the realization of a number of ideals, and have initiated numerous reforms of our societal structures. Ideological convictions have often been at the root of particular initiatives, for example in the field of economic policy. Several political strategies and interventions were, as a consequence, fiercely disputed. A few issues and ideals, however, have been of persistent and consistent concern in the political field. The 'democratization' of education is such an issue. There is, during the entire postwar period in Europe, little difference of opinion as to the relevance of public expenditures for education. Material investments and internal educational reforms have been generously sponsored by governments since the 1950s. Numerous new school buildings were constructed. Teacher education was expanded. Comprehensive educational structures were introduced, while trying to provide for equality of opportunity and to eliminate the influence of social background. Compulsory education has been extended, in most countries to the age of 16 or even 18 years. Et cetera (cf., e.g., Jonsson, Mills & Müller 1996).

This article is concerned with measuring and explaining the performance of the educational system in the postwar period. In the first part, I will present figures on the output of the educational system and on the increasing demand for education. The main part of the article is devoted to an interpretation of these figures. My interpretation will not merely focus on the particular characteristics of educational planning and educational policy. Already a brief look at society in general learns that similar evolutions are taking place elsewhere, as for example with regard to the demand for

The Performance of Social Systems
Edited by Francisco Parra-Luna, Kluwer Academic/Plenum Publishers, 2000

medical care, for legislation, or even for political decision-making. In fact, most societal systems are confronted with expanding demands. This phenomenon indicates that higher expectations and increasing participation are perhaps less related to education and educational policy as such, but that they are most of all characteristic of modern society -and thus open to general sociological analysis. In this context, I will use the systems theoretical framework developed by the German sociologist Niklas Luhmann.

1. EDUCATIONAL PARTICIPATION

To assess the performance of the educational system in one country, it is advisable to recur to clear and transparent indicators. They enable relatively easy comparisons between the performance of the educational systems in different countries, or between the performance of the educational system and other functional subsystems of society. The most transparent perspective on the evolution of educational expansion, and on the democratization of education, is provided by enrollment numbers (instead of, e.g., the number of schools, the number of teachers, the number of non-teaching staff members, or public expenditures on education). But one better does not rely on absolute data about school attendance. More relevant is to examine the historical development of the *relative participation in education*, i.e. to study the evolution of the percentage of schoolgoing youngsters within each generation. Absolute numbers are determined by the ups and downs of the birth rate. Only relative figures provide a clear indication of the changing demand for education within the population.

As the democratization of education has been such an important issue in European policy making, the performance of the educational system is calculated in terms of educational participation indices. In other words, only output indices are used to measure the performance of the educational system. With regard to education, this seems to me to be the most suitable performance indicator, given the political emphasis on ameliorating access to educational provision. (This point will be discussed later on). Of course, this particular option is also related to the fact that expenditures for the provision of educational services are fairly fixed, as about 90% of all expenditures are used to pay the loans of teachers and other personnel in schools and universities. Focussing on this input indicator would lead to the following conclusion: the less teachers earn, the more efficient the educational system. Clearly, this is hardly exceptable as an assessment of the performance of the educational system.

As an example, I will explore the evolution of educational participation in one country, viz. Belgium. Both global, overall evolutions and regional differences will be discussed -while Belgium is a federal and multicultural state. The data of which I make use to calculate these participation indices were collected during the general Censuses in Belgium. The first Census in which information was gathered on school attendance, years of schooling, educational degrees, etc. was carried out in 1961. In 1970, 1981 and 1991, this information was again collected. For my purposes, these data are better fitted than the annual figures published by the statistical services of the Departments of Education. These latter data can, elsewhere, be useful - depending on the way they are gathered. If one wants to study regional differences in educational participation, it is important that the place of residence of the pupils and students is taken into account, and not just the residence of the educational institution where the pupils and students are enrolled (cf. Vanderstraeten 1996: 15-16).

1.1 Historical trends

To present an overall picture of changes in the demand for education in Belgium, relative participation indices were calculated for each generation between 16 and 26 years. Figure 1 and figure 2 depict the evolution of these percentages, for the male and the female part of the population respectively.[12] What is most striking, is the uninterrupted increase of the different participation indices. It is only for the 25-26 year-olds that these indices have barely changed since the 1960s. An indication of the extent of the increase is, for example, the fact that the percentage of 19 year-olds attending school in 1991 is at the least as high as the percentage of 16 year-olds in 1961 (58 percent of the male population, 61 percent of the female population). Also, one out of three 21 year-olds attended school in 1991. Differences between the participation indices of the successive age groups were, on the other hand, at each Census within similar margins -which indicates that the relative amount of school leavers (with or without degree)

[12]Some problems were caused by the fact that information has not always been gathered in identical ways, and has also been processed differently. These inconsistencies influence the presentation of the results. In 1970, generations older than 20 years of age were paired (21-22 years, etc.). A total number was published for '25 years and older' in 1961; afterwards, this upper limit has gradually moved up (1991: '29 years and older'). Therefore, figures 1 and 2 and table 1 do not contain indices for the generation of 25-26 years in 1961. To enable historical comparisons, all generations above 20 years have also been paired. These inconsistencies contribute to the poor image of Belgian statistics in international perspective. Still more important is that this information is always published with a considerable delay. The 1991 data, for example, were only published in 1997.

was more or less identical in each generation. In 1991, the difference between generation X and generation X+1 lay around 10 to 15 percent.

Clearly, the increase of these participation indices is more pronounced for females. For the male population, the indices doubled between 1961 and 1991 for each generation between 17 and 24 years. For the female population, the indices also doubled for the 17 year-olds, but tripled for the 18 and 19 year-olds, and at the least quintupled for the generations between 20 and 24 years. Females have systematically caught up with the males, and even seem pulling away from them. According to figure 1 and figure 2, men outperformed women in every age group in the 1960s. In 1991, women outperformed men in the generations from 16 to 20 years; for the twenty-plus, the arrears were strongly reduced.

This overall picture can be further refined. To draw attention to some details, figure 3 and figure 4 depict the average *annual* growth of the participation indices. Two tendencies now come clearly to the fore. First: age plays a decisive and unambiguous role. The older the generation, the smaller the increase of the indices. For both the 18 and 19 year-olds, school participation annually increased with an average of 2 percent between 1981 and 1991, while the increase for the 23-24 year-olds was small (0,2 percent for the male population, 0,4 percent for the female population). Second: the average annual increase during the last decade is clearly higher than ever before. Figure 3 and figure 4 make clear that the increase grew steadily for the female population, while it leveled down in the 1970s for the male population. Therefore, the strong increase of the participation indices for the male population in the 1980s is highly remarkable. Participation in education is, in summary, higher than ever before in Belgium. *The increase of the participation indices has, moreover, never been as high as during the last decade.*

1.2 Regional differences?

The development of the demand for education in Belgium is remarkable. Belgium is, however, no longer an unitary state. It is at present a federal state, in which the major Communities -i.e. the Dutch-speaking Community in the north, and the French-speaking Community in the south- drift apart. Educational and cultural issues have been at the root of the process of federalization in Belgium. Differences were gradually institutionalized at the governmental level. Between 1961 and 1968, the national Minister of Education was 'aided' by a Deputy Minister belonging to one of the other language Communities. In 1968, two Education Ministers were appointed in Belgium. Since 1989, education is entirely dealt with at the level of the Communities; there are no longer national Ministers of Education, but only

Community Ministers who report to the regional parliaments. They take political decisions independently of each other. With regard to this development, the basic report on Belgian Education Policy by the OECD speaks of 'the diverging paths' (OECD 1991). This development raises the following question: Have these diverging political paths, which took off from a common background, given way to different patterns of educational participation?

The data relevant to answering this question are depicted in figure 5 (male population) and figure 6 (female population). There are no data available for 1961. The official division of Belgium into different language regions only dates back to 1963. Immediately before the language border was fixed, the territorial boundaries of several 'frontier' districts and municipalities were redrawn -which made it impossible to calculate these figures retrospectively, starting from the records of the Census Bureau. Inspection of figures 5 and 6 makes clear that the differences between the indices for Flanders (full line) and the Walloon Region (dotted line) are small. In the period between 1970 and 1991, the patterns which have been described in the preceding paragraph clearly reappear. There is, both in the Flemish and the Walloon Region, a continuous and spectacular increase in school participation for each age group between 16 and 24 years. Moreover, this increase is more apparent in the 1980s than in the 1970s.

The two regions are thus faced with parallel developments. Against this background, some small differences between the Flemish and the Walloon Region can nevertheless be indicated. In fact, the figures for the Walloon Region are, on the average, somewhat higher than the figures for Flanders. Although significant, this discrepancy does not allow to speak of 'diverging paths' with regard to the demand for education. Different lines of educational policy have *not* given way to different patterns of educational expansion in both parts of Belgium. This fact indicates that the demand for education depends less on the practice of policy making in education, but reflects more fundamental developments within our present-day society.

2. SOCIETAL PERSPECTIVES

In our modern society, the 'emancipated' individuals have to construe their own career and their own life. Society itself triggers them to put forward their expectations. Individuals are not just entitled to cherish individual wishes and dreams. They have to use them as probes to find out about themselves. They have to 'go for it'. On the other hand, societal subsystems do not themselves find a hold for arguments which go against the optimal fulfillment of their function. They have not developed any significant counterrationality that is able to legitimate particular limitations. The educational system (in Belgium) is no *Einzelgänger* in this regard. Science, for example, does not itself call a halt to research, even if it is feared that the results will be misused -since knowledge of dangers still is knowledge. In short, educational expansion seems to result from the structural characteristics of our modern world. An exploration of sociological theory might therefore help to clarify this pattern of expansion.

2.1 Functional differentiation

It is one of the general findings of classical sociology -which sought, at the end of the 19th century, to grasp the nature of the 'modern' in contemporary societies- that social evolution coincides with greater social complexity and increasing social differentiation (cf., e.g., Simmel 1989 [1890], Durkheim 1930 [1893], Parsons, 1951). Following Luhmann (1982: 229-254, 1997: 595-865), this statement is true but needs clarification. Societies have, according to Luhmann, first of all incorporated distinct *forms* of differentiation in the course of history. Degrees of differentiation and degrees of complexity are produced and mediated by these forms of differentiation. Accordingly, it is the form (and not just the degree) of differentiation that *accounts for the continuous reproduction of the particular characteristics of modern society*. Insight into these characteristics can be gained by closely analyzing the modern form of societal differentiation.

Following Luhmann, modern society is not characterized by the forms of *segmentation* or *stratification*, but by *functional differentiation*. Functional differentiation organizes communication processes around special functions to be fulfilled at the level of society. One can think, in this regard, of the political function of producing collectively binding decisions, the economic function of securing the satisfaction of future wants, the religious function of interpreting the incomprehensible, the scientific function of acquiring knowledge, and of course also the function of educating human beings.[13] As Luhmann (1982: 236) adds: 'Since all necessary functions have to be fulfilled and are interdependent, society cannot concede absolute primacy to any one of them [as, e.g., to the upper classes in stratiform societies]. It has to use a second level of subsystem formation to institute a primacy of specific functions limited to a special set of system/environment relations'. In a functionally differentiated society, none of the primary subsystems can

[13]Normative as well as classical functionalist reasoning implied that there 'exists' a definite set of problems at the level of society and that these problems have to be solved if society is to survive. For Luhmann, it is an empirical question whether or not a function becomes a catalytic principle for subsystem formation.

represent society in general. There is no point of view that would allow an overall perspective with universal validity. No functional subsystem can claim a position that transcends its particular function -not even the economy, the political system or religion. Instead, *functional subsystems give primacy to their particular function.*

Each functional subsystem establishes a specific, highly selective set of system/environment relations. The specificity of its function is, however, combined with universalism (Luhmann 1997: 375-376, 983, cf. Parsons 1960). The projection of universal relevance of its own function leads, as Luhmann indicates, to the *inclusion of the entire population*. In contrast with other forms of differentiation, the entire population now gains access to each of the social subsystems (e.g., through universal suffrage, or compulsory education). Access to functional subsystems becomes independent of any relation to other subsystems. Functional subsystems have to treat their environments as environments of equals, while nothing but function can justify discrimination (cf. Stichweh 1988).[14] In other words, they can tolerate indifference towards almost everything, except very special features of their respective environments. The functional subsystems do not depend on a complementary definition of their environment. Each subsystem can tolerate an open and fluctuating environment (so long as other subsystems fulfill their function). This has important advantages for the heightening of social complexity. In this sense, the form of differentiation mediates the degree of differentiation and of social complexity.

These evolutions have, as indicated, a wide-ranging scope. They do not only affect the educational system. A brief description of the medical system -which displays a particular resemblance with education (cf. infra)- may clarify this scope. At present, illness is no longer interpreted as an ill-fated relation to the world, or the consequence of disbelief or moral misbehavior. It is considered to be a specific defect in the normal physical or mental functioning of human beings, that can be identified with the help of the medical sciences, and that can be treated by specific techniques. Every individual is entitled to medical care -not just the upper classes of society. Moreover, the medical system not only strives for the healing of diseases, but focuses on health. It follows a logic that holds out unlimited prospects of complete well-being *and* of medical interventions. According to the definition of the World Health Organization, 'health is a state of complete physical, mental and social well-being and not merely the absence of disease

[14]This explains why increasing functional differentiation led to a renewed emphasis on the normative ideal of 'equality' in the eighteenth century, and ever since (Dann 1975). This normative ideal clearly has socio-structural foundations, although society still does not live up to the ideal of equality.

or infirmity'. As a consequence, the entire population is ill, and needs to be treated. As with other subsystems, the medical system hypostatizes its own function. According to its proper perspective, neither politics nor religion, neither the economy nor education, but only the medical system itself is able to claim absolute primacy in society.

Developments within one subsystem may or may not jeopardize developments elsewhere in society. On the level of the functional subsystems, there can be a simultaneous augmentation of autonomy and mutual dependence. There is no zero-sum principle; more freedom in one system does not need to imply less freedom in the others. Dependencies and independencies of functional subsystems can (under certain conditions) expand together. Every subsystem is, in as far as it gains freedom for its *own* function, dependent upon the fulfillment of the *other* functions elsewhere within society. As Luhmann writes: 'By specifying and institutionalizing functions, society *increases* its internal interdependencies. By loosening the structural complementarities of systems and environments and by providing for more and more indifference, society *decreases* internal interdependencies. It augments, in other words, internal dependencies and independencies at the same time' (Luhmann 1982: 237).[15] If the conditions of the possibility of this combination of increasing dependencies and independencies become too precarious, society might collapse (cf., e.g., Offe 1987). Modernization is certainly not synonymous with progress.

2.2 Individualization

The preceding reflections focus on structural evolutions within modern society. They describe the development of a generalized 'readiness' for expansion on the side of the functional subsystems. But they do not yet explain the increasing demand on the side of the *individuals*. They do not explain why individuals raise their expectations (and raise their expectations about the fulfillment of their expectations). The hypothesis that underpins the following analysis is that contemporary processes of self-identification, and of the production of expectations and wishes, correlate with the social structure of modernity. According to well-known sociological analyses, modernity gives way to the increasing individualization of human beings (Giddens 1991: esp. p.70-108). This evolutionary process brings about, among other things, greater differences between the individual lifeworlds, increasing disparities between individual biographies and individual

[15]Increasing interdependence, therefore, does not lead to dedifferentiation (as is sometimes argued). The dynamics of modern society resides in its pattern of functional differentiation. It is the form of functional differentiation that creates the interdependencies inside society.

strategies for dealing with or avoiding role conflicts, and a more articulate awareness of personal unicity. Not the identity of, but the difference between individuals gets accentuated in complex and differentiated societies.

Since George H. Mead (1934), it has become clear that social structures mediate the constitution and actualization of the self. The way an individual experiences herself is dependent upon the schemes which society makes available to present one's proper identity. In modern societies, schemes that allow us to think and talk about ourselves in terms of accumulated personal experiences have multiplied, and have become dominant. Important social institutions have developed that stimulate this kind of introspection, and that generate individual biographies -as, for example, diaries, psycho-analytic and therapeutical sessions, in-depth interviews in journals, magazines and on television (Hahn 1987). Individuals have to construct themselves within an authentic story of life. 'Possessing a name and a place within the social framework in the form of general categories such as age, gender, social status and profession no longer suffices [...] Rather, individual persons have to find affirmation at the level of their respective personality systems, i.e. in the difference between themselves and their environment and in the manner in which they deal with this difference -as opposed to the way others do' (Luhmann 1986: 15-16). The 'burden of proof' rests in our contemporary society on individual shoulders.

Complementary to this description of the structural determinants of individualization in modern societies, additional questions need to be be posed, viz. how exactly do individuals constitute and preserve for themselves a particular identity? Which techniques do they use? In modern society, the self-identification of individualized individuals seems to take place (to a certain degree) by means of *expectations*, and the information these expectations provide -when they are fulfilled and when they are not fulfilled. An individual uses expectations as antennas for scanning the contingency of the environment in relation to herself. Her expectations produce the difference between success and disappointment, and this difference does not depend on these expectations alone. In fact, the determinacy of her actual experience makes sure that arbitrary contrary expectations cannot be formed. Expectations enable to gather information and accumulate experiences, and thus contribute to the constitution and actualization of the self (Luhmann 1989: 149-258, 1990: 107-122, Beck 1986: 205-219, see also Bateson 1972).

Forming expectations is, as Luhmann asserts, a simple technique. It involves almost no presuppositions. An expectation can be formulated without self-knowledge and without being familiar with the world -for example, by good luck. But 'after a certain period of conscious life enriched by social experiences, completely random expectations cease to occur. [...] One is forced to orient oneself to one's own history of consciousness, however unique it may be' (Luhmann 1995: 268-269). In modern society, individuals are forced to adapt internally to the fulfillment or disappointment of their expectations. Each individual has to explain to herself the results; she needs to attribute success and disappointment to particular factors. The causes of fulfillment are, for example, attributed to oneself, and those of non-fulfillment to others (cf., e.g., Weiner et al. 1972). One can, as a consequence, adapt one's level of expectations to one's experiences, or stick to one's original expectations and fall back into resentment or look for some kind of compensation.

This way of forming and dealing with expectations and the modern principle of functional differentiation clearly affect one another. Functional subsystems seem to incite individuals to put forward their expectations and claims, in so far as these fall within their functional competence. For sure, they cannot legitimately reject these expectations. The fulfillment of their function is their 'raison d'être'. And the entire population is actively or passively built into their operational perspective (inclusion). No school or university, for example, can reject the expectations of students to receive more and better education or training. On the other hand, no individual can gain an overall view of the overcomplex operations of functional subsystems. One can only follow the course of one's expectations, and record whether what is expected really occurs. An individual's expectations enable a reduction in the 'unmanageable complexity' of society, due to 'the increasing diversification and particularization of familiarities and unfamiliarities' (Luhmann 1988: 105). They transform social complexity directly into individual fate. In this way, expectations and systems establish a kind of symbiosis. *They build a relationship that gives way to increasing growth.* Separated from each other, these elements would break down.

3. COMPARATIVE PERSPECTIVES

The previous section sketches a fairly monochrome picture. It offers a general outline of the patterns of social differentiation and individual self-actualization in modern society. This outline can be, and needs to be refined -by focusing on the development of the *educational system* within modern society. In fact, the principle of functional differentiation itself indicates ways towards the respecification of our analysis. It points to the fact that the subsystems are directed towards different primary functions of the societal system, and that they thus display (to a certain degree) distinct characteristics. Compared with each other, functional subsystems are both alike and different. Therefore, analyses are necessary that can account for the particular characteristics of education vis-à-vis other functional subsystems.[16] A similar sensitivity is needed for assessing the relevance of national data and national trends. Comparative perspectives allow to color the picture of the evolution of educational participation in Belgium.

3.1 Education

As Luhmann indicates, society consists of communications, and *only* of communications. Individuals are no element or part of the societal system. The human body and the human psyche belong to the environment of society. Communication establishes a *reality sui generis*. This does not, of course, exclude the fact that social systems and individuals mutually affect one another -as I tried to demonstrate in the previous sections. But organic, psychic and social systems consist of different networks of operations.

In the course of society's modernization, the systemic boundaries between human beings and the societal system have become outspoken. The human body and the human psyche lost much of their immediate relevance for societal communication. The media of dissemination (writing, printing, and recently also electronic devices) enabled/enable an immense extension of the scope of the communicative process. And this process clearly affected the form of social differentiation. 'A societal system that is vertically differentiated according to the principle of stratification presupposes that societal differentiation is directed by kinds of persons, by their 'quality', by their determination to live in specific castes or ranked groups. By contrast, with the transition to functional differentiation, the schematic of differentiation is chosen autonomously; it is directed only by the functional problems of the societal system itself' (Luhmann 1995: 193).[17] Functional differentiation is a differentiation of communication systems.

Most functional subsystems are, as a consequence, oriented towards problems within modern society itself (although they have, of course, an impact on society's environment). They organize networks of

[16]This implies that concepts are developed and used that are sensible to similarities and to dissimilarities. The series of monographs on functional subsystems of society, of which Luhmann was able to finish four volumes (on the economy, science, law, and art), displays exactly this sensitiveness. The series is a systematic and detailed comparison of these subsystems, that makes use of a standard repertoire of abstract conceptual tools. Luhmann's four monographs contain together about 2200 pages. In this article, I can of course only briefly point to a few characteristics of the educational system in Belgium.

[17]Another example might clarify this often disputed assumption. If, inside an organization, departments are set up around different outside groups (like customers, or circles of persons to be cared for), this strengthens the influence of these groups on the organization; they find 'their' representation in the organization. By contrast, if the structuring is chosen according to purely internal points of view, the autonomy of the organizational system increases.

communication, and depend upon 'symbolic media' (Luhmann) to fulfill their function -as, for example, politics on power, science on truth, the economy on property and money, or the family on love. The unique position of the educational system within society derives from the fact that education is *oriented towards society's environment*, viz. towards bringing about psychic and behavioral changes in human beings. It is only the medical system that occupies a similar boundary position. Each system is involved not only in the 'socialization' of human beings as an adequate environment for social systems, but also focus on human beings as such, i.e. on the development of their capacities and on their health. They are (to a certain degree) concerned with the possibilities and problems proper to society's environment. Their ideals display a clear independence from societal concerns. In the educational system, the difficult relation (both in theory and in practice) between liberal education and vocational education provides a clear illustration of the conflict between the fulfillment of education's ideal of human development and the teaching of qualifications which prepare for someone's professional life (cf., e.g., Silver 1990: 100-146, Vanderstraeten 1995: 75-141).

An important consequence of this boundary position of the educational (and the medical) system might be, at first sight, a surprise. The boundary position explains why the values of human development (and health) are hardly contested within society. Human development is one of the few values that is looked upon as an ultimate value, and that has remained largely outside of ideological controversies -in contrast with, for example, issues of (and governmental expenditures for) national defense, art collection, economic activities, religious services, or the organization of political life itself.[18] Because education does not merely involve communicative consequences, but is concerned with the *external conditions of the possibility of social systems*, there is little opposition to the ideals it advocates. Accordingly, *there are only few societal restraints on educational expansion* (cf. Vanderstraeten 1997). -Of course, this does not mean that human development deserves preference in every kind of situation. This value expresses an abstract appreciation, and loses much of its relevance when it comes to the individual development of *other human beings*. Also, individuals do not always behave as if this value is an ultimate value (e.g., prefering a holiday over a conference). The rank order only applies to the communications with which the educational system asserts itself. But its application cannot easily be disputed by other functional subsystems.

[18] One might think, for example, of the Universal Declaration of Human Rights, whose 50th anniversary has recently been celebrated. Also, most recent value surveys underpin this outcome (with regard to Belgium and its Regions, see Kerkhofs et al. 1992).

This situation has a correlate at the *individual level.* I have already indicated that society's 'great transformation' (viz. from stratification to functional differentiation) changed the conditions of inclusion within society. An individual's inclusion is no longer determined by her social origins or by birth. Involvement in a functional subsystem is dependent on the circumstances, i.e. on the way inclusion is conditioned by the subsystems themselves. Each subsystem records, as a consequence, past accomplishments. For an individual, status and prospects for future involvement come to depend upon previous events in the same system -as, for example, someone's economic position depends upon previous receipts and expenditure, or someone's scientific reputation on publications and citations. Inclusion takes, as Luhmann (1987: 188-191, Luhmann & Schorr 1988: 277-282) argues, the form of a career. The concept of 'career' should be understood in its broad sense, and not just in terms of a succession of occupational positions within an organization. It includes stagnation or demotion. It might also refer to a career of diseases, or of crimes.[19] A career is the temporal structure of the process of inclusion.

The career structure underlines the importance of decisions and outcomes in an individual's life. Henceforth, individuals feel triggered to put forward individual expectations and claims. This strategy enables them to distinguish themselves from what's common, or characterizes 'the others' (cf. Bourdieu 1979). The *accumulative structure* of a career leads, at the same time, to divergent evaluations of particular events and lifetime episodes. It enhances the relevance of the beginning of one's career, and leads to the dominance of the future over the present. Current choices and decisions get evaluated from the point of view of their future consequences, especially in one's earlier (school) years. This situation clearly stimulates individuals to invest in their education. Grades and certificates make a difference, although an educational career certainly does not neatly link up with, for example, an economic career. The hopes of manpower planning have not come true. In practice, diffuse relationships between training and selection for jobs tend to dominate (Silver & Brennan 1988: 34-52). Diploma-inflation, underemployment and overschooling have become widespread phenomena (cf., e.g., Collins 1979, Bailey 1991). Nevertheless, the social structure stimulates individuals to go for it -even when faced with counterfactual

[19] One finds a similar use of the concept of career, in relation to individual identity, in John Dewey's work: 'The human individual is himself a history, a career, and for this reason his biography can be related only as a temporal event. [...] Individuality conceived as a temporal development involves uncertainty, indeterminacy, or contingency' (1988 [1940]: 102, 111). In the field of sociology, this broad definition of career has been particularly used at the University of Chicago (A. Strauss c.s.). What Luhmann adds, is the relation between career and inclusion.

indications. The resultant process has been aptly described as 'self-generating educational expansion' (Meyer et al. 1977: 245).

3.2 National politics

The functional characteristics of the educational system have, together with the appreciation of education's value, a wide-ranging, indeed world-encompassing impact. The 'world educational revolution' is accompanied by the diffusion of standard models of school organization and curricular categories (cf. Meyer, Kamens & Benavot 1992). These convergent trends do, however, not entirely eliminate the impact of national characteristics and national politics on the evolution of educational participation within a country (cf. Archer 1979). To explain particular details of the historical trends which have been sketched in the first part of this article, reference to structural arrangements and rearrangements within Belgium (and its Communities) is necessary. I will point to two social forces, that result -partly intended, and partly unintended- from political decisions in Belgium. They provide an explanation of het accelerating increase of educational participation during the 1980s, and of the divergent evolutions within the male and the female population.

First, attention needs to be paid to the structure of secondary education. In Belgium, it was not until 1887 that a law defined this structure. Alongside the so-called Classics (where the accent was put on Latin and Greek, or on Latin and mathematics/science), modern humanities appeared (with a scientific section, and a commercial and industrial section). Also, secondary schools were founded which were intended to prepare young people for 'white collar' jobs. Until the 1930s, technical and vocational education developed independent of secondary education (OECD 1991: 53-66). The highly hierarchical, compartmentalized and pyramidal structure of secondary education particularly *disfavored the female population*. The first humanities schools for girls, for example, were only introduced towards the 1890s. It was only after the Second World War that girls really gained equal access to education. (Women gained the right to vote in legislative elections in 1948.) In the course of the democratization of education, general educational tracks for girls (or open to both sexes) were created in a systematic way. The introduction of comprehensive education (1971) further ameliorated the possibilities to move on through secondary education, and from the secondary to the higher level. This 'materially' enabled the increase of educational participation among minors, and especially among girls. And to the extent that the number of female graduates at the secondary level increased, participation at the next higher level was also likely to increase.

Secondly, the legislation about compulsory school attendance has (had) important consequences. In Belgium, compulsory education was introduced in 1914 for children between 6 and 14 years of age. This period has been extended in the 1980s. Since 1983, compulsory school attendance is full-time up to the age of 16, and part-time between 16 and 18 years. As should be clear from the data I presented in the first part of this article, the effects of the 1983-law on the participation of the minor generations has been small (cf. Vanderstraeten 1993/'94). To a large degree, this law only confirmed/confirms an existing situation. It has, however, created unintended consequences. The extension of compulsory education has a particular *push up effect*. It influences, as Dronkers (1989) and Van der Ploeg (1992) have spelled out, participation at a higher level of education, while undermining participation at a lower level (as end point of an educational career). Participation at the top is stimulated, because participation at a lower level looses its distinctive value when compulsory education is extended. We do not know the exact magnitude of this push up effect.[20] But it points, again, to the self-stimulating characteristics of the interaction between individual expectations and the educational infrastructure.

More social forces can be cited, that can account for particular details of the expansion of educational participation in Belgium (cf. Vanderstraeten 1996). This type of analysis, however, is only relevant if attention is paid to developments which characterize modern society at large, and to 'convergent and divergent trends in national educational systems' (Inkeles & Sirowy 1983). Developments in the world educational system allow us to put national trends into perspective, *and* to develop a sensitivity for national characteristics and their explanation. Patterns of accelerating growth and of the educational 'emancipation' of girls and women are not unique to the Belgian context, but this context nevertheless influences how these developments occur. The rationale of *political* decisions is adjusted in the *educational* system. Simple, linear input-output models of policy planning and policy implementation should be abandoned. Political decisions produce effects in the educational system, but they often do so in unexpected ways. Educational politics mostly has, one can say, more and less effects than presumed.

[20] This argument has been statistically 'proved', using data for the Netherlands. Thus, Van der Ploeg (1992) attributed a substantial part of the expansion of higher education to the earlier expansion of secondary education. Changes in the labor market and in families (family income and the education level of parents) correlated, on the other hand, only weakly with processes of educational expansion. These findings need, however, to be interpreted with caution. The variables that are used do not do justice to the complexity of the social context (cf. Vanderstraeten 1997).

I have, against this background, insisted on the autonomy of the educational system within society. Education is not fastened to other functional subsystems of society, such as the political system. By consequence, its development might also negatively affect its societal environment. Judging from current political initiatives, the benefits of increasing educational participation in Belgium no longer outbalance the costs. The recently introduced entrance examinations for medical and paramedical studies are intended to slow down this increase. Their use is likely to be expanded -notwithstanding both *political* and *educational* objections. But this arrangement is, from a *societal* point of view, not that objectionable. It is concerned with the coordination of the different functional subsystems within society. And it further allows individuals to put forward expectations and claims, and to probe for individual possibilities and chances. It is indeed easy to criticize this arrangement, but it is not easy to come up with better alternatives in modern society.

4. CONCLUSION

The performance of the educational system is to a large degree conditioned by general developments within modern society. As a consequence, there cannot be just one indicator of the patterns of educational expansion, and of their very complex societal determinants. But the most transparent indices are figures of relative school attendance. They display most clearly the changing demand for education. They allow to consider both the characteristics of the educational subsystem and the importance of individual careers. I have tried to indicate throughout this article that the performance of the educational system of a particular country needs to be assessed against this general background. However, if the effects of particular political interventions are to be assessed, one can recur to more precise and detailed indicators. If one aims for example to provide for equal access of minority groups, figures about the regional dispersion of educational institutions are necessary (cf. Steyn & Vanderstraeten, 1998).

But what is evaluated in this case, is the effect of particular policy initiatives, and not the performance of the educational *system*.

5. REFERENCES

Archer, M.S. (1979) Social origins of educational systems (London: Sage).

Bailey, T. (1991) Jobs of the future and the education they will require: Evidence from occupational forecasts, Educational Researcher, 20 (2), 11-20.

Bateson, G. (1972) Steps to an ecology of mind. A revolutionary approach to man's understanding of himself (New York: Ballantine).

Beck, U. (1986) Risikogesellschaft: auf dem Weg in eine andere Moderne (Frankfurt a.M.: Suhrkamp).

Bourdieu, P. (1979) La distinction. Critique sociale du jugement (Paris: Minuit).

Collins, R. (1979) The credential society: An historical sociology of education and stratification (New York: Academic Press).

Dann, O. (1975) Gleichheit, in W. Conze, O. Brunner & R. Koselleck (eds.), Geschichtliche Grundbegriffe, Band 2 (Stuttgart: Klett), 997-1046.

Dewey, J. (1988) The later works, vol. 14: 1939-1943, ed. by J.A. Boydston (Carbondale-Edwardsville: SIU Press).

Dronkers, J. (1989) Veranderingen in de samenleving en economie en groei van de onderwijsdeelname sinds 1960, Comenius, 9, 47-72.

Durkheim, E. (1930) De la division du travail social (Paris: PUF).

Giddens, A. (1991) Modernity and self-identity. Self and society in the late modern age (Stanford: Stanford University Press).

Hahn, A. (1987) Identität und Selbstthematisierung, in A. Hahn & V. Kapp (eds.), Selbstthematisierung und Selbstzeugnis: Bekenntnis und Geständnis (Frankfurt a.M.: Suhrkamp).

Inkeles, A. & Sirowy, L. (1983) Convergent and divergent trends in national education systems, Social Forces, 62, 303-333.

Jonnson, J.O., Mills, C. & Müller, W. (1996) A half century of increasing educational openness? Social class, gender and educational attainment in Sweden, Germany and Britain, in R. Erikson & J.O. Jonnson (eds.), Can education be equalized? The Swedish case in comparative perspective (Boulder: Westview Press).

Kerkhofs, J., Dobbelaere, K., Voyé, L. & Bawin-Legros, B. (1992) De versnelde ommekeer. De waarden van Vlamingen, Walen en Brusselaars in de jaren negentig (Tielt: Lannoo).

Luhmann, N. (1982) The differentiation of society (New York: Columbia University Press).

Luhmann, N. (1986) Love as passion (Cambridge: Polity).

Luhmann, N. (1987) Soziologische Aufklärung 4. Beiträge zur funktionalen Differenzierung der Gesellschaft (Opladen: Westdeutscher Verlag).

Luhmann, N. (1988) Familiarity, confidence, trust: problems and alternatives, in D. Gambetta (ed.), Trust: Making and breaking cooperative relations (Oxford: Basil Blackwell), 94-107.

Luhmann, N. (1989) Individuum, Individualität, Individualismus, in N. Luhmann, Gesellschaftsstruktur und Semantik: Studien zur Wissenssoziologie der modernen Gesellschaft, Band 3 (Frankfurt a.M.: Suhrkamp), 149-258.

Luhmann, N. (1990). Essays on self-reference (New York: Columbia University Press).

Luhmann, N. (1995) Social systems (Stanford: Stanford University Press).

Luhmann, N. (1997) Die Gesellschaft der Gesellschaft (Frankfurt a.M.: Suhrkamp).

Luhmann, N. & Schorr, K.E. (1988) Reflexionsprobleme im Erziehungssystem (Frankfurt a.M.: Suhrkamp).
Mead, G.H. (1934) Mind, Self, & Society, from the standpoint of a social behaviorist, ed. Charles W. Morris (Chicago: University of Chicago Press).
Meyer, J.W., Kamens, D.H. & Benavot, A. (1992) School knowledge for the masses. World models and national primary curricular categories in the twentieth century (Washington D.C.: Falmer).
Meyer, J.W., Ramirez, F.O., Rubinson, R. & Boli-Bennett, J. (1977) The world educational revolution, 1950-1970, Sociology of Education, 50, 242-258.
OECD (1991) Education in Belgium: the diverging paths (Brussels: Departement Onderwijs).
Offe, C. (1987) The utopia of the zero-option. Modernity and modernization as normative political criteria, Praxis International, 7, 1-24.
Parsons, T. (1951) The social system (Glencoe: Free Press).
Parsons, T. (1960) Pattern variables revisited: a response to Robert Dubin, American Sociological Review, 25, 467-483.
Silver, H. (1990) Education, change and the policy process (London: Falmer Press).
Silver, H. & Brennan, J. (1987) A liberal vocationalism (London: Methuen).
Simmel, G. (1989) Gesamtausgabe, Band 2: Aufsätze 1887-1890. Über soziale Differenzierung, ed. by O. Rammstedt (Frankfurt a.M.: Suhrkamp).
Steyn, H.J. & Vanderstraeten, R. (1998) Minority group education in South Africa: a comparative perspective, South African Journal of Education, 18, 89-96.
Stichweh, R. (1988) Inklusion in Funktionssysteme der modernen Gesellschaft, in R. Mayntz, B. Rosewitz, U. Schimank & R. Stichweh (eds.), Differenzierung und Verselbständigung. Zur Entwicklung gesellschaftlicher Teilsysteme (Frankfurt a.M.: Campus), 261-293.
van der Ploeg, S.W. (1992) Expansie van het voortgezet en hoger onderwijs: Effecten van veranderingen op de arbeidsmarkt, in gezinnen en in het onderwijssysteem, Mens en Maatschappij, 67, 156-176.
Vanderstraeten, R. (1993/'94) De gevolgen van de leerplichtverlenging voor de grootte van de schoolbevolking, Tijdschrift voor Onderwijsrecht en Onderwijsbeleid, 3, 19-23.
Vanderstraeten, R. (1995) Leren voor het leven. Opvoeding en onderwijs als maatschappelijk deelsysteem (Leuven-Apeldoorn: Garant).
Vanderstraeten R. (1996) L'évolution de la scolarisation en Belgique. Vers une nouvelle perspective sur l'expansion du système éducatif, Eduquer & Former, 2 (8), 15-27.
Vanderstraeten, R. (1997) Circularity, complexity and educational policy planning. A systems approach to the planning of school provision, Oxford Review of Education, 23, 321-332.
Weber, M. (1976) Wirtschaft und Gesellschaft (Tübingen: Mohr).
Weiner, B. et al. (1972) Perceiving the causes of success and failure, in E.E. Jones et al. (eds.) Attribution: perceiving the causes of behavior (Morristown N.J.: Central Learning Press), 95-120.

Chapter 14

Performance of health care units and the influence of management

F. TRETTER
Districto Hospital Haar, Munich

1. INTRODUCTION

The current public discussion on *functionality of health care systems* in many industrialized countries is centered around *health economics*, asking if we still can afford a quantitative and/or qualitative growth in health services. In Germany the debate aims at stabilization or even at sectorial reduction of financial input into the system with no reduction of the quantity and with a minimal reduction of quality of health service delivery. Further analysis of this paradigm leads to several problems as it is shown further on.

1.1 Systems analysis

Black box methodology as a tool of systems analysis is trying to conceptualize the system under investigation as an input-output analysis. That means at any case that input and output must be defined (Parra-Luna 1999). On a general point of view the output is the medical service delivered to the patient. The effect of this output is the obtained health status of the patient. Referring to the doctors one can say that they give an input to the patient by applying diagnostic and therapeutic techniques. The output as a consequence of the input then will be the obtained health. For the input for the patient the doctor also needs a financial and educational input which is done by the organization he is working in.

Thus it gets evident that in our case the input-output analysis has two steps:

The Performance of Social Systems
Edited by Francisco Parra-Luna, Kluwer Academic/Plenum Publishers, 2000

1. the analysis of the relation of resource input and service output of the hospital and
2. the service output as input into the patient (with the output "health") inputs and outputs have to be worked out.

1.2 Measurement and "monetarization" of health

For the measurement of health status several techniques can be used, like medical parameters and rating scales. Already at this first level of measurement there is not much consensus about the validity and reliability of the respective measures.

In other words: it is hard to construct *satisfying quantified indicators for health*.

As a consequence, there is a lack of understanding health care as a special social system functioning according to immaterial and partially subjective values like "health", "wellness", "harm reduction" etc. For a rational basis for optimizing the health care system the evaluation of the performance of the system must be based on indicators which represent the production of health in an adequate way.

On the other hand it is a rather new field of discussion to consider the health care system as an economic system.

1.3 Health economics

Conceiving health care as an economic activity requires a deeper understanding of health economic categories, constructs, laws, theories and findings. Some, but not all principles of market economics can be useful in systems analysis (and design) of health care. This must be related to a certain viewpoint like that of the insurance companies, the patients, the hospitals etc. And again: the translation of processes, states and outputs of health services into categories of economics is a reductionistic procedure which must be made very carefully and with consideration of ethical issues.

At talking about health, disease and death in terms of prices (or costs or expenditures) one looses the sight for the values of these issues:

- From a radical point of health economics only a healthy (or a dead) person is a good, cheap person.
- From a convinced medical point of view the value a person can not be expressed in monetary values because the person is representing an individual existence and value and can not be replaced.

This is the ethical dilemma between medicine and economics.

But still the dominating discussion is referring solely to the financial level of the problem.

1.4 Managers as optimizers of organizational functions

Finally one has to consider the role of *management of health care* units with regard of *optimizing the function of the system* or the "performance" of the system as it is understood in this book (input-output-relation). Activity of management largely is goal setting and regulating the action of the staff of the company. In other words, the manager tries to *increase the coherence of action of members of the organization*. By doing this he probably gets a *synergy effect* and he can increase the output of the unit without increasing the money input into the system. Speaking practically, there is a change of understanding management of health care units in Germany. Up to nowadays there was the rule that, for instance, "the best" of the doctors of a hospital will become the director of the hospital or the senior consultant of a department. In the 90s there is the increasing tendency to employ economists at management positions. This group of professionals however very often has conflicts with the medical staff. Therefore there is the wish to have leaders at top of a hospital who are trained in medicine and in economy as well.

But no matter what kind of profession is guiding the hospital the style of guidance has to be optimized. In contrast to classical style of management in public hospitals which was either very autocratic or very humanistic there seems to be a need of developing highly sophisticated forms of participatory styles of management.

Those aspects will be discussed in this paper with the aim to make these variables more precise as they are important determinants of performance of health care units.

2. OPERATIONALIZATION OF "HEALTH" AND "COSTS"

When trying to describe and to evaluate functions of the health care system or units of the system by input-output analysis one has to be concerned with the choice of the parameters and variables of the system and one has to consider the problems of quantification of those items. The first level of analysis is related to the definition of health.

One very wide definition of health according to the very early (1946) constitutional confession of the WHO is the total physical, psychic and social wellness. As medicine is only concerned with physical and psychic problems and therapeutic strategies a more narrow concept of health should be used. Of course, the persistence of a reduced health as it is the case at defects after acute diseases or at chronic illness implies a reduced social functioning. This results in social deficits which might result in depression etc. Those issues lead to the discussion of the constitution of the concept and measurement of quality of life (Salvador-Carulla 1997).

Now, when considering the clinical problems to define health and disease (or degrees of disease) first of all one has to distinguish "objective" health and "subjective" health (or illness; comp. tab. 1):

- it is possible that a patient has nearly no objectivable dysfunctions but appears and feels himself sick (e.g. functional syndromes, prodromal stage of infections, illnesses of functions of the immune system)
- it is possible that a person is very sick but still goes to work and functions socially very well (e.g. cancer)

It is wide agreement now that subjective definition of being sick is also an acceptable type (or even a component) of disease.

Table I: Objective and subjective illness

		Objective Illness	
		+	-
Subjective Illness	+	Heavy infection	Functional syndromes
	-	Early cancer	Full health

2.1 Epidemiology

At a macroscopic level *epidemiological indicators* can characterize the health of a population. For this reason health economics must be based not only at least on medicine and economy but also on epidemiology.

The main measures of the health of a population are the absolute and the relative *mortality* and the scale and structure of *morbidity*. Conceptually related is the *life expectancy* at birth (or at any other age). Life expectancy is based on recordings of mortality. On average, in industrialized countries it is increasing and right now it is at about 75 years for men. However, the group of persons being older then 65 has a high rate of sick individuals. The measurement of health of these individuals has to be supplemented by the evaluation of quality of life. Finally, the increase of life expectancy can not go to an unlimited level: there is not an extremely increasing number of 90 or 100 years old persons. There might be a biological limit probably of the cardiovascular system which at the end is the functional cause of death. This example shows some problems of choice of health indicators.

2.2 Measurement of health

When trying to obtain a more detailed measurement of health and disease one can use a lot of *rating scales* for experts and for patients. Any disease can be assessed by scales for "global clinical impression" by the trained physician. For any syndrome there are various scales for rating the severity of the disorder. Also for functional disorders (subjective disorders) scales are available which allow quantification of symptoms. This is not discussed in detail here (comp. Drummond et al. 1997, Salvador-Carulla 1997).

Although the construction and analysis of health-related indicators already has reached a high professional level the discussion in this paper will be kept rather elementary to show the principle problems.

For instance, already when considering the relevance of life expectancy as a measure of health one can see some problems: If a 50 years old person after an accident under conditions of intensive care lives additional 30 years it would be a good effect of the medical system but one has to ask for the objective and subjective quality of life. At general, it is interesting what kind of *level of social functioning* the formerly sick individual has reached or could keep one year (or longer) after discharge. This must be *related* to the *intensity of residual symptoms*. Especially, if a person looses functions and thus gets handicapped the level of social functioning must be determined. This can be measured by scales for evaluation of quality of life (Quality of life scales).

Some scales are the World Health Organization Quality of Life Scale (WHOQOL), the Nottingham Health Profile (NHP), the Quality of Life Check List (QLC). Those scales are asking for the level of adjustment in several domains of life (work, housing, family etc.) according to objective and/or subjective items.

At health economic questions modified scales or measures for quality of life are used: The Quality Adjusted Life Years (QALY), the Disability Adjusted Life Years (DALY) or the Life Years Saved (LYS; comp. Drummond et al. 1997). The principle of those scales is the evaluation of value of additional years of life which are won by the particular treatment method which should be evaluated.

2.3 Measurement of effects of treatment

Those measurement procedures and especially the scales for measurement of quality of life are important for clinical research to test if the treatment program A is better then the treatment program B. At a next step, after proofing the efficacy of a new treatment procedure one can make the comparison of the monetary input of those two programs.

For the first step, usually the *outcome* of any treatment should be measurable by an increase of health or a reduction of the disease and/or the symptoms. This can be indicated by a pre/post-comparison of relevant somatic and/or psychic *parameters* like blood pressure, heart rate, laboratory parameters or the scores of the respective *rating scales* being measured at *admission compared to discharge*. Such a ratio of post-treatment scores to pre-treatment scores indicate the *efficacy of therapy*.

However, treatment outcome also must be related to the input into treatment. This is the *time spent for treatment* measured as number of days of inpatient treatment (”*days of hospitalization*”). By this measure an indicator of *therapeutic efficiency* can be defined. However this indicator might not be sufficient as rapid techniques of treatment could imply a high relapse rate with the need of additional inpatient treatment. Therefore not only *efficacy* and speed but also *stability of treatment outcome* must be measured. Thus the ”*relapse rate*” (e.g within 6 months after discharge) or the *symptom-free interval* or the *interval to next hospitalization* after discharge can be measured.

The final step, as it was said already, is to monetarize not only the financial input as costs of treatment programmes but also their outcomes. With this aim those indicators can be expressed in monetary items by calculating the wages of personnel and prices of goods consumed for this respective health service.

However, the problem at relating health impairment of chronically ill individuals to economic indicators very soon leads to the problem of *euthanasia*.

Those aspects, for instance, get relevant when evaluating treatment strategies at *emergency medicine*: reanimation of a victim of an accident at first glance might be a good success, but if the brain damaged individual gets tested half a year later and one sees that this person has severe mental handicaps and can not live by himself one has problems – health economics probably would like to inhibit the therapeutic action of the doctor on emergency call, whereas medical ethics would say that the doctor should save life at any price. Individual ethics of relatives even could imply the wish to stop life supporting activities, saying that if the person would see his own state he would not like to stay alive anymore. Also heavily impaired individuals who want to die can not be helped at dying. Thus ethical issues can be correlated to primarily economic motives which is a danger for human values. At those questions health economics runs into ethical dilemmata: it would be cheaper for society if those people would die. But who has the right to decide over the existence of another individual, which human being can say that it is superior to the other with regard to the question to let live or let die?

This is a current controversial issue of medical ethics with remarkable international differences. This extremely complicated problem can only be solved by wide and serious discussion within the total society.

2.4 Costs of health

Switching from the level of the individual case to the health of the population one has to consider that the growth of health expenditures internationally is growing with an annual rate of 5 to 10%. From an economic point of view this can be called an increasing "over-allocation" of social resources. Obviously sooner or later this has to be stopped everywhere in the world.

In Germany, the *monetary input* into the health care system must be limited as the total macro economy can not absorb additional increase of health costs as a part of it has to be paid by the companies and because of harder conditions of economic competition for the companies. For this reason health policies try to relate the *financial input* into the health care system (Germany: 300 billions DM ; = 8,6% of Gross Domestic Product, GPD; USA: 800 billions $ = 14 % of GDP) to the quantity and quality of *output of* services. In this way the efficiency of the performance of units of the health care systems is critizised.

The health care system in the USA very often is used as a positive reference system. Some data can characterize the different systems, here comparing the US system with Germany (Zdrowoyslaw & Dürig 1997, McCornell & Brue 1995):

- In the USA 224 physicians (N=600.000) supply 100.000 inhabitants. 50% of the approximately 6500 hospitals (more then 1 mio beds) there belong to private non-profit institutions, 30 % are public hospitals and 20 % are owned by for-profit institutions. One hospital has about 154 beds.

Nearly 300 billion Dollars i.e. about 37% of the total health costs (800 billion $) are spent in the hospital sector. One bed has daily costs of about 1000.- $ (1500.- DM).

In the US 3200 DM per capita (about 4800.- DM) are spent for health costs. 14,1 % of the GDP are used for health.

- In Germany 337 physicians (N = 230.000) supply 100.000 inhabitants. Of all 2500 hospitals (630.000 beds) are about 36 % run by non-profit organizations, 39 % of the hospitals are held by public organizations and 15% belong to for-profit organizations. One hospital has about 252 beds.

Nearly 100 billion of DM were used in hospitals, this is about one third of the total health costs (about 300 billion DM).One bed costs about 450.-DM a day.

In Germany 3750.- DM per capita are spent for health.

8,6% of the GDP are used for health.

At these numbers one can see easily that in comparison to Germany in the US

- the number of physicians per capita is lower
- the health costs per capita and the percentage of the GNP are much higher

For this reason the total costs for physicians in the US are about 1,4 million $ (2,1 million DM) whereas in Germany the respective costs are 1,3 million DM.

But it is not easy to compare both systems, as in the USA

- there are less public hospitals,
- the hospitals are smaller.

This is probably also due to the fact that especially mental hospitals in Germany are larger then 800 beds. In Germany generally hospitals of this size to 90% are hold by public organizations.

At such comparisons not only the structure of the health care system but also the structure of the population has to be considered.

From an epidemiological point of view the most important thing is the group of elderly persons consuming the highest proportion of health care services and of course also the costs. Health insurance companies as the payers of health services have high costs with the elderly and impaired persons. Annual expenditures for health insurance (FRG, 1994) are ca.. 2300.-DM per Person without pensioners and ca. 5900.-DM per pensioners (Pfaff 1983, Zdrowomyslaw u. Dürig 1997).

By increase of the number and proportion of persons being older then 65 the costs of health services will further increase even if there is no increase in quality of those services. Probably even the quality has to be reduced to keep the costs under control!

Thus further analysis of the health costs leads to an ethical question like if the insurances should pay health care costs for self-caused diseases which are based on a risky lifestyle (smoking cigarettes when young, suffering from lung cancer when being old).

In Germany there is a drastic restriction of the budget for health service. This leads to a reduction of intensity of services done by physicians as there is an increasing demand by a growing number of elderly persons. Also for hospitals an economic analysis becomes important: Thus one tries to get away some expensive patients because their treatment potentially brings the health care unit into "red numbers". This is the limit of humanity, as it is criticized by lobbies of physicians.

Anyhow, the monetary input into the health care system must be compared with the output of the system and comparisons between different systems must be done. Multidimensional perspectives, however, are necessary to be applied.

3. HEALTH ECONOMICS

Health economics not yet is established in medical thinking therefore there is no agreement on the choice of indicators and on the operationalization and monetarization of the relevant variables (comp. Herder-Dorneich 1994). Physicians see the reduction of prices of their services as a result of "rationation" and they see it related to deeper economic reasoning. However, they do not know the calculatory background of the health economic approach. Health economics is a basic scientific analysis of the functionality of the health care system. It has to be constituted by both - economists and medical professional otherwise there would be a dominance of decisions of insurances to buy services of medical units offering the lowest price: In Germany 2500 hospitals have to "struggle for life" with about only a few big insurance companies. The difference in performance of health care units therefore must be operationalized in a scientific way as soon as possible. Systems analysis can help to characterize the basic structure of the controversy as it is tried to show it in this paper.

3.1 Market mechanisms

Economics is the optimum management of scarce resources. Any social unit has to exchange goods and services (services can also be classified as goods). The value of the good is the price, the price is influencing the amount of supply and of demand: the lower the price, the lower is the supply and the lower the price, the higher is the demand. The higher the price, the more is the supply, the higher the price the lower is the demand (Pindyck & Rubinfeld 1998).

At the market the price gets settled by the dynamics of interaction of demand and supply.

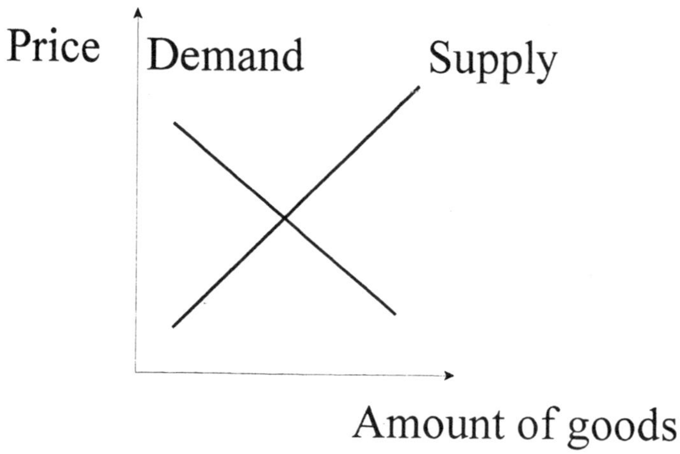

Figure 1: Demand and supply curves (s.text).

The regulation of those forces cannot be seen for health - the cheaper an operation is, the less it will be offered ("small surgery") and the more will be demanded by the many sick? The more expensive an operation will be (transplantation), the less it will be demanded by the relatively few very sick? One sees that cheap services must be made available broadly whereas expensive services have to be limited. One also can imagine that the subjective value of health is not corresponding with the price at market of medical services. The central role of insurance companies influences the market dynamics.

As it was mentioned already, the main problem of efficiency of health care is the *cost explosion*. This is caused by *factors of demand* (rising incomes, higher education, aging population, unhealthy life styles) but there are also *factors of supply* (number of physicians, medical technology, defensive medicine, asymmetry of information, medical ethics).

For regulation of the costs only a few variables can be influenced like prevention of unhealthy lifestyles, limitations at medical technology etc. Each strategy however must be evaluated in the frame of cost-effectiveness relations.

One of those cost increasing factors is the increment of income of demanders. This general relation of income and expenditures is described by Engel curves. For health there is a S-shaped curve between expenditures and income, saying that over a large scale of increment of income the health expenditures are only slowly increasing but at a very high level of income the expenditures increase strongly (s.Fig 2, comp. Pindyck & Rubinfeld 1998).

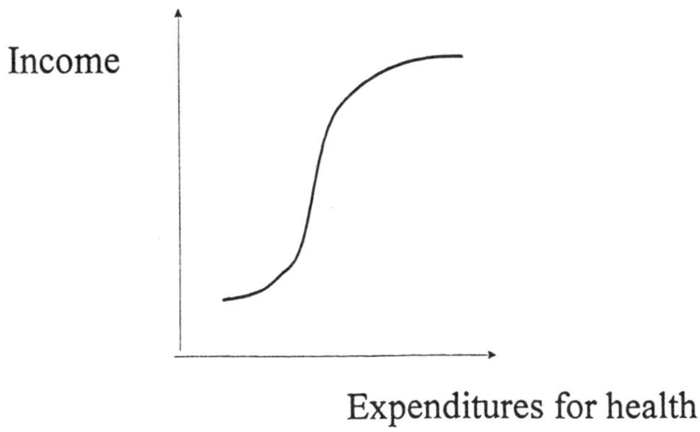

Figure 2: The relation between expenditures for health and income (Engelcurve)

3.2 Economics and the optimization function

A central task of economists is to find the maximum of outputs at a given input or the minimum of input at a given output (s. Fig. 3). Economics also tries to evaluate the efficiency of a system or of sub-units of a system.

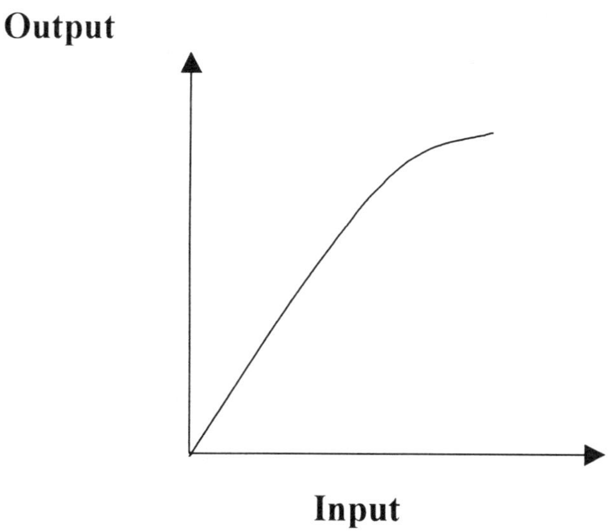

Figure 3: Empirical input-output function

From an economic view there is a limit of increment of inputs as there is a marginal increase of benefits. When reduction of input is necessary the critical level of minimal input must be found out to reach still an acceptable output.

In economics this relation is represented slightly different as production (goods, services etc.) is drawn on the abscissa whereas the costs are drawn on the ordinate (s.Fig 4).

3.3 The cost of health production

At a general level of analysis the crucial question of health economics is how to keep the standards but reduce the costs or how to stabilize the cost when expanding health services in quantity and quality. This problem can be described in economics by cost production curves. The production function tells the maximum output that can be produced with a given amount of inputs. This function will only be described here on the level of total costs. The essential proposition of the cost production curves is that there is an optimum of the ratio of production and costs: To determine the optimum economists use the indicator "marginal costs" indicating the increment of costs if the output of the firm is increased by one unit. For simplification here the theoretical relation of increment in health production to total costs are sketched (s. Fig. 4).

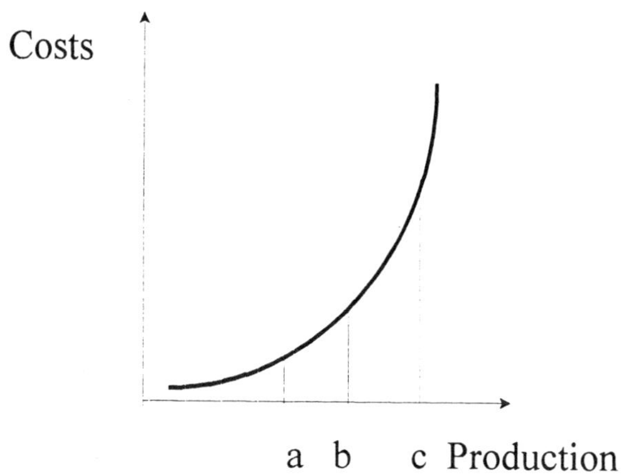

Figure 4: The health production function: The exploding costs of health services in relation to increasing production of health. Where are we now at level a, b or c?

For health economic decisions also the optimum of development of a unit must be considered. This can be done by estimating the relation of increase of utility and increase of costs of medical service per case (s. Fig. 5).

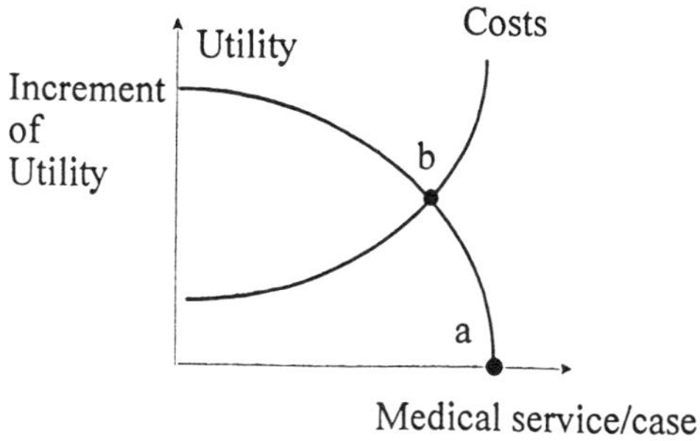

Figure 5 : Cost-utility relation:

a) The maximum of medical utility is reached when no further growth of utility is possible (zero growth). The cost curve at that point already was extremely growing and nearly implies unpayable expenses.
b) The optimal service is reached when both is realized - an already reduced increment of utility comes up if medical care is increased and the costs are not yet extremely increasing (comp. Zdrowomyslaw & Dürig 1997).

For detailed quantitative health economic analysis various analytical tools are used (comp. Drummond et al. 1997): *Cost-minimization analysis* for optimization of production of the same degree of health, *Cost-effectiveness analysis* for relating different monetary input with it´s specific degree of health effects, *cost-utility analysis* for the evaluation of monetarized health service in respect to the utility as a complex indicator (e.g. quality of life or quality-adjusted life-years; QALY), *cost-benefit analysis* for evaluation of the monetarized health effects of the health service being studied.

3.4 Examples for cost-effectiveness relations

When one wants to evaluate the global efficacy of a given health care system by using the health care indicator *life expectancy* there is evidence that the USA – although it has the highest expenditures for health in absolute numbers and per capita – internationally ranks eight. It also ranks only eleventh in "maternal mortality" and twenty second in "infant mortality" (McCornell &Brue 1996). Thus, according to those mortality indicators the US system has not a good performance. This could probably be the case because there is much emphasis on high-tech medicine which offers a socially and spatially unequal distributed service.

At a lower aggregate level of analysis one can evaluate therapeutic strategies in categories of cost effectiveness. An example where the empirical data are not complete, but where one can see the problems of health economic decisions is the treatment of alcoholics: we know at therapy of alcoholics that the abstinence rate can not be increased to hundred percent even if the patients are at an inpatient treatment programme for two years. Usually the abstinence rate is about 60% after long-term treatment for five to six months. The costs for such treatments are about 30.000 DM. On the other hand the spontaneous rate of abstinence after physical detoxification is about 10 %. This costs nothing for the insurance companies. Now, one has to establish the cost effectiveness curve for treatment (Fig. 6). If insurance pays a third of the price (10.000 DM) it will get an abstinence rate of 30 % . If it pays 5.000 DM the abstinence rate will be maybe 12 – 20 % . This result will only be acceptable if it reaches 20 %. It would not be an alternative to the relational success of an input of 10.000,- DM (30%). For this reason it is the case that only a certain threshold value of monetary input must take place to obtain a significant effect. On the other hand there will be a limit of monetary input to increase the effect significantly. This is a mathematical problem of optimization using empirical data. However, there is a lack of adequate data for such analyses yet.

In practice the efficiency of therapeutic units depends to a certain degree on efficient management of resources.Therefore management tasks have to be discussed here too.

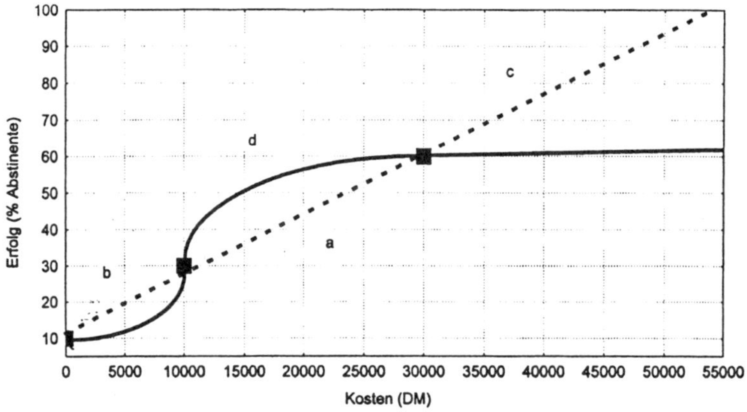

Figure. 6: Theoretical relation between costs and effectiveness at therapy of alcohol addicts (from Tretter 1998).

a) empirical curve at linear extrapolation of three data poi·nts
b) expected curve at shorter time of therapy (z.B. inpatient short-time therapy, shorter then 50 d or outpatient therapy)down to the level of spontaneous remission without therapeutic intervention

c) expected curve at long-term inpatient treatment (e.g. longer then 150 d = ca.5 M)
d) theoretical cost-effectiveness curve as a sigmoid growth function

Comments: The daily faire was calculated with ca. 200.-DM, that is about 10,000.- DM for 50 d and about 30,000.- DM for 150 d.

Abscissa: costs (DM)

Ordinate: success (rate of abstinent patients)

In economics cost effective curves usually are drawn the other way round (ordinate = costs).

Economic calculation is only one tool to evalutate the efficiency of the system. In practice, there is another way to increase health economics – the output maximizing function of management.

4. MANAGEMENT IN HOSPITALS

Management is an activity which is regulating the goals and the operations of an organization. Management of health care providing units has it's peculiarities Those can be studied best at analysis of management of hospitals.

Hospitals are very important units of health economic analyses. Even if hospitals are of the same type and therefore get the same amount of money the are differences at output. Some newspapers organize evaluationsof hospitals made by patients showing the differences. Those analyses very often are not very scientific, however they show tendencies.

s, if there are differences at hospital's output the role of management seems to be crucial. Therefore a sort excursion to the field of hospitals has to be done.

4.1 The hospital.

The hospital must be understood as a complex social organization. The output of the hospital is the service for the patients. Th quality of the ouput is not only related to the *medical service* but also to th *psychosocial service* and the *hotel function* of the hospital. This aspect gets measured by *satisfaction scales* applied to patients at discharge.

From a *systemic point of view* there is a *financial input* to the hospital namely by the insurance companies used by management to establish the *structural quality* of the hospital (staff, technology). Management, however, has to coordinate the action of the staff which can be called the "organizational input" of management into the health service. Health service itself is the output of the hospital but also is the input to the patient which is the target object (or subject) of the hospital's actions. Feedback of the patients might be a help to evaluate at least the efficacy of the hospital's actions. (Fig. 7).

The change of definition of tasks of hospitals in Germany which is taking place right now aims at the reduction of *number of beds* and at reduction of *duration of stay* in the hospital. Patients should be treated as much as possible at outpatient treatment units. Treatments longer then estimated by insurances are not payed any more!

There is, additionally, a special problem for German health care as there is a need to restructure the health care system for mentally ill persons. As in most other countries in the world also in Germany mental health services were concentrated on large hospitals with more then 1000 beds. In the 60s there was a Community Mental Health movement in the US and also in Britain, Netherlands, Sweden and other countries. The de-institutionalization was started. Complementary out-patient treatment structures were established. Because of political resistance in Germany this only came up in the late 80s and now in the 90s. In Germany in-patient psychiatric health care in is mainly offered by public hospitals which are located about 80 kilometers out of the cities. In 1974 70% of the beds were in clinics with more then 1000 beds, 70% of patients stayed longer then 2 years in the mental hospitals. One physician had to supply 50 or more patients (comp. Deutscher Bundestag 1975).

Now, in the late 90s the reformation of mental hospitals is driven by public administration: they should be split up into smaller units which are located more closely to the community. They should have outpatient treatment settings to reduce the costs (one psychiatric bed per day is only about 300.- DM).

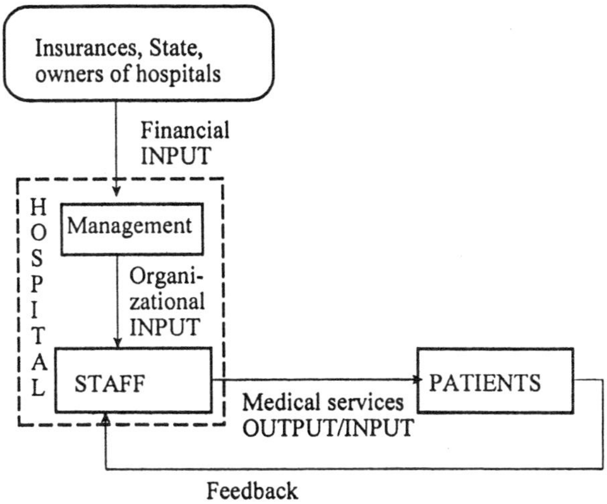

Figure 7: The flow of inputs into the hospital and outputs of the hospital as inputs to patients and the feedbacks of the patients (repeated consultation, questionnaires on satisfaction).

The state and the districts also try to establish models of privatization. The hospitals have to run their own economy then.

This process of adaptation to new health political references must be realized by management of hospitals. Thus management in public hospitals is faced with several problems:

– The Chief of hospital was a physician without being trained in management.

– Staff of (public) hospitals is thinking that the community will pay the wages no matter how the economic activity of the clinic is like.

– There has to be considered a lot of laws for personnel management as it is not so transparent at private hospitals (Law for Equalization, Law for Protection of Work Time etc.) and which is not calculated in the costs of daily hospital service. For instance, overtime has to be balanced by leisure time, there is no money left to compensate. The result is that there is a lowering of coordination and cooperation of the staff as too many individuals are at holidays , leisure time etc.

For this reason the quality of the service for the patients by far is submaximal.

There are some disadvantages of privatization for public hospitals: For instance they can not rely on deficit-spending any more. The staff has to learn that their professional activity alone can guarantee their salaries.

Little by little these issues get tackled by insurances, politicians and the public.Tasks and strategies of management for optimization of the system´s function

For the current task to optimize the health care system under restriction of financial input one only can analyze and regulate the functioning of the system at an organizational level (rationalization). *Organization* in this view means the structural aspects of functions (services). Not only financial aspects but also organizational aspects and aspects of management must be considered.

Management of a unit of the health care system, as usual, primarely has the goal to maximize the efficacy of the unit and to minimize the input to the system. Thus efficiency can be reached.

THE TASKS OF MANAGERS IN HEALTH CARE UNITS ARE EXTREMELY COMPLEX AND SHOULD BE LESS PROFIT-ORIENTED THEN IN THE FIELD OF INDUSTRY. THE WORK OF MANGERS CAN BE DESCRIBED BY THE FOLLOWING ACTIVITIES (FIEDLER & GARCIA 1987, DAFT 1997; COMP. FIG. 8):

- *Tasks* at processes of:
- planning (creating goals as "Sollwerte")
- controlling (observing the current situation, "Istwerte")
- guiding (instructing personnel)
- deciding (setting priorities)
- in *fields* of :
- corporate goals,
- personnel,
- organisational structure
- technological facilities,
- physico-spatial structures of the hospital,
- financial basis
- time structures of processes
- in *relation to* the environment ("ecological" dimensions) like:
- insurance companies

- public opinion
- policy makers
- officials
- competing corporations
- cooperating corporations
- public administrations
- politicians

The formation of a "corporate identity" is an important issue: If one assumes that all human actions are regulated by the internal concept of the self, the environment and the environmental relations of the person it is useful to extend this concept to organizations: all workers of a corporation are acting according their corporate self understanding (role of actor, function of corporation, corporate identity) and the understanding of the environment represented by the customers (e.g. patients and competitors).

In this way the self-interpretation of the hospital as an autonomous or even autopoetic system in a dynamic environment of tasks and options is adequate. In other words this is a specific version of the "social ecology of the company". The task of management is to establish this new view, otherwise the staff believes that there is a center of power somewhere which is the resource of energy (e.g. money, salaries).

Now, under the new condition that every (public) hospital has it's own responsibility for economical success an adaptation of the guidance of the hospital's staff is necessary. However, it is hard to find the appropriate style of management. Should it be

a) more autocratic or
b) more democratic?

With an autocratic pattern the speed of change could be increased but the majority of staff will disagree if it has not taken over the new view of the "social ecology of the company". Conflicts will reduce the speed of change.

Thus it is more easy to practice the democratic pattern of management by participation.

- There is a bottom-up delegation of responsibility
- The contracts with the employees can be time-limited which is not possible in state hospitals. Thus in case of changing of the functions of the hospital there is enough flexibility.

By this change the self- image of the hospitals has tor be changed : the staff of the hospital must see the world similar as the management.

4.1.1 Styles of management

First of all, there are problems of integrating medical guidance and economic guidance. The managerial decisions must be based by the medical professions of the clinic. Therefore the hospital manager must act based on expertise of physicians and nurses of the hospital. This implies a *new style of leadership* at the hospital staff: no "management by authority" any more. Also other styles of management like *management by objectives* or *management by exception* but preferentially *management by participation (management by consensus)* chould be practised (Fiedler & Garcia 1987).

For participatory styles of management the formation of collective understanding of task dynamics is essential. This means that staff must recognize the internal and external determinants of economic action of the hospital. There can be a reduction of time and energy consuming (i.e. money consuming) conflicts if the staff has a sufficiently complex mental map of the landscape of tasks and options of the hospital and it's clients. To reach this goal the management has to techniques represent the task environment of the total hospital. It has to be shown the interdependence of effects of single actions in different fields of action in the hospital and external to the hospital (politics, contacts to general practitioners, information in mass media). Some methodologies which seem to be suited for such purposes are the following (comp. Rosenhead 1997):

- Systems methodology by Checkland and others;SODA, Soft Systems Methodology etc)
- Swiss School of management in St. Gallen (comp. Malik 1989, Probst u. Gomez 1991; comp. also Hub 1994; GAMMA methodology).

Those tools of systems methodology can be used at staff to present the task environment of hospital management. Thus it can be shown that the central goal to keep the number of occupied beds at maximum is interrelated with several other variables as conditions and consequences of a successful policy of occupying beds.

This procedure can be started within a session with the staff:

At a first phase of *brain storming* were anybody can say terms which signify any aspect of the internal and external work environment of the hospital with regard to the central goal "increment of number of occupied beds". These concepts can be listed on a board. Then anybody can *estimate* the mode, strength and direction of *influence of one variable to the other*. In a next step the total network can be displayed so that one gets a better overview and finally the *functions of the total system of operations* can be discussed.

This philosophy of guidance was applied by the author with the experience that the reduction of conflicts and thus the increment of coherence was possible. There was also the impression that staff was more able to accept the complexity of management tasks for the hospital. There was a good evidence for the staff of the complexity of processes of the hospital by the visualisation. The evidence of feedback loops, the convergence of input variables and the non-linearity became evident (s. Fig.

9). Thus this technique is one practice-grounded step for systemic methodology of analyzing organizations.

There are several other advantages of this method:

- Everyday-thinking can be transformed into network thinking which also can be formalized for computer-assisted modeling.
- The satisfaction of staff is increasing as it understands the problem of managing complexity.
- Staff can be trained to solve problems of complex dynamic nonlinear systems.

By this example of guidance of staff of hospital there is evidence that guidance by consensus can have synergetic effects which help to optimize the transformation of given (monetary) inputs into medical service outputs which as a result is incresing the performance of the hospital.

This chapter completing it should be mentioned that this technique also can be applied at higher levels of health care management like in coordinative functions at public administration.

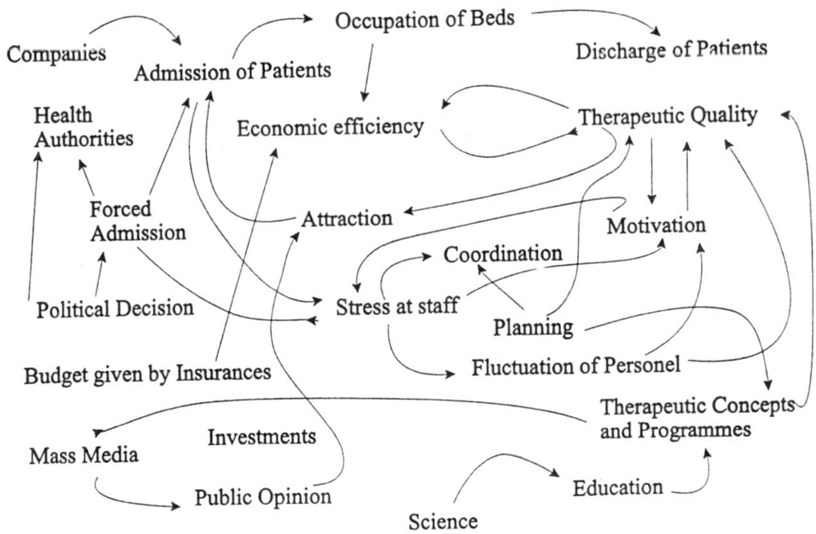

Figure 8: The map of the landscape of variables at work in hospitals.

5. CONCLUSIONS

Medical professionals need the tools of economic thinking. Determining the outcome of health services in more precise terms and their translation into the language of economics, i.e. in prices, is a way which might help to describe the functions of the health care system in a more precise way. The methodological problems at developing those indicators were described here.

Parallel to this more or less academic issue health politics, at least in Germany, is changing the total system by introducing some principles of the free market. Demand/supply concepts dominate the ideas of restructuring the health care system. The health market will be dominated by the insurance companies as demanders and payers of health services. The prices and the services will be organized by competition at cost-reduction. The customer of the insurances are only the indirect payers. They have the problem that when they sign a contract with the insurance company they are healthy and have an *asymmetry of information on risks on health and disease*. They depend on the health theory of the insurance company. Later, if they are sick they are not able to select the suppliers of medical services any more. Insurances understand themselves as advocates of their customers. Thus there might be three disparate cultures - the medical experts, the economists and the laymen. One wonders how integration can take place.

In this way the American system of *free enterprise* is the reference model for European countries and also for Germany. The Amercian model promises a sufficient or even an optimal supply with health care services. Competition will lead to the survival of the best health care service. According to this concept traditional scepticism at socialist parties and organizations however say that capitalism does not work at health care as not profit making but humanistic values (solidarity) are the supreme goals. In case of capitalistic guidelines the poor will suffer and only the rich will get the best services. Also the insurance companies will only guarantee excellent support if one pays extra fees. This inequality probably must be compensated by the state who supports the poor financially.

Market principles at health economy do not work in every country. Therefore in nearly every all countries there is the need for *public regulation* of health care systems: Quality control and quality management, spatial dispersion of services, etc. are functions of health authorities.

The central question however is how much and in which mode the regulation should be organized: *Will self regulating economy supply a good public health or do we need an authorized regulation of health care system ?*

Besides these questions, on the level of *single units* of health care there is the current everyday task to optimize the input-output ratio. In part it can be tackled by optimizing the management strategies. Considering the main task of management to motivate and to direct the staff a mapping technique showing the complexity of tasks, conditions and consequences of hospitals operations might be very useful for management by consensus. By this "soft technology" the outcome of hospitals operations might be better.

However, this theses must be tested. When one considers the problem of measurement as it was shown in the first paragraphs of this paper then it seems highly unlikely that we can expect a good recommendation for the design of management operations. Experimental organization research has a lot of methodological problems.

For this reason it might be difficult to establish golden rules for efficient management of health care units. Moreover, it seems that the introduction of conditions of a market economy in the health field will "optimize" the system in a more efficient way then scientific consideration.

6. REFERENCES

Deutscher Bundestag (Ed) (1975): Bericht über die Lage der Psychiatrie in der BRD. Drucksache 7/4200, Bonn

Daft, R.L. 1997: Organization - theory and design. South Western College Publ., Cincinnati

Drummond, M.F., O'Brien, B., Stoddart, G.L., Torrance, G. W. (1997): Methods for the economic evaluation of health care programmes. Oxford Medical Publ. Oxford

Fiedler, F.E. Garcia, J.E. (1987): New Approaches to Leadership. McGraw Hill, New York

Herder-Dorneich, P. (1994): Ökonomische Theorien des Gesundheitswesens. Nomos, Baden-Baden

Hub, H. (1994): Ganzheitliches Denken im Managment. Gabler, Wiesbaden

McConnell, C.R., Brue, S.L. (1996): Economics. McGraw Hill, New York

Parra-Luna, F (1999): Axiological systesm theory. Unpubl. Manuscr. Dep.Sociol.Univ. Madrid

Pfaff, M. (1983): Finanzierungs- und Ausgabenströme im Gesundheitssketor: eine realtypische Betrachtung fiskalischer System. Schriften des Vereins für Sozialpolitik. Bd 135, Berlin

Pindyck, R.S., Rubinfeld, D.L. (1998): Microeconomics. Prentice Hall, New Jersey

Rosenhead, S. J. (Ed) (1997): Rational analysis for a problematic world. Wiley, New York

Salvador-Carulla, L. (1997): Measuring quality of life in cost analysis – controversies and use in mental health. In: Katschnig, H., Freeman, H. , Sartorius, N. (Eds); Quality of Life in mental Disorders. Wiley, New York, pp.287-304

Samuelson, P.A. (1995): Economics. McGraw Hill, New York

Tretter, F. (1979): Medizin und Umwelt. Unpubl. Diss.Univ. Munich

Tretter, F. (1998): Die Ökologie der Sucht. Hogrefe, Göttingen

Zdrowomyslaw, N., Dürig, W. (1997): Gesundheitsökonomie. Oldenbourg, München

Chapter 15

Indicators of social performance in the heartland of the United States

DARIO MENANTEAU-HORTA
Center for Rural Sociology and Community Analysis, University of Minnesota

1. INTRODUCTION

Challenges to social development in the United States provide a fruitful arena to explore performance issues of the socio-economic system within the context of sustainable development. This exploration is urgent, as research reveals increasing inequality and a widening gap between the rich and the poor in American society.

According to a U.S. Census Bureau report, the gap between the rich and the poor has been growing since 1968. Over the last three decades, the average income of those households in the bottom 20 percent of earners increased less than 1 percent (from U.S.$7,702 to $7,762), while those in the top 20 percent had an average income growth of 44 percent (from U.S.$73,754 to $105,945).

The same government agency announced in August, 1998, that at least three in ten persons in the United States were living in poverty. The U.S. Government defines poverty according to income and considers poor a three-person family with a total annual income below $13,650.

Poverty is a social reality that paints a complex picture of inadequate performance of the economy as it affects certain groups more than others. Poverty rates differ by race, age, and family structure, and they are dramatically higher among Blacks, Hispanics, children, and households headed by single mothers.

From a social and political perspective, the search for new development strategies that take into account social performance of economic systems is needed worldwide. The responsibility upon the more industrialized nations

The Performance of Social Systems
Edited by Francisco Parra-Luna, Kluwer Academic/Plenum Publishers, 2000

to provide clear leadership in this quest seems imperative, as they control far more resources than the rest of the world. In 1990, for instance, the poorest 70 percent of the world's people earned 7 percent of its income, while the remaining 30 percent earned 93 percent. So far, it is commonly observed, the benefits of economic growth have not trickled down to the larger population of the world nor to all people in each country.

Development models that emphasize economic growth based on a *laissez-faire* ideology tend to consider poverty and inequality as temporary conditions. Kuznets (1955) suggests that economic growth may go together with a worsening of income distribution during its earliest stages, but in later phases of development the benefits of growth would spread among more people.

Research on social organization and income inequality demonstrates, however, that neither the free market nor individualistic competition has been able to bring a balanced and sustained development to all members of society. This situation, particularly acute in developing countries, is also persistent in the United States.

Growing concerns about income inequality and social problems in the U.S. suggest that the overall rates of economic growth are no longer sufficient to evaluate the performance of a society characterized by uneven development among economic sectors and communities. Unbalanced economic growth and unequal distribution of resources between metropolitan centers and rural areas have become a serious developmental issue (Seligson, 1984:3-7). The gap between urban and rural areas within states or counties (Fox, 1978) and the discrepancy of individual earnings between regions has widened since the 1970's (Hirschman and Blankenship 1981).

The ability of rural communities to provide for the basic needs of their people has dramatically declined despite economic growth in other sectors of the national economy (Wilkinson, 1991).

The post-war prosperity in the United States made possible an increase of about 40 percent in the median family income in the 1950's. This economic expansion was followed by a similar rate of growth in the 1960's, but with disturbing uncertainty and decline during the 1970's, 1980's, and 1990's.

Myrdal (1975) captured in a highly critical observation some problems of social performance in the U.S. His statement clearly points out a typical example of successful economic growth without a fair process of social development. Myrdal remarks:

> *"It is not a tenable position for the United States to be that country among the rich countries that has the most and the worst slums, the highest rate of unemployed and unemployables, and the least developed health services, and that is the most niggardly toward its old people and its poor children who are so many, as well as being the country that*

leads the whole Western world in violence, crime and corruption in high places."

Today, individuals in the U.S. who are homeless, unemployed, living in poverty, and without access to basic health services continue to be a serious social, economic, and political problem. In the mid-1980's, family income was almost equivalent to the levels registered in the country in 1969, and the proportion of income shared by those at the bottom 40 percent of wage-earners was lower than at any time since the end of WWII (Danzinger and Gottshalk, 1988). As Braun (1991) remarks, "The real issue is not whether income has become more unequally distributed within the United States, but rather just how bad the situation is."

The growing gap between rich and poor not only denotes a form of distorted development and poor performance of economic institutions, but it also tends to nurture conditions of social conflict, disorganization, and violence. This situation became prominent in rural areas during the farm crisis of the 1980's and 1990's as well as in urban centers with the Los Angeles riots of the 1990s. As a result, the problems of poverty and inequality have taken center stage of social and political debates.

Poverty, therefore, may be considered a function of both availability of resources and the way resources are distributed in society. While the rational utilization of resources is an indicator of economic efficiency, the distribution of them is a measurement of fair management and social performance.

This analysis focuses on the meaning of the concept of performance and its possible measurement. For this purpose, a brief discussion about various perspectives regarding poverty is followed by indicators of income distribution and social well-being estimated for counties of a midwestern state in the United Sates. The observed relationships between income inequality and social well-being provide some empirical measurements of social performance.

2. THE MEANING OF PERFORMANCE

The concept of performance is commonly defined as a process of carrying out an action and denotes aspects of effectiveness and accomplishment. Performance applied to social institutions and human behavior usually involves the notion of implementation or fulfillment of an objective or claim, as well as the pursuit of efficiency as the capacity to achieve what is desired.

Peterson (1995) discusses performance criteria for economic systems based on two elements: Efficiency, defined in terms of rational individual choice and market equilibrium; and Equity, which refers to income distribution inspired in the principles of social fairness. While traditional economics advocated efficiency as the most important criterion separated from social values, the new institutional economics recognizes axiological issues in the analysis of how societies organize themselves to secure goods and services to satisfy needs.

This author underlined a broader view of performance, saying that "narrow market-based notions of efficiency are replaced with concerns about social inadequacy and sustainability (with the question) -how well are the material needs of society being met?" He argues that efficiency (defined in terms of non-wasteful resource use) and equity both contribute to the success of society in sustaining and reproducing itself.

The question of system performance is politically relevant as it relates to social institutions, public policy, and the sustainability of the entire system. From a political perspective, what is being pursued is a better society for all of its members. Thus, the level of performance of social institutions, including the economy, could be measured by the results in the satisfaction of individual and social needs.

Social performance, therefore, is not only the "output" of an event that makes people feel good or happy but the results of a process attempting to make the entire social system better.

Parra-Luna (1997) calls attention to the importance of values and the contributions of axiological analysis in assessing the global efficiency of political systems. He remarks that global axiological considerations based on accepted international values(i.e., overcome inequities between countries, acceptance of democratic decisions, etc.), are essential in order to achieve worldwide sustainable development.

Following this line of thought, it is reasonable to argue that the definition of social performance not only denotes a measurable output or a concrete accomplishment, but it also includes interests and values. Individuals operate within a social and cultural context usually influenced by an axiological dimension which needs to be considered in determining the functioning and performance of social institutions. Accordingly, social performance may also be seen in terms of its implications and consequences for future policies and behavior.

This relationship between performance and public policies can be observed in the way a society deals with poverty. The definition of poverty according to certain arbitrarily determined income levels may dramatically alter the formulation of policy and the political consequences of poverty alleviation programs. By changing the definition of poverty, it is possible, therefore, to develop policies and programs that may be more or less inclusive of the population in need of services. If that is the case, the definition of performance will also require adjustment.

In the area of education, the measurements of performance include the accomplishment of learning objectives in relation to student outcomes (Mehrens and Lehmann, 1984). Performance-based evaluation is rapidly becoming the preferred assessment approach in modern education.

In the field of public health, the concept of performance is considered related to assessment of health outcomes. Perrin and Koshel (1997) suggest that performance in public health involves three parts: (a) health status outcome, in terms of change (or lack of it) in mental or physical status; (b) social functioning, which refers to the ability of an individual to function in society; and (c) consumer satisfaction, determined by patients' and clients' positive responses to health services and programs.

Miringoff and Miringoff (1995) discuss an Index of Social Health for the U.S. and the need to monitor social performance and the strength of key social institutions. They report that the overall pattern of U.S. social health, measured by 16 indicators, has declined from 74/100 in 1970 to 41/100 in 1992. They state that such significant "decline in the nation's social performance may well help to explain the sense of unease felt by many Americans today. Although social health, as a whole, is rarely discussed in government or in the media, overshadowed as it is by economics and politics, it clearly requires the kind of sustained attention that we give to these other areas of national concern."

An important contribution to the measuring of social performance at the national and international levels was made by the United Nations Development Program (1990) with the Human Development Index(HDI). This index is a composite indicator that calculates the GDP based on real purchasing power (not exchange rates), literacy rates, education, and life expectancy.

3. ON POVERTY

What are the causes of poverty?

Why do some social units and/or groups have more poor people than others?

Can poverty be eliminated by changing the poor's values, attitudes, and motivations?

Is poverty a result of the lack of economic performance?

Despite many efforts to provide satisfactory answers to questions concerning poverty, most arguments still remain inconclusive.

Explanations of poverty may be grouped under three broad categories: (A) individualistic perspective, (B) structural approach, and (C) limited resources theories.

3.1 Individualistic Perspective

The individualistic perspective emphasizes the importance of personal traits, whether genetically inherited or learned, in determining poverty. Four theories within this perspective may be identified. They all assume working ability of individuals as the main mechanism for economic failure or success.

3.1.1 Social Darwinism.

Social Darwinists argue that people become poor simply because they are biologically inferior which, in turn, leads to low working ability and, consequently, low earning capacity. It is also assumed that biological inferiority is largely genetically inherited from one generation to another (Herrnestein, 1971). According to this viewpoint, the existing poverty rates are inevitable conditions of societies.

3.1.2 Functionalism.

Functionalists see the stratification system as an inevitable feature of society (Davis and Moore, 1945). By implication, they perceive poverty as a result of some people occupying functionally unimportant or low-paid jobs that can be easily filled. Functionalists, therefore, tend to view the system of differential rewards as normal and suggest that limited personal qualification and low performance are the causes of poverty.

3.1.3 Human Capital.

The arguments of the human capital theorists are similar to those of the functionalists except they do not start explaining poverty from the stratification system. They assume that the poor—-people with less talent and/or less resources—-are usually unable and/or unwilling to invest in human capital, (i.e., education or vocational training), and this results in weak bargaining power in the labor market and low income (Mincer, 1958; Becker, 1964).

3.1.4 Culture of Poverty.

According to this viewpoint, poverty would be more likely to exist in a cultural setting where individuals have negative attitudes towards work, are unable to defer gratification, maintain strong beliefs in fatalism, and lack motivation to achieve (Lerner, 1958; Lewis, 1959, 1966; McClelland, 1961; Hagen, 1962; Inkles and Smith, 1974).

3.2 Structural Approach

Opponents of the individualistic perspective argue that structural factors (i.e., class, wealth, power, and performance of the social system) rather than personal traits are the main determinants of poverty. Structuralists conceive poverty as a result of the process of discrimination, institutional barriers, and lack of opportunities for some segments of society (Merton, 1957; Duncan, 1969; Miller and Roby, 1969; Rossi and Blum, 1969). According to this perspective, poverty is considered dysfunctional, and the best way to reduce it is to change the reward system and the class structure in society.

Some authors suggest that the acquisition of personal traits is not only the result of the internalization process but also a response to the outside environment such as the reward-punishment system (Valentine, 1968; Pareek, 1971; Galbraith, 1979). They argue that to improve the poor's working ability or to change their values and behavior, partial programs in training and education are not sufficient. It is also necessary that the whole social structure be changed to improve its performance at the same time.

3.3 Limited Resources Theories

Some scholars perceive the problem of poverty as a result of either the lack of or the uneven distribution of resources. Representatives of these viewpoints are found among the following three perspectives:

3.3.1 Trickle-Down Theory.

Some economists argue that poverty is largely due to limited resources (i.e., low level of economic development) in society. For some of these researchers, raising the level of economic development, defined in terms of the creation of more goods and services, is the best way to eliminate poverty. They assume that, at the mature stage of development, the fruits of economic growth will automatically "trickle down" to the population, including the impoverished (Rostow, 1952; Kuznets, 1955; Hirschman, 1958).

3.3.2 Uneven Distribution Theory.

Others who contest the trickle-down thesis contend that, based on the development experience around the world during the past few decades, economic growth alone does not guarantee benefits for the poor (Lipton,

1977). They insist that the most effective way to improve the living standard of the poor is to change the distributive system during the time of economic growth (Anderson, 1964; Griffin and James, 1981).

3.3.3 Dependency Approach.

"Dependentistas" have argued that development and distribution of social and economic resources are strongly related to internal hierarchical positions within a society and its external relations with other units in the world system (Frank, 1967; Dos Santos, 1970). Since poverty is assumed to be a function of the level of economic development and resource distribution, by implication, then, the level of poverty in a social unit is also determined by the unit's hierarchical position and its relations with other units within a larger system. This point becomes especially important when poverty rates are compared not only between nations but also among regions and communities within a country.

Summariz.ng these various perspectives described here, it may be said that, in general, theories that approach the problem of poverty from an individual level usually assume that the poor tend to have insufficient education and are more likely to express deviant cultural values, limited working attitudes, and low achievement motivation. An opposite viewpoint is one that considers poverty as a function of the availability and distribution of resources among members of society (Germani, 1980:85).

4. ON INCOME INEQUALITY

From the classical and neoclassical economic perspectives to the Marxist school of thought, income distribution is viewed as the outcome of processes that are primarily "economic." According to this viewpoint, the processes of production, competition, markets, and profit maximization are seen as the main determinants of performance affecting income distribution.

In recent years, however, more attention has been given to social factors, particularly in the study of development, poverty, and inequality.

The dramatic impact of income inequality on poverty levels is well documented by Londono and Szekely (1997) in reference to Latin America. The interaction between inequality and poverty clearly shows that greater income inequality is associated with higher poverty levels. According to these authors:

"Poverty in Latin America increased dramatically in the past decade because of the tremendous increase in income inequality, which the 1990's have failed to correct. Had income inequality remained unchanged, the crisis of the 1980's would have generated poverty for only half the number of people than it actually did."

Despite the fact that the question of income distribution is one of the most important issues in economics and also in the organization of contemporary society, the explanations given about changes in distribution and income inequality are neither entirely convincing nor conclusive.

Domar (1947) and Lewis (1955) suggest some internal conditions of society such as capital, technology, and skilled labor as important factors for growth. Huntington (1968) emphasizes the type of institutional arrangements. Hoselitz (1960) refers to the impact of the cultural system, while Lerner (1958) and McClelland (1961) brought forward behavioral and motivational factors in development. The implicit assumption in most of these works coincides with the assumption advanced by Williamson (1965) and Paukert (1973) that the higher the level of economic growth, the lower the degree of income inequality.

Some researchers, however, find that there is no significant relationship between economic growth and income inequality (Chase-Dunn 1975). Studies provide contradictory information as to whether a higher level of economic growth actually results in lower degrees of income inequality. The thesis of growth with more equality is confirmed by Nelson (1984) but rejected by Gartrell (1981). Tsui (1987) reports that states with higher levels of economic growth had lower degrees of income inequality in 1950, but data for 1970 and 1980 show the opposite to be true. Tsui's conclusion is that higher levels of economic growth in the United States appear associated to higher degrees of income inequality.

Danziger and P. Gottshalk (1988) describe several conditions which have been proposed as the main causes of inequality. Although there is insufficient evidence to prove the single validity of each one of these conditions in explaining income inequality, it is more likely that they may have a collective influence upon inequality. Overall, the conditions suggested by these authors combine quite well the economic factors such as cyclical unemployment, economic growth, and inflation and the social processes such as demographic changes, welfare programs, industrial changes, and family structure.

Birdsall and Londono (1997) suggest even closer linkages between the socio-political factors and the economic elements as essential ingredients of a development strategy to reduce inequality and poverty. They explain that greater access by the poor to productive assets, education, and credit can contribute to higher economic growth, but the poor must have also "access, with voice and choice, to decision-making mechanisms." They concluded that "inequality in its different dimensions has to take on greater prominence in discussions on poverty and economic and social development."

5. MEASUREMENT OF INCOME INEQUALITY AND SOCIAL WELL-BEING.

5.1 The Gini Coefficient.

Income inequality was measured by estimates of the Gini coefficient for all 87 counties of a midwestern state located in the heartland of the United States. The Gini Index of income inequality varies from 1 (perfect inequality) to 0 (perfect equality). Gini coefficients are based on family income and measure the degree of income inequality in each county.

Data on number of households classified in nine income categories and total household income for all counties was obtained from the State of Minnesota 1990 Population Census. For this study, counties were divided into metropolitan(7), urban(16), and rural(64) counties using the classification of standard metropolitan statistical areas (SMSA) reported by the Census.

5.2 Indicators of Social Well-being

A general measurement of social well-being for the same counties was constructed by estimating four sub-indexes with the following variables:

1) **a socioeconomic index**, measured by median family income, education, poverty and housing;

2) **a health status index**, including infant death rate, total death rate, and death rate from pneumonia and influenza);

3) **a family status index**, registering two-parent household and percent of female-headed households;

4) **an alienation index**, measured by the suicide rate and death rate from chronic liver disease.

These variables represent several dimensions related to the standard of living of the population in each county. They portray social and economic conditions as well as health and household characteristics. The last two indicators, the death rate from chronic liver disease and cirrhosis (usually considered related to alcoholism) and the suicide rate are both measurements of alienation and social disengagement. Similar indicators have been used by other studies of social well-being in the United States (Ross, P.J., H. Bluestone, and F.K. Hines, 1979). Data sources for these indicators were the 1990 Census of Population and statistical publications of the Minnesota State Department of Health.

The general index was calculated applying the following formula:

$$I_k = \square_I Z_{ik}$$

Where: I_k = the value of the general index for county k;
X_{ik} = the value of indicator i for county k;
M_i = the mean value of indicator I for all counties;
S_i = the standard deviation of indicator I for all counties; and
Z_{ik} = the standardized Z-scores of each indicator of each county estimated as $Z_{ik} = (X_{ik} - M_i)/S_i$.

The standardized Z-scores permit comparison of the scales of the different indicators utilized in this analysis.

6. RESEARCH FINDINGS

6.1 Income Distribution Patterns

Data indicate a disturbing pattern of income inequality in the heartland of the United States. The following are some of the major findings:

* At least half of all Minnesota counties have high indexes of income inequality with Gini coefficients equal to or above .4000. This is higher than the estimated Gini value for the United States as a whole, which is .3359.

* Urban and metropolitan counties tend to have lower levels of income inequality than do rural counties.

* Five counties with the lowest Gini coefficients are urban, and four of them are also metropolitan counties.

* The two largest urban counties with the two main cities of the state show higher levels of income inequality than do other metropolitan counties. In fact, the level of inequality found in the county with the largest city (.4291) is the highest of all counties in that state.

* Urban counties with relatively high income inequality portray very high levels of income discrepancies similar to levels found in rural counties.

* More than twenty rural Minnesota counties with Gini coefficients above 4118 show higher levels of income inequality than in developing countries such as Egypt (.3694), India (.4022), and Sri Lanka (.4113).

Table 1. Income Inequality in Localities of North Central United States and Selected Countries

Lower inequality		**Lower inequality**	
Type of US locality	Gini coefficient	Country	Gini coefficient
Metropolitan	.3070	Netherlands	.2664
Metropolitan 2	.3465	Belgium	.2722
Urban	.3466	Japan	.2773

Lower inequality		**Lower inequality**	
Metropolitan	.3479	Switzerland	.3002
Higher inequality		**Higher inequality**	
US Locality			
Rural 1	.4216	El Salvador	.3957
Rural 2	.4219	Egypt	.3964
Rural 3	.4222	India	.4022
Rural 4	.4223	Sri Lanka	.4114
Rural 5	.4248	Portugal	.4141
Rural 6	.4252	Thailand	.4180
Rural 7	.4258	Indonesia	.4242
Midwestern state	.3414	United States	.3359

Concentration of economic and human resources in urban and metropolitan areas may be a consequence of an "extractive" type of growth. "Extractive" industries include agriculture, lumber, and minerals, which, through time, have removed resources from rural communities to cities and international markets. Human capital has followed a similar pattern with migration of workers from rural areas to urban centers.

This unbalanced development appears to be closely related to the model of economic growth chosen by many industrial societies and less developed nations. In both cases, they help create income inequality, which affects not only poor societies of the so-called "third world" but also the U.S.

Once income inequality is established, the mechanisms by which it is sustained are largely related to efforts to maximize profits, by downsizing the labor force while raising other costs such as monetary and other rewards for owners, investors, and a few at the top of managerial functions.

6.2 Patterns of Social Well-Being

The general index of social well-being indicates better conditions of social and economic achievement and performance for urban counties than for rural areas.

The overall range of scores for Minnesota counties in 1990 was from -9.744 to 12.751. Seven of the sixteen urban counties had the best scores for social well-being in the state. The urban counties that had the lowest scores of social well-being are distantly located from the metropolitan area and the two major cities.

The lowest social well-being scores were found in rural counties. Although these counties do not follow a strict regional pattern, it is apparent, when comparing other low-score counties, that large areas of western and northwestern regions score significantly lower than do the south, central, and northeast regions where the metropolitan and urban counties are located.

Metropolitan counties had a higher level of median family income, ranging from $39,926 to $48,098, than did the non-metro counties with a range of $20,406 to $43,196. The percent of people 25 years or older with bachelor's degrees in the metropolitan counties was 28%, compared with only 15% in the non-metropolitan areas. In the non-metropolitan counties, 8.4% of the families were below poverty level, compared with 5.8% in metropolitan counties.

While there are differences between metro and non-metro areas, there are also differences between urban and rural counties. Urban counties, for instance, showed higher family income than did rural counties. The median family income in urban counties ranged from $28,373 to $48,098, while in rural areas this family income varied from $20,406 to $36,521.

Also, 26% of people 25 years or older have a bachelor's degree in urban counties, compared with 13% in the rural counties.

In rural counties, 8.8% of all families are below poverty level, while in urban counties the proportion of families in poverty is 6.2%.

Finally, when the five counties with the highest scores in the index of social well-being are compared to the five counties with the lowest scores, additional differences emerge. First, the scores vary from 9.17 to 12.75 on the high end and from -7.07 to -9.74 on the low end. The median family income in counties with higher scores ranges from $43,196 to $48,098, while the median family income in counties with lower scores ranges from $20,406 to $26,413.

Educational attainment is higher for counties with higher well-being scores, with 39.7% of people 25 years or older having bachelor's degrees, compared with 9.8% in counties with lower scores. In this last group of counties, 14.5% of the families live below the poverty level, compared with 3.6% in counties with higher scores.

Research findings revealed a high correlation between the Gini values of income inequality and the general index of social well-being. Results suggest that higher levels of income inequality, apart from concentrating more income in fewer hands, are also associated with infant mortality rates, restricted access to medical services, and other public health problems. Inequality, however, tends to diminish as educational attainment of the population improves and the median family income increases.

7. CONCLUSIONS

In a market-driven economy, income distribution represents the most visible mechanism by which a social system shares its resources, benefits, and rewards among its members. Income distribution is, therefore, a powerful indicator of socio-economic performance for the satisfaction of social needs.

Problems of income inequality and poverty are net results of uneven development among economic sectors and a lack of systemic efficiency. Even in industrialized nations, including the United States, the "magic" of modernization and the "promises of development" have become elusive and distant from vast segments of the population. Moreover, the gap between the rich and the poor has widened over the last decades.

This study illustrates the extent to which performance of the social system can be evaluated by measuring and analyzing income inequality and social well-being in a society. In this case, estimates of income inequality for localities (counties) and measurements of social well-being in the state of Minnesota are proposed as performance indicators of the social and economic system in the same localities.

Data confirm that development in the United States, usually defined in terms of economic expansion, has resulted in uneven growth, increased poverty, and greater inequality.

8. ACKNOWLEDGMENTS

The author expresses his gratitude to

Dr. Christos T. Papadas, currently a member of the faculty of The Queen's University of Belfast, for his valuable assistance in early stages of this research at the University of Minnesota. His contribution in estimating measurements of income inequality and the index of social well-being utilized in this report are highly appreciated.

9. REFERENCES

Anderson, W. H. (1964). "Trickling down: The Relationship Between Economic Growth and the Extent of Poverty Among American Families.": *The Quarterly Journal of Economics* 78(4): 511-524.

Beach, C. (1988). "The Vanishing Middle Class? Evidence and Explanations," Institute for Research on Poverty, *Discussion Paper*, 1988.

Becker, G. S. (1964). *Human Capital*. New York: National Bureau of Economic Research.

Birdsall, N. and Londono, J. L. (1997). "Asset Inequality has Obstructed Reductions in Poverty." IDB: Development Policy.

Bluestone, B.,and Harrison, B. "The Great American Job Machine: The Proliferation of Low Wage Employment in the U.S. Economy," Joint Economic Committee of the U.S. Congress, 1986.

Braun, D. (1991). *The Rich Get Richer, The Rise of Income Inequality in the United States and the World*. Chicago: Nelson-Hall Publishers.

Chase-Dunn, Ch. (1975). "The Effects of international economic dependence on development and inequality: a cross-national study." *American Sociological Review* 40 (December):720-38.

Danziger, S. and Gottschalk, P. (1988-89). "Increasing inequality in the United States: what we know and what we don't," *Journal of Post_Keynesian Economics*, 11.2, 174-195, winter.

Davis, K. and Moore, W. (1945). "Some Principles of Stratification." *American Sociological Review* 10: 242-249.

Domar, E.D. (1947). "Expansion and Employment." *American Economic Review* 37 (1): 34-55.

Dos Santos, T. (1970). "The structure of dependence." *American_Economic Review* 60(2): 231-236.

Duncan, O. D. (1969). "Inheritance of poverty or inheritance of race?" pp. 85-110 in D.P. Moynihan (ed.), *On Understanding Poverty*. New York: Basic Books, Inc.

Fox, K. (1978). "Uneven regional development in the United States." *Review of Radical Political Economies* 10(3):68-86, 87-93.

Frank, A. G. (1967). *Capitalism and Underdevelopment in Latin America*. New York: Monthly Review Press.

Galbraith, J. K. (1979). *The Nature of Mass Poverty*. Massachusetts: Harvard University Press.

Gartrell, J. (1981). "Inequality within Rural Communities of India." *American Sociological Review* 46:768-82.

Griffin, K. & James, J. (1981). *The Transition to Egalitarian Development*. New York: St. Martin's Press.

Hagen, E. (1962). *On the Theory of Social Change*. Homewood, Illinois: The Dorsey Press.

Herrnstein, R. (1971). "I. Q. in the meritocracy." *Atlantic Monthly*. September 1971:44-64.

Hirschman, A. O. (1958). *The Strategy of Economic Development*. New Haven: Yale University Press.

Hirschman, C. and K. Blankenship (1981). "The North-South earning gap: Changes during the 1960s and 1970s." *American Journal of Sociology* 87(2): 388-403.

Hoselitz, B.F. (1960). "A Sociological Approach to Economic Development," in Pp. 53-83, *Sociological Aspects of Economic_Growth*. Illinois: The Free Press.

Huntington, S. (1968). *Political Order in Changing Societies*. New Haven, CT: Yale University Press.

Inkles, A. and Smith, D. (1974). *Becoming Modern: Individual Change in_Six Developing Countries*. Massachusetts: Harvard University Press.

Kuznets, S. (1955). Economic Growth and Income Inequality." *American_Economic Review*, 45 (March): 1-28.

Lerner, D. (1958). *The Passing of Traditional Society*. Glencoe, Illinois: The Free Press.

Lewis, O. (1959). *Five Families: Mexican Case Studies in the Culture of Poverty*. New York: Basic Books.

Lewis, O. (1966). "The Culture of Poverty." *Scientific American* 215(4): 19-25.

Lipton, M. (1977). *Why Poor People Stay Poor: Urban Bias in World_Development*. Massachusetts: Harvard University Press.

Londono, J. L. and Szekely, M. (1997). "Distributional Surprises after a Decade of Reforms in Latin America," Washington, D.C., Inter American Development Bank.

McClelland, D. C. (1961). *The Achieving Society*. New Jersey: Van Nostrand.

Mehrens, W.A. and Lehmann, I.J. (1984). *Measurement an evaluation in_education and psychology (3rd ed.)* Chicago: Holt, Rinehart, and Winston, Inc.

Merton, R. K. (1957). *Social Theory and Social Structure*. Glencoe, Illinois: The Free Press.

Miller, S. M. and Roby, P. (1969). "Poverty: changing social stratification." Pp. 64-84, in D. P. Moynihan (ed.), *On Understanding Poverty*. New York: Basic Books, Inc.

Mincer, J. (1958). "Investment in human capital and personal income." *Journal of Political Economy* 66(4): 281-302.

Miringoff, M., and Miringoff, M.L. (1995). America's social health: The nations need to know. *Challenge*, 38, pp 19-24.

Myrdal, G. (1975). *Against the Stream, Critical Essays in Economics*, New York: Vintage Books (p.284).

Nelson, J. (1984). "Income Inequality: The American States." Social Science Quarterly, 65, 3:854-60.

Pareck, U. (1971). "Poverty and motivation: figure and ground." Pp. 300-317, in V. L. Allen (ed.), *Psychological Factors in Poverty*. Chicago: Markham Publishing Co.

Parra-Luna, F. (1997). "Hacia la Medida del Desarrollo Sostenible." in *Sustainable Development, edited by Elohim,* Parra-Luna and Stuhler. University Complutense of Madrid, Spain (Vol. 2)

Perrin, E.B. and Koshel, J.J. (Eds.) (1997). *Assessment of performance measures for public health, substance abuse, and mental health.* National Research Council. Washington, D.C.: National Academy Press.

Peterson, J. (1995). For whom? Institutional economics and distributional issues in the economics classroom. *Journal of Economic Issues, 29 (2)*, pp 567-574.

Rossie, P. H. and Blum, F. D. (1969). "Class, status, and poverty." Pp. 36-63, in D. P. Moynihan (ed.), *On Understanding Poverty*. New York: Basic Books, Inc.

Rostow, W. W. (1952). *The Process of Economic Growth*. New York: Norton.

Rubinson, R. (1976). "The world-economy and the distribution of income within states: a cross-national study." *American Sociological Review* 41:638-659.

Seligson, M. (1984). *The Gap Between Rich and Poor: Contending Perspectives on the Political Economy of Development*. Boulder, Co: Westview Press.

Tsui, Ch. (1987). *Modernization and Occupational Income Inequality:The Case of the U.S.,1050-1980*. Ph.D. dissertation, University of Minnesota.

United Nations Development Program (1990). *Human development report*,1990, New York.

Valentine, C. A. (1968). *Culture and Poverty*. Chicago: the University of Chicago Press.

Williamson, J.G. (1965). "Regional Inequality and the Process of National Development." *Economic Development and Cultural Change*, 13:3-39.

Wilkinson, K.E. (1991). "The Future of the Community in Rural Areas." in *The Future of Rural America,K.E.* Pigg (editor). Boulder, CO.: Westview Press.

Chapter 16

Global modeling and international performance

MICHAEL P. BYRON
Adjunct Faculty, Palomar Community College, California

1. MODEL CONCEPTUAL AND EMPIRICAL FOUNDATIONS.

The modern international system began to coalesce around five centuries ago. Technological, demographic, political, economic, and organizational factors account for this occurrence (Levy; 1983; Chirot, 1985; Modelski, 1987). Its constituent entities interact via the mechanisms of both trade and warfare. The research questions, which I will investigate for this paper, are: 1) Does the international system "learn". Specifically, does it constitute a complex adaptive, "learning" system. I will utilize real world warfare frequency data (Levy, 1983) to evaluate this proposition. 2) What type of international system (realist/neo-realist type system or liberal/neo-liberal/idealist) most nearly corresponds to selected observed performance characteristics of the international system. I will utilize variants of my computer model in which either military power alone determines the outcome of interactions between nations (realism-type paradigm), or, military power and trade determine outcomes for international interactions (liberalism-type paradigm). I evaluate the performance of these two classes of systemic variant. By definition, performance (T) is equal to outputs (Y) divided by inputs (X). Throughout this essay my general indicator for performance shall be adaptation. These two variants in essence represent between-group conflict only, versus a combination of between group-conflict and between-group co-operation. In doing this I am explicitly positing that trade and war are inversely related. There is broad prior theoretical and empirical support for this proposition. For example,

The Performance of Social Systems
Edited by Francisco Parra-Luna, Kluwer Academic/Plenum Publishers, 2000

Mansfield (1994, pp. 31) notes: "...moreover there is considerable evidence of an inverse relationship between trade and commerce." I should note that he interprets these findings from a neo-realist perspective, which is not wholly consistent with that which I shall develop below.

In conducting this inquiry, I will articulate and evaluate a model of the international system, which learns with respect to time. I hypothesize that systemic learning, in the context of declining frequencies of interstate warfare, has occurred in response to increasingly rapid gains in total systemic power. These power increases are hypothesized to have caused the adverse consequences of warfare, in general, and great-power warfare, in particular, to evince a corresponding increase. This creates 'selection pressure' upon the system to 'learn' to avoid this ever-increasing danger to its well being, and indeed, to its existence. Systems, which are capable of evincing such adaptive learning, are referred to as complex adaptive systems.

With respect to international trade, I predicate this inquiry upon the explicit assumption that, unlike warfare, which is at best zero-sum, trade is non-zero sum. This occurs because of what is commonly known as Ricardo's Law of Comparative Advantage. Stated as generally as possible, it asserts that given two (or more) groups each of which produces various goods and/or services with differing degrees of efficiency (cost), trade the groups is always to the mutual advantage of each group, even if one group produces all of the goods and services which another does, and does so with greater efficiency.

As one researcher observes:

> *Its [trade] invention represents one of the very few moments in evolution when Homo sapiens stumbled upon some competitive ecological advantage over other species that was truly unique. There simply is no other animal that exploits the law of comparative advantage between-groups. Within groups, as we have seen, the ants, the mole rats, the Huia birds beautifully exploit the division of labor. But not between groups. David Ricardo explained a trick that our ancestors had invented many, many, years before. The Law of Comparative Advantage is one of the ecological aces that our species holds.* (Ridley, 1996, pp. 210).

Thus, while between-group conflict is a paleontologically ancient resource allocation mechanism for all living organisms, between-group co-operation, as exemplified by trade, is a *unique* discovery of humanity. As we shall see, this leads to some interesting predicted consequences for the global system.

1.1 The Crisis-Driven Evolutionary Learning Model.

I predicate my model of crisis-driven learning upon the hypothesis that fundamental systemic level learning occurs principally in response to crisis. Crisis, in this context is defined as any situation in which the application of previously encoded knowledge fails to resolve some situation in a manner in which harm, or the potential for harm, is averted (Byron, 1996, 1997). I assume that the probability of learning is proportionate to the perceived probability of harm arising from not resolving the situation. For the purposes of this study, which is predicated upon an analysis of relative warfare frequencies, this probability is assumed to be unity.

Crisis-driven learning occurs at the systemic level when states are unable to resolve some problem as a result of their individual, state-level and below, efforts. This provides them with strong incentives to attempt to cooperate with other states to resolve the common problem that threatens them with increasing "harm" (loss of power in the context of the model). If successful, these efforts may lead to the creation of new behavioral norms, and/or international regimes. If unsuccessful, war may result.

I have published specific details regarding this systemic learning model elsewhere (Byron, 1996, 1997). Very succinctly, I theorize that systemic learning begins in the minds of individual decision-makers. When some problem, or difficulty, arises which the application of pre-existent inter-state adjudicative, or behavioral algorithms such as international regimes, consultative bodies (e.g. the G-8), or generalized international norms, are unable to resolve, the system experiences a *crisis*. If the direct, or anticipated, consequences of inaction exceed those of implementing change, then inter-decision-maker agreement is sought. If consensus can be attained, then new behavioral algorithms are incorporated into the structure of the international system, as noted above, in the form of new norms, consultative bodies, or regimes. If this process fails, then conflict between states becomes more likely, as attempts are made to force new behaviors which are most beneficial to a given state's interests upon other, possibly unwilling, members of the international system.

Broadly speaking, the reality of systemic learning is increasingly accepted. For example, Modelski (1994, pp. 324) states, "Successful learning produces structural change." What is at most at issue is the nature, and significance of such learning.

Levy (1994) in his magisterial survey of learning and foreign policy distinguishes sharply between individual level learning and collective/systemic learning. He defines individual level learning as comprising: "a change in beliefs at the individual cognitive level..." Levy, 1994, pp. 278). Regarding group/systemic level learning he states:

> *the reification of learning at the collective level is not analytically viable. Organizations do not literally learn in the same sense that individuals do. They only learn through individuals who serve in those organizations, by encoding individually learned inferences from experience into organizational routines.* (Levy, 1994, pp. 287).

These definitions appear accurate enough. Where they are inconsistent with my crisis-driven evolutionary learning model is in failing to recognize that individual and group level learning, in the context of the international system, are inseparably bound together. They are aspects of a seamless web of interactions. They collectively constitute different stages of one, holistic, process. Restated, they constitute a unitary complex adaptive system. Furthermore, the difficulties inherent in incorporation of new system level learning, which Levy (1994) articulates, explain why systemic learning *tends* strongly to be crisis-driven.

The international system thus, can be viewed holistically as comprising a structured linkage between many individual human minds, which is temporally extended. These minds do not exist in isolation, but rather exist in the context of a plethora of social structures, which, in turn, influence decision-making. Hence, learning for these minds, is substantially influenced by the context in which it occurs. This has two important effects: First, decisions are often less than optimal as the ability to interpret information is circumscribed. Thus learning may be both adaptive, as well as maladaptive. A more precise, objective, definition of the terms adaptive and maladaptive in the sense of relative *fitness* will be utilized subsequently in this study. Second, as individual level learning must occur only in this context, all prior individual learning, which has subsequently been incorporated into the structure of the system affects all current learning. Information flows throughout the international system are characterized both by feedback, as well as, feedforward, channels. Adaptation is not necessarily a linear process. For example, unrestrained development may lead to ecological problems. These problems then create a crisis which leads to changed developmental strategies. Given this reality it is reasonable to consider that all learning, at any possible level of analysis, takes place within a unified "design space" as articulated recently by the noted philosopher of science Daniel Dennet (Dennet, 1995). As will be discussed below, a system so constituted, is defined as comprising a complex adaptive system. This is the approach that I shall utilize. It's the explanatory viability of this model, which shall be empirically tested below.

Regarding warfare, which occurs with increased probability, in the context of my model, when attempts at decision-maker agreement fail; a given state's propensity towards engaging in warfare is empirically related to its level of democratization (Doyle, 1983, 1986). I hypothesize that the underlying variable is the state's degree of inter-state reciprocity (cf.

Axelrod, 1984). Reciprocity correlates with democracy because individual actors possess *innate* goals. These goals include health, wealth, security, knowledge, liberty, distributive justice, ecological well-being, personal fulfillment, and self-development. States, which are constituted internally so as to maximize these goals, necessarily possess high degrees of individual political rights and civil liberties. This is because individuals must deal reciprocally with one another to maintain and perpetuate at the group level, these individual benefits. As democracies are predicated upon individual political and civil rights, they are highly reciprocal in structure. When they encounter similarly constituted states these encounters are characterized by the easy development of multiple, reciprocal, crosscutting ties between a profusion of state (and non-state) actors. This easy growth of dense, reciprocal, ties makes conflict difficult (Axelrod, 1984).

Conversely, interactions between non-democracies, or between a democracy and a non-democracy, would not logically be anticipated to develop such dense, crosscutting, reciprocal, ties. Relatively few inter-state actors would characterize non-democracies, with the state possessing final, summary, decisional authority in all cases. Reciprocity would continue only so long as the ultimate state authority determined it to be useful. Thus all such ties are few, and fragile. War is much more probable in this case, than it is in the latter one.

Culture affects state actor behaviors (Eckstein, 1997). While much cultural variance is nominal, those aspects of culture which determine propensity towards democratic/reciprocal behaviors must, logically, determine, ultimately, a given states propensity towards war-proneness. War-proneness is certainly a non-nominal variable. Thus, culture determines reciprocity, and reciprocity determines war-proneness. War-proneness, in the context of a crisis-driven evolutionary learning system acts as a driver of systemic level learning via the mechanism of adaptation/maladaptation. Adaptive strategies result from the diminishing payoffs attained by war initiators, over time, in the context of historically increasing levels of systemic power.

Increasing systemic power generates the crisis, which the contemporary world system is experiencing. Power derives ultimately, from the amount of information (i.e. "knowledge") available to system actors at any given time. Given that this amount of information has been growing with ever increasing rapidity throughout the approximately 500 year existence of the modern world system, the amount of power available to system actors for inter-actor interactions, such as warfare, has shown a corresponding increase. This increasingly asymptotic rise in the potential for harm to the world system's constituent nation state elements, and consequently, to the viability of the entire system, constitutes the systemic crisis upon which this study is predicated.

1.2 Complex Adaptive Systems.

I hypothesize, as noted above, that the global system comprises a unitary *learning* system. Holistic, multi-level, learning systems, as I've articulated above, *are*, complex adaptive systems. There is no universally accepted definition for a complex adaptive system. A particularly succinct definition is: "A complex adaptive system can be defined as an adaptive network exhibiting aggregate properties that emerge from the local interaction among many agents mutually constituting their own environment." (Tetlock & Belkin, 1996, 260). Still another definition of complex adaptive systems involves, "...a medium-sized number of intelligent, adaptive agents acting on the basis of local information." (Casti, 1996, x). The study's computer model will constitute just such an adaptive network. It will manifest emergent, "bottom-up" properties arising from local interaction among agents (nation-states) which constitute their own environment. There will be a medium sized number of these adaptive agents (10), which will act on the basis of locally available information. Accordingly, the model system will comprise a complex adaptive system by definition. This system will learn to adapt to an endogenously generated selection pressure caused by increasing systemic power, engendering crisis-driven evolutionary learning in the context of dyadic warfare between system elements.

To address the question of how the model generated data compare with those of the real world system, we need to look at the behavior of the real world system. Levy's real world findings may be succinctly summarized by his observation that:

> *The results are not perfectly congruent across all of these indicators, but some overall patterns emerge. In general, interstate war involving the Great-powers has been diminishing over time. There has been a strong decline in the frequency of [interstate] war, particularly in the frequency of Great-power war.* (Levy, 1983, 135).

Thus, the real world system appears to be learning how to deal with the crisis, caused by warfare in the context of rapidly increasing systemic power, which threatens its continued existence, by increasingly seeking out non-warfare solutions to inter-state conflicts. In effect, it appears to evince a trajectory through its learning space manifold towards a no-war equilibrium attractor, (a point towards which the apparently random motion of the model system's trajectory converges) driven by selection pressure arising from this endogenously generated crisis. Via this process of adaptation performance becomes, on average, over time, more optimal. These real world, empirical findings for the 1495 to 1975 period will serve as a basis for empirical falsification of the model's findings.

2. METHODOLOGY.

2.1 Construction of the Model.

My investigative strategy was to utilize MS QuickBasic to create a model international system program. The model system consists of ten mutually interacting actor-elements, which represent the nation-state constituents of the world political system. This value was selected for several reasons. Levy (1983) in his study of great power warfare between 1495 and 1975 identifies a cumulative total of fourteen great powers. Given the assumption that the 10 model system nations represent only those nations which possess sufficient power to significantly influence the trajectory of the entire model system through learning space, and that such model nations are functionally equivalent to Levy's great powers, the model is consistent with phenomenological reality. Also, 10 is a large enough number of model states to allow for collective interactive behaviors to emerge, without being so large a number as to be either too unwieldy to keep track of, or impractical for simulation.

My computer model of crisis-driven learning is predicated, as I've articulated above, upon the hypothesis that systemic level learning occurs principally in response to crisis. In this context, I define crisis as any situation in which the application of previously encoded knowledge fails to resolve some situation in a manner in which harm, or the potential for harm, is averted. In the context of the model, culture "K", represents a common meta-schema for a national polity. Thus, it follows that crisis-induced cultural change will likely occur discontinuously, by discrete increments, which represent replacement of less "fit" schematic elements. This change will occur with maximum probability when it is 'forced', that is, in a crisis situation.

The model consists of four global, "top-down", difference-equation-driven variables along with another variable, which sets an upper limit upon one of these variables. These variables are modified in value via local, "bottom-up" interactions among system actors, by operation of a WAR subroutine, and a CRISIS procedure, in conjunction, where applicable, with the model's TRADE subroutine.

The learning space manifold within which these various interactions occur is configured as a virtual 10-dimensional array. Each "dimension" corresponds uniquely with one of the model's 10 constituent nation states. Each time the program is initiated a unique model system is created de novo and positioned in an arbitrary location in this space, consistent with a moderately low level of learning.

Recalling that each of the 10 dimensions is, in actuality, a range of discrete values provides direction in this space. Motion along this range towards a no-war equilibrium solution (that is towards an attractor which represents a discrete solution to the problem of attaining this condition) corresponds to motion along the C-axis towards its end-point, 0. Thus, lower C values, and correspondingly, higher amounts of learning, (a closer approach to a no-war equilibrium problem attractor- solution) correspond to motion "deeper" into the learning array.

This visualization also allows for the geometry of the learning array to be comprehended: As the end-point, or vertex, for C is common to all 10 nations, then this vertex is the common vertex of learning space. All 10 dimensions therefore meet at this one point. At this unique point, and only at this point, (which corresponds to perfect systemic reciprocity) they coincide.

As I visualize it, the learning space manifold looks somewhat like an inverted 10-sided step-polygon, which steps 'downwards' towards this singularity in increments of .05 K. Each of these points is located exactly .05 K units from all adjacent points. Its 10 sides, or dimensions, as noted above, correspond to the 10 nations that comprise the model system. Topographically, the model's learning space, which lies within the interior volume of this 10-sided inverted step-polygon, is contorted by the presence of attractors, which represent unique solutions to a system's no-war, crisis-driven, equilibrium search.

It is important to note that this learning space manifold is *endogenously generated*. This means that the learning space manifold is not some pre-existent metric, into which a model system is placed; rather it exists because a given system exists. Within the top-down parameters allowed by the modeling software, it is free to vary as a function of instant, unique, systemic characteristics. Thus, even the 'location' of attractors *can* change as a given system evolves.

In order to dynamically model (simulate) its selected aspects of phenomenological reality, the model incorporates five variables, four subroutines, and five parameters. A brief description of each follows. For detailed information please refer to the technical appendix (appendix one).

Model Variables.

Relation. This is a measure of affect between states. It is scaled between +1.0 and -1.0.

Reciprocity. This is a measure of the tendency of any state "I" to engage in reciprocal "tit-for-tat" interactions with another state "j".

Power. This is an absolute measure of the resources available to a given state "I".

Culture. This is an indicator of the behavioral characteristics of a given state "I". It serves as an indicator for learning.

Coupling. This is a measure of the "openness" of a given state "I" to another state "j" in its interaction environment.

Model Subroutines.

WAR subroutine. This is the algorithm which determines whether two states "I" and "j" shall engage in dyadic warfare. It also determines stochastically, based upon objective characteristics determined by the instant values of the other variables and parameters used for the model, stochastically, what the outcome of the war, if it occurs, shall be.

CRISIS subroutine. Under specified circumstances this algorithmic process determines, in a manner strictly in accord with theory, what adaptive (or maladaptive) learning shall occur for a given state 'I".

TRADE subroutine. This subroutine, analogously to the WAR subroutine, determines whether two nations "I" and "j" shall engage in dyadic trade.

COLLAPSE subroutine. This is the procedure by which a combination of deterministic and stochastic factors are weighted to compute a non-zero probability that the international system will collapse due to the persistence of inter-state warfare in the context of rapidly increasing systemic power.

3) Model Paramenters

PI. This is the ratio of instant aggregate systemic power to initial systemic power.

C. This is a measure of a given model system's "position" within the learning space manifold.

CYCLE. This is the indicator for the number of program iterations which have elapsed.

D. This is a nominal indicator for effective distance between any dyad of states "I" and "j" within the context of the model system.

G. This is a systemic metric parameter which varies depending on the relative frequency of trade and war inter-state interactions between model system states at any given time. Figure 1, below, displays this information schematically.

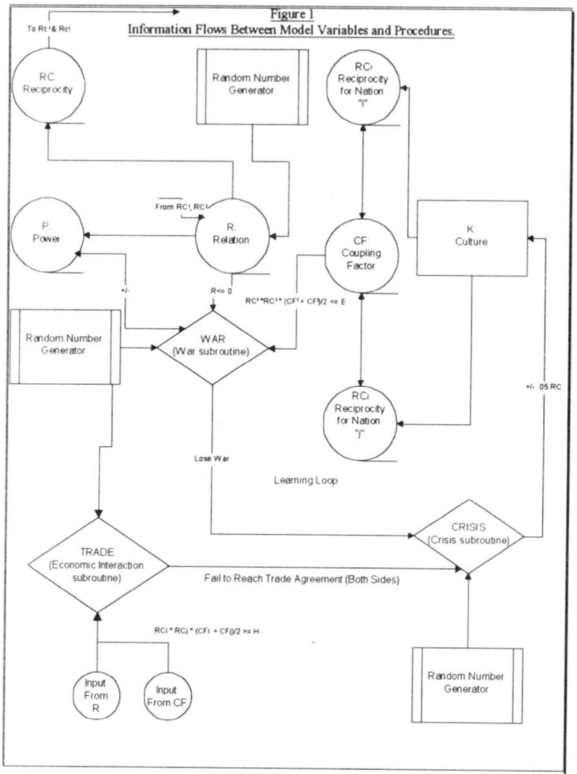

Figure 1. Information flows between model variables and procedures

3. ANALYSIS AND EVALUATION OF THE MODEL

The study's hypothesis is that the international system constitutes an adaptive learning system. This occurs via the crisis-driven adaptive, performance optimizing, mechanism described above. It intends to determine whether this learning system is best characterized as being dominated by between-group conflict (realism-type paradigm, or whether between-group co-operation exerts a significant influence over systemic processes (liberalism-type paradigm). To begin to evaluate these propositions I created five variants of the model. The first of these referred to as "PWar" is identical to the next three, except that it does not utilize the model's TRADE module. It corresponds to a pure, realist (e.g. Gilpin, 1981) international system paradigm. As such it is the control, or baseline, systemic model against which the other variants are to be compared.

Subsequently, I generated three variants of the model, which were identical to the above-described "PWar" system, except that they employed the TRADE module. These model systems utilized a fixed threshold value to determine algorithmically whether, or not, conditions were appropriate for a given nation "I" to engage in trade with another nation "j". The purpose of this is to investigate the sensitivity of the outcomes for trading interactions to arbitrarily preset parameters.

These three model systems are labeled: 1) "hEWar" (i.e. "high TRADE module enabled War), 2) "mEWar" (where "m" represents "medium"), 3) "lEWar" ("l" for "low"). They meet the conditions for triggering the TRADE subroutine (and hence international trade) specified by: IF $RC^{ij}_{(t)}$ * $((CF^i + CF^j) / 2) \geq$ H AND $R^{ij}_{(t)} > 0$ THEN GOSUB TRADE, where H is equal to .35, .30 and .25 respectively.

In the real world system there is no analogue for these fixed threshold values. It is a self-regulating, complex adaptive system, as discussed above. Accordingly, I created a fifth, and final, version of the model, termed "ArWar" (Auto-Regulating War). This class of model system algorithmically calculates its own triggering threshold values in accordance with a simple formula based upon the instant systemic value of reciprocity (RC): IF $RC^{ij}_{(t)}$ * $((CF^i + CF^j) / 2) \geq (1 - (.85$ * MEANRC)) AND $R^{ij}_{(t)} > 0$ THEN GOSUB TRADE. Here the term (1- (.85 * MEANRC)) replaces the fixed constant "H" used for the preceding three system variants. "MEANRC" designates the instant value of mean systemic reciprocity.

Thus, the instant values of the TRADE module's triggering threshold depends critically upon the temporally corresponding values of mean systemic reciprocity. This is *non-arbitrarily* consistent with the model's underlying assumption that levels of international trade are dependent upon the relative absence of warfare from the international system: more warfare engenders less trade, until a threshold is reached at which trade ceases. As, in the context of the model, reciprocity and war are inversely correlated, it follows logically that a system characterized by higher levels of reciprocity ought to be characterized by higher levels of international trade.

The simple 1 - (.85 * MEANRC) threshold-setting algorithm serves to maintain logical consistency with core theoretic assumptions. All model systems begin with an initial mean systemic RC of .5, thus, the initial TRADE threshold is .575. This means that the early model international system is highly anarchic, and war-prone, making trade difficult. As crisis-driven learning occurs the system gradually becomes less anarchic, and hence more conducive, for trade. Thus, the triggering threshold for trade must evince a corresponding decrease. At a mean systemic RC level of approximately .882 the triggering threshold for international trade becomes equal (i.e. .25) to that for interstate war. At higher levels of mean systemic reciprocity it actually becomes very slightly less. Relatively few model system runs attain such high levels of mean systemic reciprocity without either reaching a no-war equilibrium, (undergoing a cessation of inter-state warfare) or collapsing due to continuing warfare. However, those that do are generally characterized by the presence of one, or two, states that still engage in inter-state war, while being surrounded by an otherwise peaceful system of trading states. In such a systemic configuration, and only in such a configuration, the model allows two states, which go to war, to also engage in trade, during the same program cycle. The reasoning here is that this type of system has become predominately "trade-like", as opposed to "war-like". Accordingly, it evinces different behaviors when stressed. Looked at from another perspective, trade could be considered to be a "weapon" to integrate an aberrant member of the otherwise pacific international system.

I ran each of these five systemic variants 500 times in order to produce a statistically meaningful N of cases. The resulting data files were read into EXCEL, and sorted by whether, or not, they had collapsed before inter-state warfare had ceased. Subsequently, I used the system variant with the fewest valid runs, (PWar) with 347 valid, or non-collapsed, cases, to establish the N for the study. All valid cases numbered above 347 for the other four system variants were discarded. Elimination of these excess cases was purely random, involving no bias on the part of the author, other than accepting Excel's numbering of them arising from the above described sorting procedure. These data are displayed in table 1 below. Values for PI are rounded to the closest whole number.

Table 1. Data for Systemic Variants.

System	N	Cycle	PI	C	RC	CF	Collapse %
PWar	347	92.91	273,256	.245	.816	.773	.306
hEWar	347	56.78	556,130	.247	.878	.832	.272
mEWar	347	33.52	196,902	.278	.885	.838	.276
lEWar	347	24.94	183,861	.316	.883	.836	.140
ArWar	347	86.64	123,687	.246	.859	.814	.194

It is clear that selection of any of the several fixed values for triggering the TRADE subroutine produce systemic outcomes that are 'preferable' to those of the baseline PWar type system in that the probability of collapse is lower, while the corresponding values of systemic reciprocity and coupling are greater. Values for C do tend to be greater for the fixed TRADE subroutine triggered cases. However, as learning continues via trade interactions for this class of system, this distinction is not critical. [All comparisons are significant at the .05 level, or better.]

Of course, it is not intuitively reasonable to assume that trade interactions in the real world system possess a fixed triggering threshold. The ArWar model systems avoid this difficulty, as described above. A quick glance at figure one should serve to make clear that ArWar type systems are substantially more adaptable, overall than are any of the other system variants. Terminal C values for ArWar and PWar systems did not evince a statistically significant difference.

Overall, it is clear that systems, which are influenced by economic interactions, learn with greater rapidity. This conclusion is buttressed by noting that all trading systems studied were able to seek out no-war equilibrium attractor-solutions more rapidly than was the case for the non-trade dominated systems. In all cases, their probability of systemic collapse was substantially reduced. Reciprocity was greater for this class of system. Additionally, they were substantially more highly coupled. Learning occurred at much lower levels of net mean systemic power (PI) than was the case for the realist paradigm PWar systems. No statistically significant difference was found between the mean terminal C values of the two classes of systems. Each is thus equally 'stable', ceteris paribus. However, as noted in the discussion above, learning continues after the cessation of inter-state warfare via the mechanism of interstate trade crises. Thus, even in this category, the trading systems exhibit greater viability. As noted above, this provides compelling tribute to the power of Ricardo's Law of Comparative Advantage. Given these findings, I incorporated the ArWar TRADE algorithm in to the baseline model. Doing so, in the context of the above assessment, is a-priori reasonable in that it brings the model into closer correspondence with phenomenological reality, is consistent with crisis-driven evolutionary learning (as it renders the system more 'adaptive'), and does not represent arbitrarily selecting some optimal value in order to make the model 'work.'

Building upon this, I next altered the manner in which the WAR subroutine's triggering algorithm is calculated. Initially, this value had been a simple constant value of .25. This value has been chosen as it was intuitively reasonable. Recall the WAR subroutine triggering algorithm: IF $RC^{ij}_{(t)} * ((CF^{i} + CF^{j}) / 2) \leq .25$ AND $R^{ij}_{(t)} \leq 0$ THEN GOSUB WAR. Thus, any two nations "I" and "j" which possessed a *mean* reciprocity of .65, or *less*, and which had a signed relational value which was *negative*, could, potentially, go to war with one another. This corresponds with the real world observations that democracies do not go to war with one another as two high reciprocity (RC values of above.65 per algorithmic definition) model nations would not meet the dyadic war criteria. Consistent with this, two low RC nations would potentially, meet these criteria. Furthermore, as per empirical observation, war between a "democratic" (defined as a model system nation possessing a RC value in excess of .65) and a "non-democratic" model system nation would be, potentially, possible. The problem with this fixed threshold is that it is predicated upon an unrealistic assumption: that wars occur, in all times, and under all circumstances, under invariant conditions. Indeed, if this assumption were to be true, then learning derived-changes in the structure of the international system would have no effect upon the conditions under which wars occur, only upon their frequency. It appears much more logical to me to posit that the threshold for war occurrence is lower when the international system is more anarchic, and that, conversely, it is significantly greater when the system is characterized by trading interactions among high reciprocity nations.

To operationalize this assumption, while preserving the intuitive and observationally derived warfare assumptions embodied by the .25 war-triggering threshold value, I developed a simple algorithm to produce this dynamic triggering value: $\overline{RC} - (\overline{RC}/1.65)$. This dynamic triggering algorithm produced a *mean* triggering value of very nearly .25, thus preserving the benefits of the fixed value threshold, yet it varied as a function of mean RC, thus allowing warfare triggering conditions to vary in an intuitively reasonable manner, consistent with theory. For example, in a 500 N run test, its value, averaged over the 500 runs, was .2544 for the PWar class of model systems, and .2489 for the ArWar class of model system.

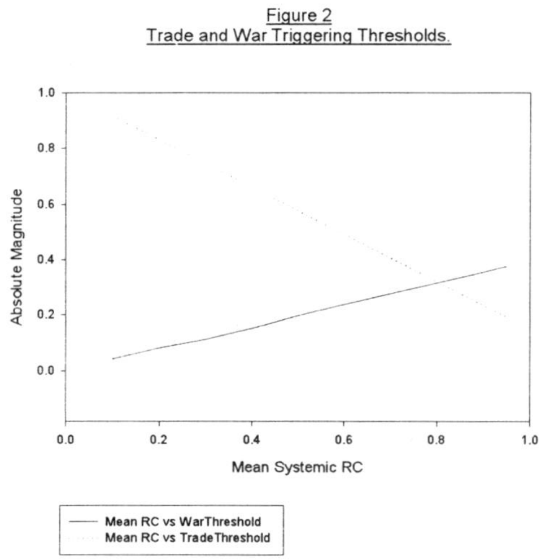

Figure 2. Trade and war triggering thresholds.

Figure 2 above displays the resulting slope of the WAR-triggering function, in conjunction with that for the above-discussed TRADE-triggering function.

Note that: A) The slope for the TRADE-triggering algorithm is *steeper* than is that for the WAR-triggering algorithm. This depicts the operationalization of my assumption that trade can, generally, occur more easily, and under a broader range of systemic conditions, than can war. B) At a value of mean systemic RC of about .81 the trade and war algorithm slopes intersect. For higher mean systemic values than this, a given "ij" dyad can engage in *both* war *and* trade during a *single* program cycle. I explain this by noting that for high RC systems, war, instigated by one, or two, low RC nations, is an anomaly. Under these conditions, for such a "tradelike" system, trading with war-prone nations becomes a type of defensive "weapon" designed to promote the integration of the rouge nation into the otherwise pacific, tradelike, international system. Table 2 below presents the data for the relationship between various fixed values of the war-triggering threshold and several systemic variables and parameters. For comparison, corresponding values for ArWar and PWar systems are also provided. These systems are fully self-regulating with respect to determining their war and trade thresholds. The methodology employed is identical to that utilized above in Table 1, except that an N of 604 cases were required (rather than 500) to obtain the 226 non-collapsed cases for the .50 triggering threshold data. This is because I made a decision that a study N of at least 200 cases would be required to allow for significance testing. An unadjusted N of 100 cases was generated for triggering values of .60 and .70. This was done as their collapse rates exceeded 90 percent. This makes attaining an N of 200 uncollapsed cases infeasible. Consequently, these data must be considered as being less reliable than the rest of the data in Table 2. Note that values for PI are rounded to the nearest whole number.

Table 2. Selected Systemic Measures for Varying WAR-Triggering Threshold Values.

Thresh-old	N	Cycle	Collapse %	PI	C	RC	CF
.10	226	47.33	0.4	156	.520	.686	.650
.20	226	75.22	8.4	179,911	.314	.810	.768
.25	226	83.01	14.8	105,090	.261	.867	.822
.30	226	88.80	22.0	88,494	.216	.894	.847
.40	226	118.79	39.9	404,727	.176	.915	.867
.50	226	154.62	61.9	75,038,199	.130	.939	.793
.60	100	211.85	93.0	161,426,525	.105	.837	.793
.70	100	270.63	98.0	4,449,233	.100	.812	.792
PWar	244	109.57	51.2	338,258	.188	.836	.792
ArWar	244	92.29	14.8	67,952	.227	.891	.845

Per table 2 the systemic parameters of Cycle, Collapse, and C, along with (probably) PI, vary linearly with respect to value of WAR subroutine triggering threshold. Conversely, RC and CFevince a clear, curvilinear, variance with respect to this threshold. Their values are maximal between triggering values of .40 and .50.

As explicated above the WAR triggering threshold is *intended* to produce a mean threshold value of around .25. This is predicated logical grounds, in the context of my crisis-driven evolutionary learning theory. As this value is less than .40—.50, this selection may appear substandard. However, I would note that we can now see from table 2 that this lower value also corresponds to lower values for Cycle, and Collapse. Hence when the totality of all systemic parameters are taken into account, this lower value appears to more nearly represent the composite optimal triggering value.

Which of these systemic classes best correspond to real world data? Recall Levy's (1983) comments:

> *.... some overall patterns emerge. In general, interstate war involving the Great-powers has been diminishing over time. There has been a strong decline in the frequency of war, particularly in the frequency of Great-power war.* (Levy, 1983, 135).

Clearly, in both classes of model system inter-state warfare is observed to decline with respect to system time. What about declines in the relative frequencies of war for 'great powers' vis-à-vis 'minor powers'? Table 1 makes it clear that a decrease in warfare frequencies, both absolutely, and as a function of increasing power, occurs. This is consistent with theory.

It must be noted that the mean terminal systemic values presented in table 2 above are statistical composites. No given system actually possesses these values for more than one, or possibly two, of its several terminal systemic parameter values. The below data were selected with this in mind. These data are presented below in table 3. Note that values for PI are rounded to the nearest whole number.

Table 3. Terminal Systemic Values

System #	CYCLE	PI	C	RC
ArWar 1	84	31,202	.12	.84
ArWar 2	88	1,007	.23	.90
PWar 1	125	1,228	.25	.81
PWar 2	95	7,561	.09	.88

Following Levy's (1985) operationalization procedure, model system nations were grouped into two categories: great powers and minor powers. Any nation which possessed at least 15% of total systemic power during a given program cycle was assigned to the great power category for that cycle. As relative power between model nations changed, membership in these groupings also varied. The three categories employed in figure 2 above represent, respectively, all wars (Wars), wars involving at least one great power (GPWar), and wars between great powers (GP2x). Per Levy's real world empirical findings, we would expect wars between great powers to decline most precipitously, with respect to system time, wars involving at least one great power to decline somewhat less rapidly, with all wars, including minor power vs. minor power wars to decline with greater relative slowness. In both cases, those of the power-driven model, as well as for the representative trade-enabled systems, warfare frequencies are found to decrease, on average, over time, both absolutely, and as a function of increasing nation-state power. Figures 3-d below visually depict the data presented in Table 3.

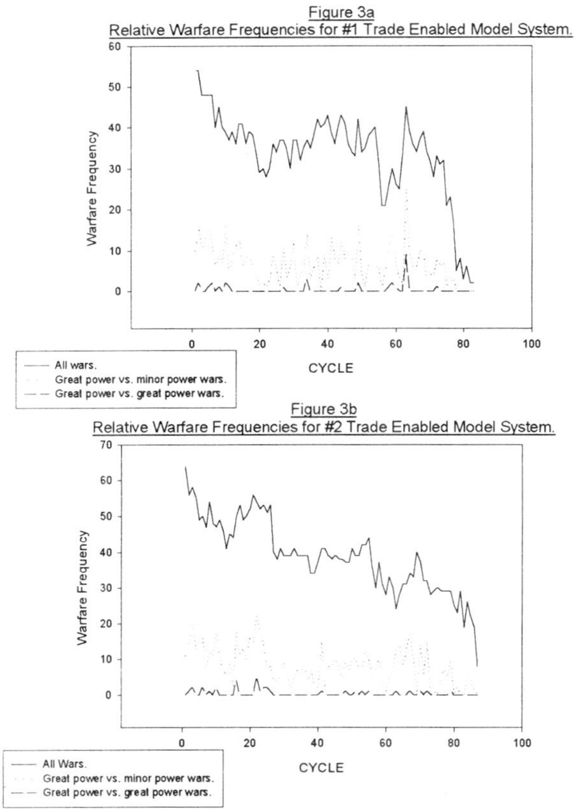

*Figure 3. a:*Relative warfare frequencies for #1 trade-enabled model system. *b.* Relative warfare frequencies for #2 trade-enabled model system.

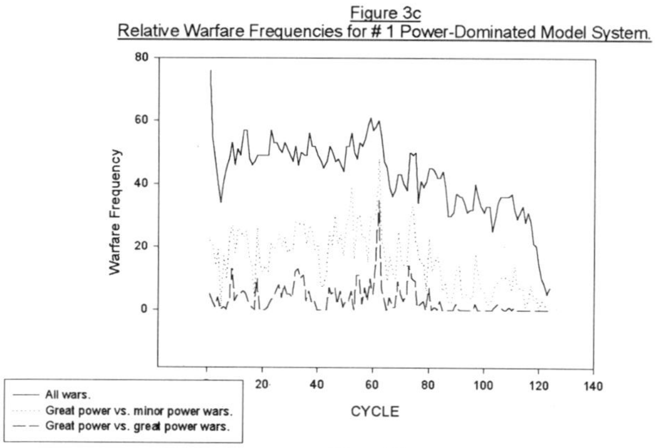

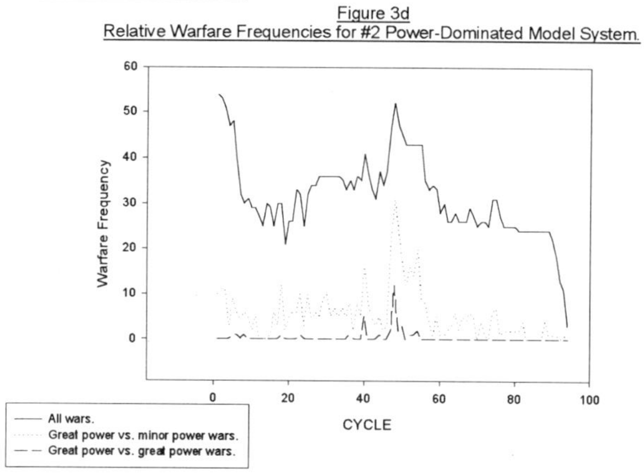

Figure 3. c. Relatve warfare frequencies for #1 power-dominated model system. *d.* Relative warfare frequencies for #2 power-dominated model system.

It seems clear that Levy's findings fail to falsify either hypothetical systemic configuration. Nonetheless, as both model systems are variants of a complex adaptive learning systemic model, their close correspondence with Levy's real world empirical findings does evince strong support for the core hypothesis that the real world is indeed such a learning system. The remaining issue concerns the role of economic interactions in determining the temporal evolution of the actual world system.

At this point it will be informative to evaluate the trajectories and topologies of representative examples of each class of sample model system. Figures 4a through 4h, presented just below, accomplish this. The set of model systems referenced in table 3, above, are utilized for this purpose. Sigma Plot's 3-d modeling software was employed to create these graphs.

This type of data display portrayed by the several figures presented below, is knows as a 'mesh plot.' System time and warfare frequencies form its x and y-axes. These two dimensions, in effect display the area of the manifold in which inter-state warfare occurs.

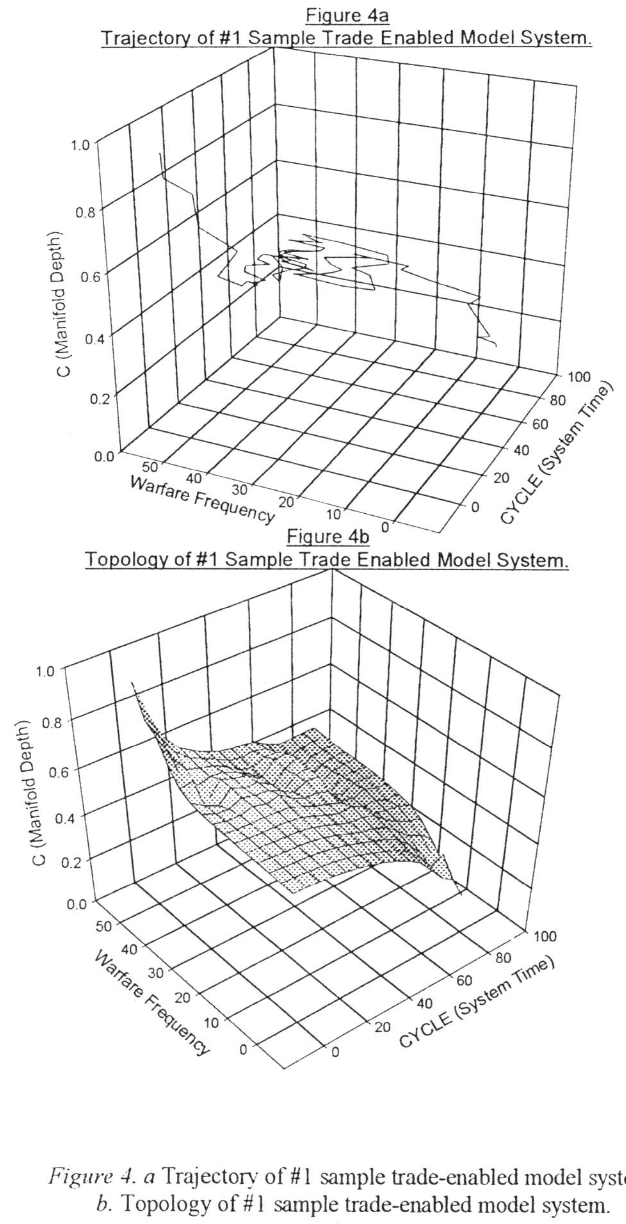

Figure 4. a Trajectory of #1 sample trade-enabled model system.
b. Topology of #1 sample trade-enabled model system.

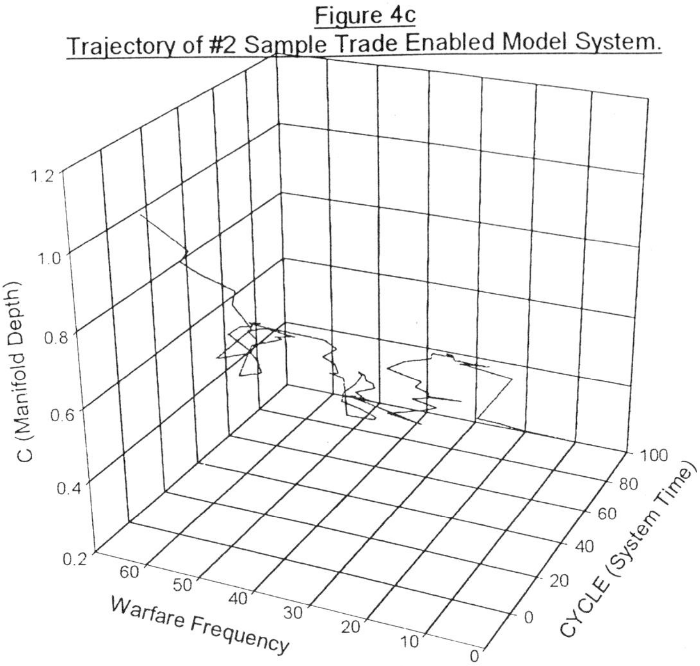

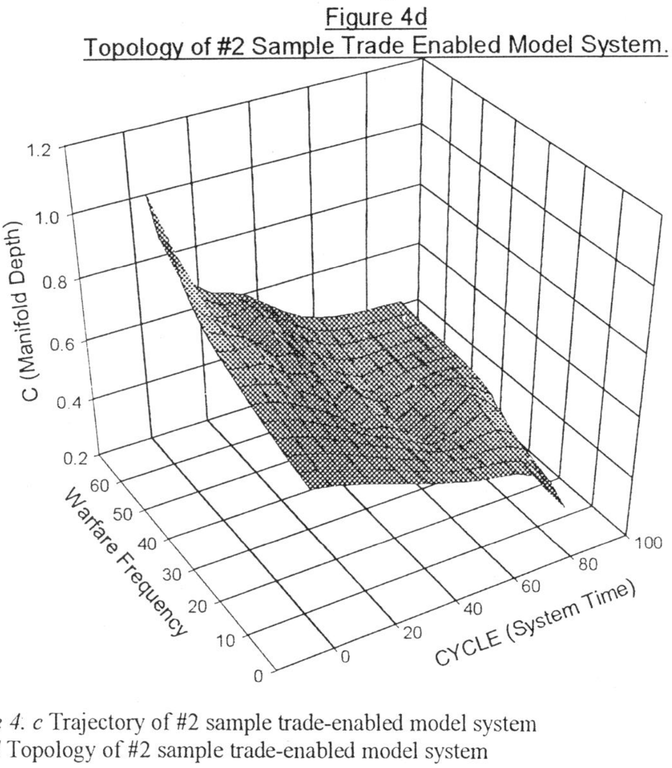

Figure 4. c Trajectory of #2 sample trade-enabled model system
d Topology of #2 sample trade-enabled model system

Figure 4e
Trajectory of #1 Sample Power-Dominated Model System Within Learning Manifold.

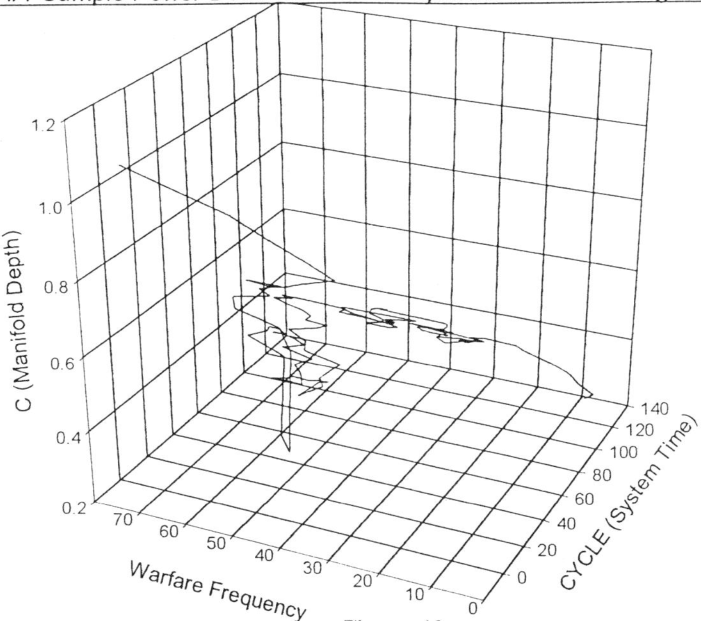

Figure 4f
Topology of #1 Sample Power-Dominated Model System.

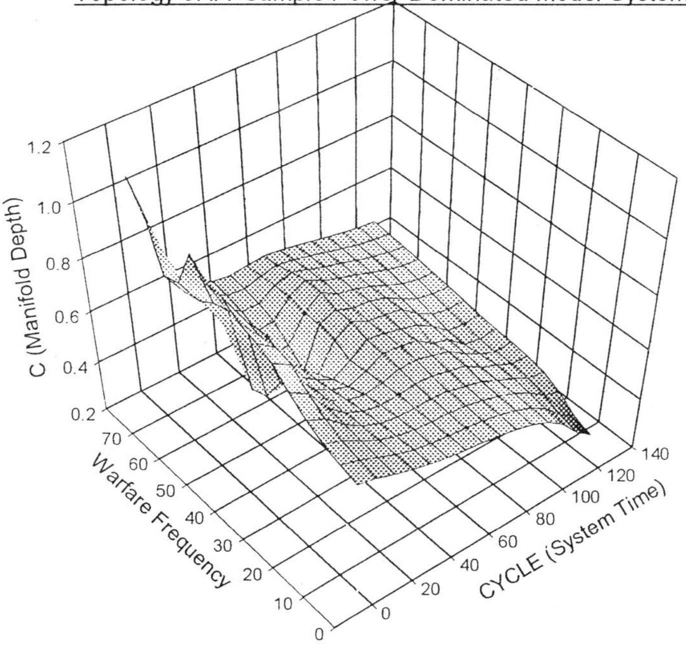

Figvre 4. e Trajectory of #1 sample power-dominated model system.
f Topology of # sample power-dominated model system.

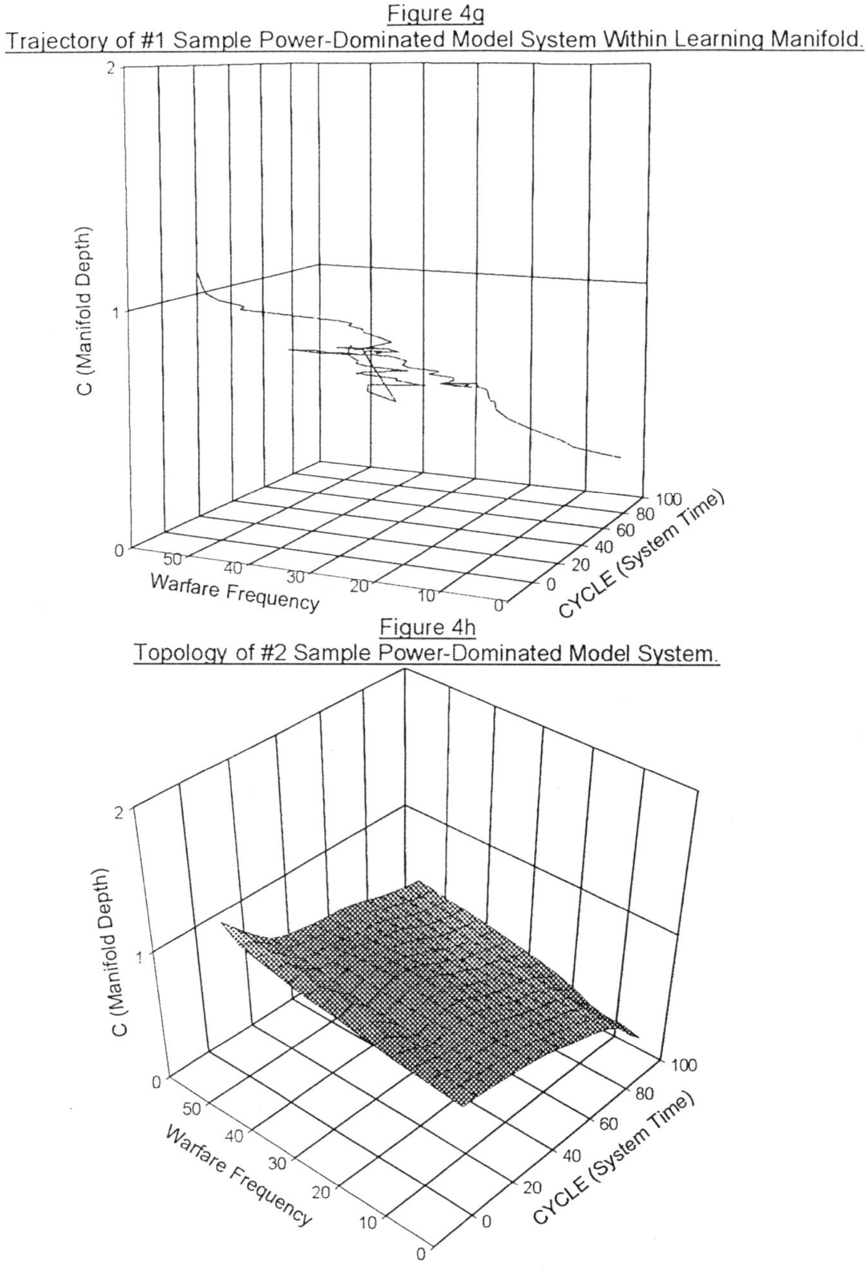

Figure 4. g Trajectory of #1 sample power-dominated model system within learning manifold.
h Topology of #2 power-dominated model system.

Depth, in the form of a z-axis, is provided by the systemic measure "C" as described above. By recalling that C provides an (logarithmically adjusted with a range between *approximately* 1.0 and 0.0) indicator for the model system's instant position, or "depth" in the learning space manifold, the rationale for its selection as the z-axis, or depth information variable, should be clear.

Thus, each mesh plot graph depicts third-dimensional, topological contortions, of a 2-dimensional surface, which represents warfare frequencies with respect to system time. Recall that each model system is experiencing endogenously generated selection pressures, in the context of increasing systemic power, to find a solution to its crisis of continuing warfare. Also recall that motion 'deeper' into the learning space manifold corresponds to greater systemic 'learning' (in the form of higher mean K and RC values, for example). Now, because this learning space is endogenously generated, with each individual model system being unique, it must follow that there are unique locations, or points within this manifold which correspond to precisely those systemic values which a given system requires in order to 'learn' to eliminate inter-state warfare. The deeper such a point lies in the manifold, the more 'stable' it is. [Think of a marble rolling into a shallow depression—it can easily roll out again. As the depression becomes deeper, this outcome becomes progressively less probable.] Putting all of this together results in the mesh plot graphs being interpreted as representing *fitness landscapes* for each model system. In the context of these landscapes, increasing *depth* corresponds to increasing *fitness* (where fitness represents decreasing war-proneness). The depressions can now be seen as representing the location of no-war equilibrium *attractors*. Via the above described mechanism of crisis-driven evolutionary learning, adaptively increasing fitness corresponds to increasingly optimal performance. Deeper attractors are more stable than those located higher up on this fitness landscape. Thus the system can be considered as 'wandering' randomly across this fitness landscape, driven by selection pressure to seek out a *stable* attractor, which represents a valid solution to its no-war equilibrium problem. Figures 4a, 4c, 4e, and 4g, represent exactly this. As such, they are 3-d graphs of their respective system's trajectory across the corresponding fitness landscape. Taken together, each set of figures graphically represents both the trajectory of each model system within its learning space manifold, as well as the contortions of that learning space manifold which are indicative of the presence of these point attractor-solutions.

Power-dominated systems *on average* were found to possess lower dimensionality, and hence appear as being more nearly 2-dimensional, or 'flat' surfaces. Conversely, the economic enabled model systems *generally* evince substantially higher dimensionality, and thus appear to be more fully 3-dimensional, or 'crumpled.' This indicates that the trade-enabled systems are significantly more volatile. They adapt more often, and with greater rapidity, on average, than do the power-dominated systems. Empirical corroboration of this observation is provided by recalling that "depth" information is provided by the system parameter "C." If the hypothesis that ArWar systems, exhibit, *on average*, greater dimensionality (technically, greater *fractal* dimensionality) than do corresponding PWar systems, were valid, then I would logically anticipate that corresponding values of C for each model system variant would also exhibit relative differences in dimensionality. To test this hypothesis, I evaluated an N of 25,969 consecutive cyclic values of C for each systemic variant. Mean kurtosis values for ArWar systems were found to be -.319. The corresponding values for PWar systems were -.394. These differences were statistically significant at better than the .001 level. A second test of the hypothesis utilizing an N of 24,774 other, randomly generated cases, yielded corresponding values of -.248 and -.439, respectively. Kurtosis is a measure for how much a given distribution varies from normal. Distributions, which are negative, are termed *platykurtic*. Such distributions are, on average, "flatter" than is a standard normal distribution. Both the Ar and P distributions fall into this category. However, the Ar distribution is substantially less "flat" than is the P distribution. That is, it exhibits substantially higher (relative) dimensionality than does the P distribution. This finding is exactly what would be logically anticipated to occur if the hypothesis that ArWar systems show greater fractal dimensionality, and hence volatility, and hence adaptiveness, than do PWar systems were, indeed, true.

4. DISCUSSION

Clearly, both types of model systems produce similar outcomes with respect to Levy's empirically observed real world phenomena of warfare frequencies, which decline as both a function of time and power. This is almost certainly due to their both seeking out no-war attractor-solutions in their endogenously generated learning space manifolds.

Equally clearly, the trading systemic variants exhibit higher dimensionality. Corollary with this, they exhibit greater (and more rapid) adaptiveness. My conclusion is that this visually represents the operationalization of the law of comparative advantage upon this class of system's trajectory and topology. Recall Ridley's (1996, pp. 210) comments:

> *There simply is no other animal that exploits the law of comparative advantage between groups. Within groups, as we have seen, the division of labor is beautifully exploited by the ants, the mole rats, and the Huia birds. But not between-groups...The Law of Comparative Advantage is one of the ecological aces that our species holds.*

This is the source of the volatility, the adaptiveness, and the dimensionality, observed above. So which systemic variant best corresponds to the real world system? I think that the question is misleading. Here's why: Both system variants are crisis-driven, complex adaptive learning systems. As such, both operationalize in such a manner as to produce results consistent with corresponding real world empirical data.

Given these conclusions, it follows that such a world system must become more complex, in the sense that it incorporates ever more information into the system's structure (Byron, 1997; Modelski, 1987). This incorporation takes the form of a proliferation of international regimes, global and regional free-trade agreements, software piracy "understandings", new norms such as nuclear non-proliferation, communications agreements, military alliances refashioned to protect against "chaos" (i.e. NATO), and so forth. Reciprocal trading, informational, decisional, and communications pathways multiply. Epistemic communities begin to flourish (e.g. Haas, 1990).

What is happening is that the global system is becoming ever more sensitive to violent disruption, ever more vulnerable, to serious harm, or to actual collapse. In tandem with this trend, the amount of military power available to inflict just such harm is growing exponentially. Simultaneous with these trends, the global system strives to maintain its dynamic equilibrium. For example, increasing wealth in the context of increasing inter-state inequality generates new crises. These crises must lead to increasing learning, represented at the systemic level by a substantial lessening of this crisis-engendering inequality if sowing the seeds for future interstate warfare--at ever greater levels of systemic power is to be avoided. This creates circumstances in which rapid adaptation, which incorporates a 'memory' for previous learning, is essential to systemic survival. If there were ever a time, place and circumstances in which crisis-driven learning would be anticipated to occur, this would surely be it. If there was ever a time when a new "trick" which would enhance survival would be strongly selected for this must be it. Recall Ridley's "...ecological ace ..." comments above.

In terms of survival, trade-mediated systems (liberalism-paradigm) are vastly more viable than are (realist paradigm) systems in which interactions between nations are mediated *primarily* by power contests. Their performance is more optimal. Systems which emphasize between-group co-operation unquestionably possess a clear advantage over those which rely on between-group competition, with co-operation reserved to within-group interactions. At this point I'd like to pose a related question: If this is so then why hasn't nature yet exploited it? Why don't, for example, ant colonies, trade? The answer to this question provides the answer to which model system variant most closely corresponds to the real world: All previous interaction systems, whether human, animal, or insect, have been *low power* systems. It is only at *very high levels of power* that the advantages of between-group co-operation become manifest. As, until very recently, the world system had been a (relatively) low power system, these advantages were not manifest. Thus, it closely approximated a pure realist paradigm, power-mediated system. Only recently have ever more rapid increases in systemic power begun to alter this prior reality. Recall that the metric of the system's learning space is endogenously generated. Power-driven systems must, therefore, possess quantifiably different metrics than do trade-driven systems. As discussed above, this is indeed the case. Both model variants seem, despite their differently operationalize assumptions, to be trying to tell a consistent story. The core elements of this story are that the world system is indeed, a complex adaptive learning system. In this context the system evinces substantially different attributes for high power configurations, than it does for low power ones. Much of the contemporary debate between realists, in all of their variants, and liberals/idealists, in all of their variants, can be attributed to a failure to comprehend the reality, and consequent implications of, this dynamic continuum of systemic adaptiveness.

5. CONCLUSION

In summation, the world system is a complex adaptive learning system. It is presently engaged in a rapid race between learning and collapse. Under conditions of temporally ever more swiftly increasing power growth it is experiencing increasingly strong 'selection pressure' to reconfigure itself in such a manner as to take advantage of the survival opportunities offered by the more optimal performance of between-group co-operation predicated systemic configurations. This is the 'crisis' that my study has focused upon. It is worthwhile to recall that the Chinese ideogram for crisis is created by conjoining the otherwise separate characters for 'danger' and 'opportunity.' As such high-power systemically generated conditions have never previously existed, we are located in a unique moment in history—not only human history, but indeed the entire history of life on Earth. This is our contemporary danger—as well as our opportunity.

6. APPENDIX ONE. TECHNICAL APPENDIX

6.1 Study model variables

As noted above, the four algorithmically defined variables are 1) Relation 2) Reciprocity 3) Power and, 4) Coupling. Each is difference equation defined. As such each imparts global, 'top-down' properties upon the model international system. Appendix One lists these several equations.

a) <u>Relation.</u>

Relation, R: This variable has a range of between +1.0 and -1.0, for each of nation i's relations with other actors comprising the model international system. It represents the instant, net, relationship between any two model systemic nations, "I" and "j". Relation is a zero-sum variable. In effect, each nation possesses a fixed quantity of relational capital. This occurs as the relational difference equation 'assumes' that nations (or more specifically individual/group level nation-state decision-makers) have limited amounts of attentional capacity. Consequently, as more attention (i.e. a stronger signed relationship) is focused upon certain inter-state interactions, the attention available to be focused upon other state interactions diminishes (cf. Jervis, 1976; Simon & Kaplan, 1991; Allport, 1991). For purposes of the model, it is employed primarily to allow for logical consistency. Thus, nations, which engage in warfare with one another, possess a mutually negative interrelationship, which exceeds some threshold value. Conversely, nations, which engage in trade, possess a mutually positive interrelationship, which exceeds another threshold value. The difference equation for relation is:

1) Relation: $R^{ij}_{(t+1)} = R^{ij}_{(t)} + b$;

where $b = (.25 * ((RC^{i} * RC^{j})/2)) * \frac{(R^{ij}_{(t)} - (R^{i+j}))}{(N - 1)}$

Note: the fixed value .25 above is replaced by the value $\overline{RC}$ - $(\overline{RC}/1.65)$ in certain variants of the model.

b) <u>Reciprocity.</u>

Reciprocity, RC: This variable represents propensity to engage in reciprocal interactions between any two model systemic nations, "I" and "j". It's the model's indicator for level of inter-democracy lack of war-proneness, in a limited, Axelrodian, sense.

According to my theory of crisis-driven learning, increasing levels of systemic learning are characterized by an increasing profusion of international regimes, in the context of an ever broader and deeper diffusion of international norms. This has the effect of increasing mean values of systemic reciprocity. Therefore, mean systemic reciprocity will serve as a measure of net systemic learning for any arbitrary time "t." The difference equation for this variable is:

2) Reciprocity: $RC^{I}_{(t+1)} = 2 * \frac{(\sum {}^{+}R^{i}_{(t)})}{(\sum |R^{i}_{(t)}|)}$

c) <u>Power.</u>

Power, P: This variable represents total power which a given state, "I", possesses at any time t. Power, in the context of the model, may be defined as the ability of a given state "I" to potentially influence, or affect, the behavior of some other state "j", while possessing a commensurate ability to resist being involuntarily influenced by state "j".

The power equation 'assumes' that propensity towards reciprocal behavior is conducive to increasing a given state "i's" relative share of total systemic power. Conversely, non-reciprocal behavior decreases "i's" relative power. My reasoning is that co-operation allows limited resources for state "I" to be 'productively' utilized. Low reciprocity states, on the other hand, choose to 'squander' some portion of their power by engaging in conflict. This (culturally determined) behavioral strategy becomes increasingly less likely to yield net benefits (i.e. an increase in the power of war-initiator states) as systemic power increases ever more rapidly. Thus, their relative share of total systemic power tends to diminish, on average, over time. This ultimately leads to increased 'selection pressure' upon them.

The model is predicated upon an assumption that net systemic power is increasing over time, (as a function of, generally available, increasing knowledge) concomitantly generating a crisis of system survival, in the context of a war-dominated international system. Consistent with this assumption, the power difference equation incorporates a term for incrementing power from program iteration to program iteration, somewhat like compound interest. The difference equation for power is:

3) Power: $P^{i'}_{(t+1)} = P^{i}_{(t)} + a$,

$$\text{where } a = \frac{\left|(\Sigma^{+}R^{i}_{(t)})\right|}{\left|(\Sigma R^{i}_{(t)})\right|} + \frac{\left|[Q * (\Sigma^{+}R_{(t)})]\right|}{\left|(\Sigma R_{(t)})\right|}$$

$$\text{where } Q = \left|P^{I}\right| + [P^{total}/(45 - (5 * RC^{total})]$$

d) <u>Culture.</u>

A variable, "K" represents culture. K has a potential range of 0.00 to 1.00. Initial values of K are arbitrarily assigned with 5 nations receiving "high bound" culture K values of .75, and the other five receiving "low bound" culture values of .25. This yields a mean initial K value of exactly .5. I selected this value in accord with the explicit assumption that it represents the lowest mean K value at which systemic coalescence, in a manner analogous to that of the modern world system circa 1500 AD can occur. K simply sets an upper (though not a lower) limit for the potential range of state "i's" RC values. RC values are free to vary within the upper bound set by this limit. K varies by discrete increments of .05 because cognitive, or behavioral, change is assumed to occur discretely as discussed above.

e) <u>Coupling</u>

Finally, the model incorporates a coupling parameter, "CF". CFprovides a measure of how closely coupled any two nations "I" and "j" are at any time t. In effect, it determines the extent to which any nation "I" may interact with, and thus algorithmically influence, the behavior of a given other nation, "j". Evaluated for the system as a whole (by computing its mean value) it provides a measure of the degree of 'openness' of the system. As such it is closely related to reciprocity, as a perfectly reciprocal system would be anticipated to be a highly coupled system. Conversely, a perfectly non-coupled system would be anticipated to exhibit little, if any, reciprocity. Thus, reciprocal systems would be expected to exhibit numerous national and sub-national, bi-directional interaction or communications channels. Correspondingly, these relational pathways would be expected to decline in tandem with diminishing coupling. Its difference equation 'assumes' that perfect coupling can be closely approached, but never actually reached. Taken together, these four difference equation determined variables provide the model system with a coherent "top-down" structure. That is, they constitute global, deterministic, transformation procedures that determine how the model system evolves with respect to system time. The difference equations defining this variable are:

4) Coupling: If $RC^i < MRC$: $CF^i = MRC - ((RC^i - MRC)/2)$
If $RC^i > MRC$: $CF^i = MRC + ((RC^i - MRC)/2)$
If $RC^i = MRC$: $CF^i = MRC$
where $MRC = RC^{total}/(N - 1)$

6.2 Study model processes

a) <u>WAR Subroutine.</u>

The model's WAR subroutine is triggered when the following criteria are met: IF $RC^{ij}_{(t)} * ((CF^i + CF^j) / 2) \leq E$ AND $R^{ij}_{(t)} \leq 0$ THEN GOSUB (i.e. go to subroutine) WAR. "E" is an algorithmically determined, positive, triggering threshold value that will be described in a subsequent section of this paper. The WAR subroutine produces, in most cases, a "winner", as well as a "loser." Winners have their power increased, while losers suffer a corresponding decrease. About one out of ten times no winner/loser is selected. This reflects the ambiguities inherent to warfare. Losers also undergo the CRISIS procedure. This subroutine provides a "bottom up" stochastic, process which allows for non-deterministic evolution and learning to occur and influence the trajectory of the model system. My thinking here is that warfare "winners" having successfully increased their power via application of existent behavioral algorithms (represented by their culture or "K" value) will experience little, or no, pressure to modify their behavior. Conversely, the application of these existent behavioral algorithms has resulted in grievous harm, in the form of diminution of power, to the war loser. Per my theory of crisis-driven learning, this situation ought to constitute a crisis, as I've defined it above. This actor perceived crisis, per theoretic assumptions, leads to an adjustment of the actor's behavioral algorithms.

b) <u>CRISIS Subroutine.</u>

Nations which lose at WAR experience two outcomes, the first, an algorithmic reduction in their power, was mentioned above. The second is that they undergo the CRISIS subroutine. Here, their K values are randomly raised, or lowered, by an increment of .05. The probabilities for each outcome are equal. This discrete value is selected in conformity with the model's assumption that cultural change occurs adaptively/maladaptively, by discrete intervals, corresponding to schematic/memetic replacement, of less "fit" behavioral schemas in response to crisis. Continuing inter-state warfare in the context of increasing systemic power, as discussed above, engenders crisis.

In essence, via operation of the model's WAR subroutine, and its CRISIS subroutine, each nation, sequentially, compares itself dyadically, with all other nations. This comparison determines whether the two nations will engage in war, and if so, which one will initiate it. The effects of this process include substantial adjustments in power and, for losers at war, culture, and hence, reciprocity. It is this local, interactive process, which allow for the system's adaptive learning behavior to emerge locally within the context of its difference equation specified global "top down" structure. This subroutine allows for additional, non-algorithmic, "bottom up" learning to take place in the context of non-warfare interaction among systemic actors. It is here, in the context of the model, that adaptation occurs. The result is an additional, stochastic, input, affecting the model system's trajectory through learning space.

c) <u>TRADE Subroutine.</u>

This subroutine operates similarly to that of the WAR subroutine. Its triggering threshold is represented by the variable "H." The principle differences are that relations between nations "I" and "j" must be positive, (i.e. RC^{ij}) as opposed to negative, for the WAR subroutine. The triggering values ("H"), as we shall see below, are, as is the case for the WAR subroutine, an algorithmically determined variable. Also, for purposes of this inquiry, if failure to reach a trade agreement occurs, BOTH actors undergo the CRISIS procedure, and hence learning, as opposed to only the loser doing so in the in WAR subroutine.

My reasoning is that if crisis is modeled to be the primary driver of systemic learning, then any class of conditions, the failure of which would likely be perceived by system actors as creating harm to their well being, ought to be a generator of crisis. This is because of the fact that as both actors are similarly deprived of their potential gains both similarly experience crisis. Simply put, both, by being deprived of their potential winnings, become "losers". This triggers the CRISIS module.

I should note that the CRISIS procedure is algorithmically triggered in this manner with between about one-fifth, to one-tenth, the frequency which it is triggered by the corresponding algorithm located in the WAR subroutine. Both algorithms are in part deterministic, relying upon instant values of system variables, and in part stochastic, relying upon input from a random number generator. For high levels of reciprocity trade-instigated CRISIS decreases rapidly, diminishing to zero at very high RC levels. This reflects my assumption that crises are generally perceived by state actors as offering less potential for harm than is the case for defeat in war.

The TRADE subroutine is predicated upon the explicit assumption that all trade interactions are mutually beneficial, (recall the discussion above concerning Ricardo's Law of Comparative Advantage) in terms of increased power accruing to each party. At relatively low levels of mutual reciprocity (RC^{ij}) these benefits accrue disproportionately to the more powerful of the two trading entities as a function of the ratio of their respective power. However, as RC^{ij} increases, the distribution of benefits becomes progressively more equitable. For high values of RC^{ij} (as I shall explicate just below) it is actually *possible* for the less powerful nation to benefit disproportionately. In so doing, the TRADE algorithm reflects my interpretation of the application of Ricardo's Law of Comparative Advantage to this class of interactions.

6.3 Study Model Parameters

Operation of this theory-predicated model creates a learning space manifold in which systemic learning occurs. Evaluation of system behavior in the context of this manifold requires the introduction of several new systemic measures.

The first of these is termed "PI." PI represents the value of the ratio of instant systemic power to initial systemic power. It provides an indicator of the model system's absolute magnitude of systemic power increase. As there is a very great range within which power may vary between the inception and termination of each model system, the resultant PI values for two or more model system's can vary by many orders of magnitude.

A second is "C". C varies uniquely for each system. It has a maximum value of *approximately* 1.0, (each model system is unique) and a minimum value of 0.0. It provides an indicator of the model system's trajectory within the learning space matrix. It does this because instant systemic power and systemic position within the manifold are complementary concepts. It is calculated according to the formula: $C = \left| 2 * (1/ P_{total}{}^{.2}) \right|$. Because it is logarithmically derived from the instant value of systemic power, it provides an (logarithmically scaled) indication of systemic 'depth.' This scaling methodology allows for easy visualization of changes in systemic power. As these changes can occur over many orders of magnitude, direct viewing is wholly infeasible. When C is evaluated with respect to system time, it provides an indicator for systemic trajectory. Were it to be displayed three-dimensionally, it would offer a glimpse of the topology of the learning space manifold itself.

Yet another parameter is labeled "CYCLE". It designates the number of program iterations, which have transpired. As such it provides the measure for system time. By definition CYCLE represents the time in which one iteration of the model system program occurs.

There is also a nominal distance parameter, "D." It possesses a potential range between approaching zero, to approaching one. It is lowest for "far apart", low net reciprocity, nations. Conversely, it is greatest for "nearby" high reciprocity nations. "Near" and "far" in this context are nominally determined by the absolute value of the difference between each model nation's pre-assigned identity number. The ten model nations are arbitrarily numbered as nations 1 through nation 10. Thus, nation 1 and nation 5 are separated by an absolute value of 4, for example. This is their "distance." As this distance may be 'bridged' with varying degrees of efficiency depending upon internal conditions within nations I and j, its *effective* value is modified somewhat by their unique, instant, values of reciprocity. I assume that more highly reciprocal nations are better able to 'bridge' distances.

D is utilized, in conjunction with reciprocity, in determining which nation in a given ij dyad will "initiate" trade or war. This is significant as initiators of both trade and war are treated differently, in terms of calculating power for the given dyadic interaction, than are non-initiators. This asymmetry tends to "favor" the initiators of trade, while correspondingly "penalizing" (in terms of net power change arising from the transaction) the initiators of war. The magnitude of this biasing ranges from negligible/non-existent at very low levels of systemic power, to being very pronounced for high levels of systemic power. It is also utilized in calculating the relative power which each member of the ij dyad brings into the trade or war interaction. Overall, D serves to allow the model's assumption that continuing warfare, in the context of rapidly increasing systemic power represents a threat to the existence of the system, along with the well established corollary to this that trade and warfare are inversely related (e.g. Mansfield 1994), to incorporate nominal distances between model nations into the model's ij dyadic relationship calculations. Simply put, the farther 'apart' any two nations are, *ceteris paribus*, the less power that they can bring to bear for any mutual interaction.

My model system program operates by sequentially and dyadically evaluating the relationships which exist between a given model nation "I", and the other nine model nations. This includes determining whether or not "I" should engage in war or trade interactions with another nation "j", and if so, whether stochastic interaction factors determine that it undergo crisis-driven learning. Once all 10 nations have undergone this process, the program repeats these procedures, thus beginning a new program iteration, or cycle.

The TRADE subroutine also incorporates a systemic metric parameter "G." This is defined as being 7.75 – total systemic reciprocity. As total RC can possess any value within the range 0.0 to 10.0, this systemic parameter decreases in value as a given model system becomes more interactively reciprocal. For very high RC systems, it actually becomes negative. The effect of G is to lessen the structural inequality which otherwise exists between relatively more and less powerful trading nations. For very high values of Rc, it allows for substantially less powerful nations to, potentially, benefit in an absolute sense from trading interactions. My reasoning here is that high reciprocity systems are qualitatively different from low reciprocity ones. This is because between-group co-operative interactions (trade) come increasingly to be favored over between-group competitive interactions (war). This has the effect of increasingly reorganizing the structure of the international system in such a way as to 'learn' to incorporate Ricardo's Law of Comparative Advantage into its structure. As these effects do not exist for a power-based system, G is utilized only in the TRADE subroutine. When G is negative, trading interactions may constitute an actual transfer of power (wealth), in an absolute sense, between the more, and less powerful members of a given ij dyad. This occurs not due to an unrealistic assumption of altruism, but rather because once the "rules" of the system come to fully incorporate the law of comparative advantage, asymmetric power based relationships become relatively less beneficial to the longer term interests of both parties. Both will benefit more if their interactions occur on the basis of relative, functional equality. Thus my reasoning here simply operationalizes the logic of increasingly integrating the law of comparative advantage into the structure of the international system via the mechanism of crisis-driven evolutionary learning.

COLLAPSE Subroutine.

A final systemic parameter is generated by operation of the collapse subroutine. This module algorithmically determines the probability that a warfare-prone system will collapse as a result of continuing warfare in the context of rapidly rising power levels. As there is insufficient empirical data to determine actual, real world historical probabilities, its values are, necessarily, somewhat arbitrary. However, they do provide a standard metric against which the effects of varying one or more systemic parameters can be evaluated. Succinctly, the probability of systemic collapse is very low at low levels of systemic power. As power increases, so does the probability of collapse. For very high levels of systemic power it becomes unity. The collapse probability increases occur discontinuously; as a given power threshold is attained, the collapse probability rises correspondingly, in a stepwise manner. At this point we are now able to proceed with the analytical portion of the study.

Comprehensive details concerning the operationalization of these several difference equations, why particular values were selected for constants, and so forth are available in Byron, 1996. All equations were developed non-arbitrarily, from, and in the context of, pre-existent theory. This information may also be accessed at either: http://www.geocities.com/~mpbyron/hdiss.htm, http://www.geocities.com/Athens/Acropolis/3956/DissCh4.rtf.

Comprehensive details regarding the extension of the model to incorporate trading interactions, as well as the modification of the Q term in equation 3 above to its present configuration may be accessed at: http://www.geocities.com/~mpbyron/JSESArt.htm OR: http://www.geocities.com/~mpbyron/JSESArt.rtf

The MS QUICKBASIC 4.5 program utilized for this simulation may be perused (or downloaded) at: http://www.geocities.com/Athens/Acropolis/3956/Warnew.bas.htm.

Note that program code lines of greater than 80 characters are split onto two (or more) lines. They must be rejoined for the program to run.

7. REFERENCES

Allport, Allan, (1991) Visual Attention; in *Foundations of Cognitive Science*, Posner, Michael I., Ed., MIT Press, Cambridge MA.

Axelrod, R., (1984) *The Evolution of Co-operation*. Basic Books, New York, NY.

Baars, Bernard J., (1997) *In the Theater of Consciousness the Workshops of the Mind.* Oxford University Press, New York, NY.

Byron, Michael P., (1996) *Crisis-Driven Evolutionary Learning: Conceptual Foundations and Systemic Modeling*, Ph.D. Dissertation, UMI, Ann Arbor MI.

_____, ________ _., (1997) Crisis-Driven Evolutionary Learning: A Procedural Mechanism for Systemic Learning, *Journal of Social and Evolutionary Systems*, 20(2): 179-83.

Casti, John I., (1996) *Would Be Worlds How Simulation is Changing the Frontiers of Science*. John Wiley & Sons Inc. New York, NY.

Chirot, Daniel, (1985) The Rise of the West, *American Sociological Review*, 50: 181-94.

Dennet, Daniel, (1995) *Darwin's Dangerous Idea Evolution and the Meaning of Life*, Simon and Schuster. New York, NY.

Doyle, Michael W., (1983) Kant, Liberal Legacies and Foreign Affairs, Part I, *Philosophy and Public Affairs*, 12(3): 205-35.

_____, ________ _., (1986) Liberalism in World Politics, *American Political Science Review*, 80(4): 1151-69.

Eckstein, Harry, (1997) Culture as a Foundation Concept for the Social Sciences, *Journal of Theoretical Politics*, 8(40): 471-97.

Gilpin, Robert, (1981) *War and Change in World Politics*, Cambridge University Press, New York, NY.

Jervis, Robert, (1976) *Perception and Misperception in International Politics*, Princeton University Press, Princeton, NJ.

Levy, Jack, S., (1983) *War in the Modern Great Power System 1495-1975*. The University Press of Kentucky. Lexington, KY.

____, ____, _, (1994) Learning and Foreign Policy: Sweeping a Conceptual Minefield, *International Organization*, 48(20): 279-312.

Haas, Ernst, (1990) *When Knowledge is Power*. University of California Press, Berkley, CA.

Mansfield, Edward, D., (1994) *Power, Trade, and War*, Princeton University Press, Princeton, NJ.

Modelski, George, (1987) Is World Politics Evolutionary Learning? *International Organization*, 44(1): 1-23.

________, ______, (1996), Evolutionary Paradigm for Global Politics. *International Studies Quarterly*, 40: 321-342.

Ridley, Matthew, (1996) *The Origins of Virtue Human Instincts and the Evolution of Co-operation*. Penguin Books USA Inc., New York, NY.

Rumelhart, David E., (1991) The Architecture of the Mind: a Connectionist Approach; in *Foundations of Cognitive Science,* Posner, Michael I., Ed., MIT Press, Cambridge MA., 133-60.

Schater, Daniel, (1991) Memory; in *Foundations of Cognitive Science*, Posner, Michael I., Ed., MIT Press, Cambridge MA, 683-726.

Simon, Herbert A., (1985) Human Nature in Politics; The Dialogue of Psychology With Political Science, *American Political Science Review,* 82: 293-304.

Simon, Herbert A., & Kaplan, Craig A., (1991) Foundations of Cognitive Science; in *Foundations of Cognitive Science*, Posner, Michael I., Ed., MIT Press, Cambridge MA, 1-47.

Tetlock, Philip E., & Belkin, Aaron, (1996) Counterfactual Thought Experiments in World Politics Logical, Methodological, and Psychological Perspectives. Princeton University Press, Princeton NJ.

Chapter 17

The performance of the world system: a critical viewpoint

Let us learn to organize the performance of the global system

ELOHIMJL (ELOHIM JIMÉNEZ-LÓPEZ)
Retired Professor, Instituto Politécnico Nacional (México)

1. THOUGH TO INTEND TO CONSTITUTE A PERFECT SOCIETY IS EVIDENTLY UNREALIZABLE.

Many of us still prefer to see human society as an **ideal organization**, as if it were the outcome of right and timely actions that would aim, by definition, at improving systematically and continuously, the human way of life or in other words at pushing ahead tireleesly a certain **social progress**. An organization of this kind is an impossible idea, basically because it should be managed necessarily by idealized beings willing permanently to act eagerly, on their own but also cooperatively, in order to become connected completely and harmoniously among them while trying to become engaged in endeavours that would aim at pushing forward the development of the society they belong to. It should be altogether the outcome of voluntary participation aiming inevitably to get involved forever in some of the various sorts of societal systems (families, groups, circles, associations, communities,...) already in existence, carrying out actions as needed for improving their way of life. Though most people have noticed that <u>working cooperatively</u>, they could deal more effectively and consistently, with all their common concern in order to improve their way of life or at least increase their survival chances, very often many of the same people have got involved in cooperative actions that aimed at annihilating other groups or communities. <u>Apparently rather few humans are motivated to work for</u>

The Performance of Social Systems
Edited by Francisco Parra-Luna, Kluwer Academic/Plenum Publishers, 2000

improving the perspective of the human society as a whole; for making its evolvement more humane and humanitarian.

At present, numerous volunteers made possible several organizations to accomplish their tasks. For example the United Nations Development Program (UNDP) offered from 1971 to 1995 such option to 10,000 people from developing and industrialized nations. They have been providing services (monitoring, health care, social conditions, equity and youth, agricultural and industrial development) and assistance in economic and political development, acting as volunteers during two weeks to two years in over 130 countries. They were motivated quite probably according to values determined peculiarly under the influence of 'united-nations-criteria', in actions for helping other people to find out how to improve their way of life or perhaps only for increasing their survival chances. [SPSSI, April 1995, p. 12]. This volunteering is basically an idealized human behaviour for dealing with some basic human concerns related to theories, plans, projects and prescriptions of 'aid and development', which point to tackle poverty. It sounds reasonable to believe that the values that motivate such kind of volunteering are intrinsically humane. After all "Responsibility is not a duty; it is a voluntary act. It is response to the needs of the other. To be responsible means to be able and ready to respond" requiring in addition the explicit accompaniment of respect which "...provides a corrective to keep responsibility from deteriorating into domination and possessiveness" [Barnes, 1994, p. 46]. What causes then, too many inhuman and inhumane performance of human beings? Anyhow volunteering is a difficult choice compared with academic research encouraged by awards.

In fact, until today, the number of volunteers has been relatively very small compared to the need of help and support; while most efforts organized, according to the present concept of aid, which was born inside the Marshall Plan (1948), have been usually either restricted exercises for attenuating symptoms of non-tackled problems or biased actions intending to favour the manifestation and evolvement of something else, but obstructing in practice the need of recognizing the causes of such problems, which remain ignored, while the economic and financial intentions and interests of companies and governments acting behind are masked: "It is not a package to aid the Mexican people, nor the Mexican economy . It's a package to bolster the failing policies implemented in our country for the last 15 years, and to prop up the regime." said Carlos Heredia in 1994. "Purely economic assistance from foreign-aid agencies doesn't resolve the crucial question - the concentration of power in Brazil." said Gilmar Mauro in 1984. "What is the use of aid if it is counterbalanced by capital flight." said Charles Abugre who worked for ISODEC in Ghana. "Major problems are solved not through charitable action, but through political action." said Alex de Waal, who worked for African Rights. "Aid will only work if it is a genuine contract."

said Hans Singer who was present at the Bretton Woods Conference in 1994. [NI, November 1996]. Responsibility and respect are two values that may motivate volunteers to devise proper circumstances for tackling the real causes of the problems, and consequently to create a humanitarian ambiance. But the number of actions organized under the name of 'humanitarian aid' and their qualities do not help remedy the effects of too many policies conceived by the economic fundamentalits who seem to be fascinated by the 'economic order' that prevails, whose values continue causing more poverty, explotation and hunger.

Could the utilization of one or another of the thousands of methodological approaches, invented with goodwill by many researchers, improve such 'humanitarian aid'? Every researcher on methodological procedures, believes that his/her meticulous routine, though conceived and designed for an idealized society, may help inevitably, after being adapted to the real conditions of one particular society, promote intelligently the fair evolvement of the multifarious features of that society. They are quite convinced that, under proper conditions, it is possible to use idealized apreciations for inducing the basic knowledge needed for pushing ahead the evolvement of real affairs.

It would mean that an idealized society may be seen as a theoretical hypothesis, though necessarily simplified, that would help, for example to create an axiological approach for finding out proper answers to: 'why society?' or '...yes a human society but, what for?' in order to determine what might be our raison d'etre as humans. However, it may be argued that an idealized social system will help axiologically: <u>determine</u> real 'person-roles', <u>design</u> consistently 'group-roles', <u>identify</u> actual 'needs', <u>organize</u> a genuine 'satisfaction' of such needs', if and only if such aims are implemented by means of self-organizing actions in the domain of a particular society in order to create conditions for allowing first hundreds, and later thousands and millions of people to develop gradually their respective personalities, each one learning to implement "the person-role, as a subject potentially able to be aware of what occurs in the system; able to feel specific but changing, and possibly contradictory needs; able to carry out what is-called 'historic action' and modify the system and, if necessary, the system environment, at will." [Parra Luna].

Then, it can be assumed that intellectual speculations derived from idealized interpretations or generalized knowledge, under proper circumstances, may become useful means for determining how a society my evolve, though an idealized view of social interactions could not offer effective recipes for improving automatically the evolvement of any real society, because inside every particular community most human beings perform diversely and fuzzily according to their own specific interests, which are determined individually or inside small groups, while foreign's interests are disregarded or frankly rejected. Due to accumulated experiences most people do not express explicity even their personal motivations, which usually are determined unconsciously through the peculiar cultural framework that every individual has managed to integrate, though unconsciously, in his/her individual mind. The complication increases because usually most people get involved in one or another societal system without knowing how different are their appreciations of a single concern. Today, it is urgently necessary to build a bridge between the analysis of idealized views and the many diverse peculiar factors that make real social problems so puzzling to cope with. But the need of 'humanitarian aid' increases qualitatively and quantitatively faster and faster, while many experts and researchers continue simplifying their appreciations instead of learning systematically to examine thoroughly what causes everything to become more and more complicated.

All along the last decades the human situation and its perspective have become every day more controversial, because the main characteristic of our civilization, inherited from previous civilizations, is the unconscious assumption that the evolvement of any civilized intention makes necessary, in advance, to homogenize increasingly the human performances. But homogeneous situations and perspectives demotivate most people from participating courageously and harmoniously in the creation of new options, causing in addition in most minds the emergence and stimulation of a selfish individualism. In practice, the societal systems constituted in the domain of most civilizations have compelled billions of people to act almost exclusively for surviving individually, learning necessarily to develop opportunistic attitudes, because most people are encouraged to expand their personalities in unified ambiences, while competing for reaching goals whose diversity has been systematically restricted in order to impede on purpose their heterogeneous manifestation.

The dynamics of our homogeneous civilization, imposed gradually and efficiently on the whole planet during the last centuries, tends nowadays to generate more and larger difficulties precisely because most organizations and institutions do not recognize that every individual behaves and evolves differently from every other. "Alfred Binet conceived and designed an intelligence test as a way of identifying public students who needed help in learning. In the schools of Maria Montessori, teachers get out of the way and students learn at their own rate." These two intellectual figures recognized that every human has the right to be a unique one, and to develop his/her potentiality on his/her own "The core of human solidarity is the realisation of the importance of human personality" "We must support as much as possible the child's desires for activity; not wait on him, but educate him to be independent." [AMI, 1970]. There is no point in wondering why these two advanced contributions have been reinterpreted in accordance with the homogenizing trends that prevail. The Binet tests have been and still are used, more often than not for selecting few people for leadership positions without identifying the wide range of human capabilities that all the others have potentially. [Time, March 19, 1999]. Nowadays, Montessori education is often an expensive education for the elite's children, while most educational organizations keep trying to form homogeneous characters, train uniformly brains and hands, in order to produce useful human instruments that may be exchanged as commodities, as they are demanded by the civilization's trends.

Another dramatic and even tragic result produced by the civilizing tendencies is offered by the practice of the official medicine: the allopathy, which intends to develop medical treatement against diseases, without even trying to identify properly the particular reasons that have caused a peculiar patient to fell ill. Usually similar symptoms of different unhealthy disorders are treated with the same medicine, previously commercialized by means of homogenizing procedures. The homoeopathic cure - discovered and developed by Samuel Hahnemann at the beginning of the 19th century and developed by him and his followers since then - approaches the specific problem that a particular person faces when he/she needs to recuperate a relatively healthy state, putting his/her whole self into a unique self-organizing task generated by a medicament, that is specific for that individual. [Lockie & Geddes, 1995]. The point of depart of Hahnemann is the assumption that the physiological features of each person are different from anyone else's.

The process of homogenization, being now deeply rooted as one of the 'cultural' values that characterizes the civilization's dynamics, is in fact one of the main causes of the 'illness' that have been disturbing increasingly the performance of our species. Homo Sapiens Sapiens (HoSaSa) has become an endangered species because the human civilizations have impeded most humans to find out on their own a possible way for allowing each one to contribute with his/her performance to push ahead the evolutionary process of the whole species. Many human actions conceived under the influence of a generalized homogeneization have become sources of degenerative phenomena. Under such circumstances, its 'successful' (?) proliferation during the last decades, will cause suddenly our annihilation.

Meanwhile, it has become acceptable and justifiable to believe that "man has been making wider and deeper his control over nature by means of science and technology. During the last millenia he has dared to explore all kinds of terrestrial and outer space circumstances making continuously discoveries of natural secrets and inventions of all kind of tools, machines, devices and processes that have increase his dominion on the Earth.". Comments of this kind are not innocuously wrong. They are instead dangerously biased as they intend implicitly to convince the reader and the listener that this 'man' could be any of all the humans (men and women) that have been born on the planet since several millenia ago. They help to avoid any particular man or woman to grumble about the way he or she has been discriminated in one or another particular society because, after all - it is argued with convincing accent - "With each discovery the condition of the human race has changed in some way for the better, as new understanding has brought more enlightened modes of thought and action, and new techniques have enhanced the material quality of life" [Burke, 1985, p. 303]. But most institutions keep up their operation assuming that the evolvement of civilized achievements has been until today a continuation of the intrinsic human features that made possible the manifestation of those synergistic endeavours that helped evolutionary forces to conform intelligent humans. It is now urgently indispensable to examine why our immediate ancestors and most of us continue 'pedalling' in wrong directions. "Today progress is sought for, too much and too exclusively through adult qualities. Thus civilization is based on the triumph of force, on violent conquest, on adaptation, on the struggle for existence and the survival of conquerors. The sad consequences of this development show themselves in the religious and moral field, in social economics and in international politics (Maria Montessori)." [AMI, 1970, p. 31].

It is evidently false to diffuse individual intentions, experiences and achievements as if they were the outcome of efforts carried out by the human species. Being it equivalent to idealize and believe that some individual endeavours will improve absolutely and 'automatically' the conditions and perspectives of billions of humans. To continue arguing in this way has become increasingly risky because any complain of any individual that could demand to be taken into account by the magnificent evolvement of the so-called civilization is considered either the reaction of a maladjusted human being unable to become a civilized individual, or a subversive attempt against the successful trajectory of humankind. Our splendorous civilization has become a disastrous fallacy, though in the sight of billions of 'second and third category' humans it is still a dream. It is indeed necessary to eradicate the processes of idealization and homogenization that are deeply rooted in the civilized way of thinking because they are used for justifying some of the inhumane and irrational sides of its paradoxical evolvement. To find out how to create a feasible process for modifying radically this tendency makes necessary to be better and better aware of how, at present, this idealized thinking is disrupting the values in the mind of experts, scholars, scientists, engineers, educators and many other people, who have been educated in practice according to an extended opportunism. Let us believe that it is still valid to assume that many humans remain potentially capable and willing to participate at any time in discovering either gradually or even suddenly how HoSaSa must start a self-organized treatement for organizing its curative power on its own.But, we cannot deny the 'successful' emergence of the Western Civilization

Though its evolvement has produced quite **paradoxical** and even **contradictory circumstances**, which have become a growing **source of degenerative forces** . By all appearances, the human way of life that prevails today, all over the world, either because it has been already established or because it is effusively admired or because it is strongly desired, is a manner of living determined by the 'historical weight' of one civilization, the Western one 'evolving' as a culmination of several previous civilized adventures. This kind of life looks as if it had been gradually though unconsciously 'structured', according to cultural values originated mainly in the teleological thinking, seemingly conceived and developed in a large measure through the Jewish way of dealing with customary affairs in the real world. Was it the only option that could emerge for the evolvement of the human world after the fall of the Roman Empire?, or was it the most feasible outcome of human experiences accumulated in those years? There is not point to speculate about other possible options that could have arisen in the past. What is advisable is to examine how should we conduct now the evolvement of the 'new', but still old and primitive civilization, which appeared violently in the European atmosphere, and evolved under the influence of inssufficiently and often wrongly oriented motivations of members of Christian communities, who were fighting against many other communities, in order to increase their domain presumably for aasuring their survival. Anyhow the Jewish community has survived and the various Christian communities have proliferated 'successfully' supported by controversial assumptions and conflicting beliefs which have generated numerous fights that have aimed until today, at searching how to survive individually while defeating & annihilating enemies, who had similar aims.

Though these assertions are only simplified conjectures, it couldn't be denied that, all along the last centuries, a relatively new way of thinking became gradually the source of actions that made possible the gradual constitution of relatively viable perspectives for the development of human societies in the future. A thinking that could help many humans to constitute his/her performance in accordance with conceptual values that allowed each one to transcend his/her appreciations of the reality at present, towards desirable and necessary situations, which presumably would be created in a 'feasible' future. It is the **teleological** thinking, whose significant possibilities are better appreciated when it is confronted to the typical thoughts expressed by wise men of the ancient Greece and Rome, which were characterized by the lack of interest in the past events and the lack of creative imagination for helping to conceive what should be done in the future. "Much of their life (the people's life) was led in a kind of perpetual present: their knowledge of the past was limited to memories of personal experience, and they have little interest in the future. Time as we know it had no meaning" [Burke, 1984, p. 91].

Anyhow, it just could happen during the Renaissance in Europe, that humans started learning to develop the notion about the need to elaborate in the human mind **'images sufficiently coherent of the past as needed for organizing human performances for dealing better with the needs at present'**, which led towards **'images sufficiently consistent of the present for being employed as a guide for helping to improve gradually the organization of feasible actions in a future, being it relatively uncertain'.**

It has been indeed a new way of thinking, derived from a better comprehension of the terrestrial circumstances that has helped people to create the Western Civilization. It has been a thinking that could generate key advanced issues that became the basis of transcendent changes which we couldn't deny: "the arrival of Aristotelian logic from medieval Arab Spain, the rediscovery of perspective geometry in Renaissance Florence, the development of the printing press, the study of trajectories that found the law of gravity, the religious and agricultural causes of the Industrial Revolution, the pairing of medicine with statistics in revolutionary France, the volcanic origins of evolution,..." [Burke, 1984]. These events, in spite of the many deaths caused by numerous and longer wars organized by elites fighting in order to get power and those produced massively by the Black Death in 1347 and the Indian cholera in 1817 allowed larger number of people in Europe to conceive new conceptual values that helped determine new perspectives. "the desire for balance and symmetry;... the rule of law, that encourages and protects acts of innovation with patent legislation, ...(giving) an individual the confidence to explore, to risk, to venture into the unknown, in the knowledge that he, as an innovator, will be protected by society; the right of each person to express his individuality;... we expect that tomorrow will be better than today..." [Burke, 1984] which altogether helped to consolidate the modern Europe.

"From the fifteenth century on the dominant spirit in Western Europe was that of expansion. People were no longer content to remain within their walls: walls of class, walls of occupation, walls of fixed duties and obligations, walls of cities and territories. They had a sense that a new world lay outside their self-imposed boundaries,..." [Mumford, 1944, p. 231]. Then, the Western thinking evolved pretending to justify its selfish values, making necessary to motivate leaders and decision makers to organize explicitly actions that aimed:

* the land expansion which has been taking place massively since 1500, causing necessarily extensive fights against indigenuous peoples, the unjust conquest of them and their achievements, and very often their total annihilation;

* the industrial expansion that started effectively since around 1830, which helped to conceive how to humanize (?) the slavery in order to continue the explotation of human resources;
* the economic expansion, organized by means of research and development policies, implemented since the 1950s, which is helping now to conceive how to humanize (?) the explotation of workers or to substitute them by automatic devices in order to take advantage more efficiently of every terrestrial situation.

Since only few centuries ago this expansionism has engendered new circumstances that have created and consolidated the power of the elites, making feasible for them the explicit and effective utilization of empirical, heuristic and scientific knowledge in the generation of more and more technological innovations, which have been inserted gradually and extensively in most societal systems, in order to: (1) 'modernize' the production processes, increasing their efficiency, productivity, quality, and consequently also the profits; (2) 'modernize' the standard of life of loyal individuals who, supported by opportunistic attitudes, should help them to play the role of organizers, supervisors, overseers, foremen, stewards; (3) 'modernize' the economic relations, increasing the profits of the international trade and exchange of natural products and manufactures; (4) 'modernize' the social facilities, implementing veiled policies conceived for helping millions of people to accept the perspective of being always potential human resources; (5) 'modernize' the public entertainment, distracting effectively people and deviating the attention that they might give to inconveniences generated by the expansionism... It has been a modernization well instrumented because soloists and choirs have been included for claiming loudly, when it has been considered convenient, that it --this modernization-- will satisfy inevitably, sooner or later but no immediately, the needs of all those willing to participate.

"Indigenous societies often share a set of values, that are in marked contrast to Western priorities. ...they are all connected through their profound relationship with the land." First peoples seek to retain that respect for nature through a sustainable and co-operative way of life." Why the Inuit, Saami, Kayapo, Maori, Hmong, Cree, Quechua, Santal and many others have been very often reluctant to co-operate? "Nothing has been so destructive to indigenous peoples as what we call progress. Mines, dams, roads, colonization schemes, plantations, cattle ranches and other expressions of 'economic development' have forced indigenous peoples from lands they have occupied for centuries and severely damaged local environments. Deforestation, desertification,and degradation of fragile, marginal lands first affect indigenous peoples, the traditional inhabitants. When they seek to stop a project, there are wider issues at stake." [Burger, 1990, p. 17 & 75].

The westernization of the human way of life has been worldwide implemented causing firstly the conquest and domination of many indigenous peoples, who had been learning, though hardly, to maintain sustainable the utilization of natural possibilities and secondly the exhaustive explotation of minerals, forests, lands, animals and people. "European colonialism has been one of the most destructive processes in human history...Within 400 years, (after 1492), successive waves of expansion had pushed much of the world's original peoples off their lands and decimated their numbers." "The colonizers came with a sword in one hand and a Bible in the other, 'to bring light to those in darkness, and also to get rich', as the soldier-chronicler Bernal Diaz del Castillo put it (in Tenochtitlan). [Burger, 1990, p. 76].

The explosive expansion of the human population since the 30s seems to be essentially a side effect generated unconsciously by the consolidation of the Western values which were reinforced by the defeat of Hitler who didn't realized that it was not possible any more the land expansion, that he had thought was indispensable for the progress of Germany. But the world wide situation in 1945, which was created with a lot of violence, made necessary pretty soon to conceive a new procedure for continue making feasible the expansion of power. Since then a Third World War implemented with economic and financial tools as weapons, is still taking place --though it was not explicitly declared-- between rich countries (the winners) and the poor peoples (the providers of all kind of resources) [Elohimjl, 1998 (1)]. During very recent years this thinking still aspire to legitimize not only the human supremacy confronted to all the other living species, but also the right of the elites --the 'more intelligent humans' (by definition)-- to use as human resources billions of impoverished people

No doubt, the process of westernization has been carried out 'successfully', but only in the sight of those relatively few humans who have always played the role of colonizers, after they confirmed their abilities as conquerors and robbers. It has been argued that it could not happen other way because "Western man's aggresive attitude toward nature and toward primitive peoples was... a diversion from the assaults he had long carried on against his immediate neighbors" [Mumford, 1944, p. 345]. Being they, by all appearences, unable to learn how to sustain certain kinds of life in harmony with many other terrestrial circumstances, when they looked for new options they didn't have other choice than performing as colonizers. When they had already emigrated, they obliged selves to engage in 'transfering' forcibly cultural values to other minds, without asking anybody why and how to act because they had to settle down in new lands as masters because they couldn't subordinate themselves unconditionally to other customs, traditions and routines. Consequently they did not respect the existence of many other different anthropological, ethnical, religious values that indigenous minds had developed and assimilated under very diverse circumstances. They even killed many people, thinking that it was justified to exterminate the primitives. "But man's attitude toward the primitive was ambivalent: one side of him was ashamed of his own greed and hard ambition: it led him to turn with loving interest to the very object he attacked. ...so the exploration of the earth was followed by a new sense of awe and wonder over all the forms of life." [Mumford, 1944, p. 345].

Due to historical circumstances the Western civilization has become the culmination of the historical trajectory of humankind (though it has become the main cause of its possible collapse in the near future), being it oriented by Jewish and Christian ways of thinking, supported by means of decisions 'scientifically and technologically minded' "By a score of different channels the interest in nature widened... (also) in the vast variety of living organisms with which man's own life is interwined (and as well) in their nature and development and transformation" [Mumford, 1944, p. 345]. But, most of these decisions have been conceived for being implemented selfishly, narrowly and shortsightedly under the assumption that everything on the Earth ought to be seen with utilitarian eyes, which means to employ them for making profitable the operation of firms and companies in order to benefit necessarily and exclusively the owners of land, machines, terrestrial circumstances, resources, etc. Today a widespread ownership makes necessary to learn how to handle properties by means of resource management according to values determined through a generalized marketing, employing all kinds of simulations carried out by means of computer technology. Every year more managers, practitioners, scientists, engineers and technologists, trained in schools and universities are trying to sustain the evolvement of the surrounding civilization: 'a well fabricated procedure that has systematically devalued on purpose the basic needs of people and other living beings.

After all, the contribution of Jewish and Christian ideas in structuring systems for assuring the civilization's evolvement have been always well assimilated by the elites, whose members keep learning very efficiently and also from time to time dishonestly - by neccesity - though almost always according to the law. The elites have also shown that they are clever enough for learning on their own how to give themselves new extended privileges, making possible gradually and consistently to maintain in equilibrium the 'troubling' evolvement of many of the systems that they create and transform in order to assure that the benefits for them continue growing.

Anyhow, despite many difficulties this civilization has evolved in accordance with values determined by leaders taking into account rather scarcely the values of millions of people who aim simply at surviving. The confrontation is potentially a very serious crisis. "What happened to Greece, Rome, China, or India has no parallel in the world today: when those civilizations collapsed, they were surrounded by neighbors that had reached nearly equal levels of culture, whereas if Western civilization should continue its downward course, it will spread ruin to every part of the planet; and its going will consume the very forces and ideas within its own tradition that might have given a start to its successor." [Mumford, 1944, p. 392].

The problem grows now qualitatively causing the 'survival chances of the human species'.to become a huge and enormously complicated dilemma. But, the features of this civilization have become de facto guidelines for more humans, in spite of being increasingly evident its ethical inconsistence and its ecological incongruity. "While Western civilizations have been spearheading the intellectual justification of conquest of nature, most peoples advocate it. Certain religious and philosophical sytems of the Western world may deny the supremacy of humanity over nature in theory. But when the opportunity arises, peoples such as the Indians, the Chinese, and the Japanese are usually just as willing as Europeans or Americans to despoil the Earth." [Ferkiss, 1993, p. 215].

Besides it is this westernized mind what motivates and helps many well intended human beings, interested in the evolvement of the Human World and the progress of the human society, to determine their respective performances, either as scientists searching out the causes of many phenomena in their surroundings or as engineers or technologists searching scientifically or empirically to comprehend real things and phenomena for finding out how to transform terrestrial elements and circumstances in facilities for being sold to other humans, while arguing that in this way their terrestrial situation will improve. Most educational processes are now systematically updated for accomplishing better such a purpose. However, more often than not most professional performances and educational curricula disregard and even ignore any ethical, ecological and ethological commitment, assuming that such concerns 'if necessary' ought to be exclusively attended through religious or other sorts of ideological membership.

2. HOMO SAPIENS SAPIENS IS FEELING QUITE ASHAMED

After being confronted to the irrefutable evidence of being a living species constituted by men and women who may perform cleverly and skillfully but quite diversely because each individual thinks and acts peculiarly, causing altogether different situations to arise and grow, which very often become contradictory appreciations of the same reality, when two or more persons neither respect the other's views nor negotiate intelligently enough the variety of their impressions. Many human performances of such a kind seem to be the source of serious, not well attended, controversial situations in the evolvement of the actual civilization, whose features become inevitably: (1) more and more inconsistent with the species' needs; (2) increasingly incongruous with the biodiversity's requirements in every terrestrial circumstance and (3) largely inconvenient for the planetary attributes that make possible our presence in the time and in the space. It means in the first case that most humans, everywhere, are now searching desperately to survive but playing passive roles in the pretended evolvement of civilized circumstances, meanwhile active minorities are creating a world of luxury and entertainment for their exclusive enjoyment, ignoring how poverty and misery are growing in their surroundings. In the second case a tragic perspective has been generated for the biodiversity that is necessary in this planet. Meanwhile in the third case the ozone diplomacy cannot implement what is needed for safeguarding the planet. The phenomenon known as civilization has in fact obstructed the evolutionary potentiality of the human species.

However, many humans still believe that our higher intelligence will help unavoidably to organize properly and better the survival chances of our species. For them it has become uncomprehensible to realize that many clever performances organized for the civilization sake:

1) obstruct the healthy evolvement of Gaia, our living Earth, causing the degeneration of those evolutionary circumstances that made possible this planet to become the only one suitable enough for the kind of life that could culminate with the emergence of human beings. [At present Gaia.is seen and treated by most humans as if it were simply an infinite source of minerals, plants and animals for being employed or being transformed by intelligent humans in useful things for human purposes while assuming that she, as a living being, will take care of herself, which means that it is not noticed that she is a particular kind of living being, fighting for surviving in the Universe, but being obliged in addition, since around 6,000 years ago, to support the influence of civilizations, that in her view have been almost always inconsistent with the evolutionary process on the Earth].
2) negate in practice the rights of all the other living species to exist according to their features and needs,

i.e. sharing harmoniusly the terrestrial circumstances; consequently such negation is causing the extinction of many species. [Even the survival needs of every other living species are disregarded, as it is assumed that the humans may take freely any living being for utilitarian purposes and recently for the sake of free trade and business, no matter it may lead gradually to the extinction of the whole species. When one or another living species is considered a pest, it is condemned unilaterally to annihilation].

3) humiliate billions of human individuals, as they have been condemned by means of unfair treatement to be only objects, though alive ones. [Needs, values and motivations of most people are in practice very scarcely taken into account, because some minorities, constituded in powerful elites, have decided that most people are only objects, commodities or resources for being trained, employed as workers or employees or no more than servants, serfs or slaves, and eventually for being put aside as unemployed fellows or expelled anywhere or annihilated when they are not needed and they are too many people to feed].

When searching to know how it could happen that the intelligent performance of HoSaSa has become a set of actions against the healthy evolvement of many living circumstances, HoSaSa included, it was necessary to distinguish the attributes of those elements of the species or groups of people that constitute HoSaSa:

Peoples: the vast majority of human beings, who have been trying to enjoy their lives, though they have become increasingly impoverished; who used to live exclusively for surviving physically, physiologically and emotionally, though they accept to get involved in all kinds of entertainment, which no doubt help them to survive psychologically. It means that usually they have been and still are motivated to live according to available circumstances, though they do not seem to comprehend what might be their role on the Earth, neither to identify their raison d'etre, neither to grasp what is happening on the planet nor to examine or analyse why so many contradictory situations are generated so cleverly and skillfully.

Elites: relatively very small groups of powerful human beings who, along the last millenia, have performed quite 'skillfully' and successfully as groups of individuals learning effectively how to become self-privileged and powerful entities in every civilized domain, which apparently have been created by groups of this kind. They tend to believe that they are the cleverest people and act accordingly in the area where they could find out how to exploit successfully the performances of many other people who were living in the surroundings and also those who had emigrated from other regions and suddenly became fascinated by civilized 'facilities'. In addition it is convenient: first to recognize that most elites have been fighting against each other, defending each one its dominion or to expand it; and second to notice that sometimes elites that were defeated while fighting have lost their power and even were annihilated.

Decision makers: members of the elites engaged in finding out how it might be better to organize the evolvement of every elite. Though they are relatively rather few, they locate themselves in powerful positions in the hierarchy of the existing societies, performing busily and cleverly for making monetary and/or financial profits through business conceived under the assumption that many natural circumstances ought to be transformed according to civilized intentions; organizing the land expansion, the industrial expansion, the economic expansion, the financial expansion... by means of exchanges, trade and wars, in order to expand the power of the elite they belong to. They are quite convinced that their role on the Earth is precisely the kind of actions they have been dealing with.

Accomplices of the elites: privileged employees or servants of decision makers, who are quite a large number of individuals, well trained for playing roles imposed by civilization's trends. Though, they are not conscious enough about the reasons that condition their performances usually they choose to accomplish their work efficiently, engaged in specific tasks as researchers, engineers, technologists, artists, educators,... really as workers searching to determine professionally or not, but intelligently in their view, how to transform more aspects of the natural world into new features of civilized environments. They tend to work very hard for sustaining the elitism, and even intend to implement actions similar to those used in the past by those who could reach an elitist position, but soon they realized that such means have been forbiden by the elites already constituted. Anyhow they keep trying to perform better, assuming that sooner or later an invitation for being accepted as a member of one or another of the existent elites will arrive.

Rapacious and avaricious specimens: ambicious and courageous human beings, who make up their minds to be rich and powerful whatever the means, but instead of trying to perform as efficient accomplices, they prefer to behave cleverer searching how to develop new skills and abilities that may open new domains of power - in one or another society - related with the expansion of innovative circumstances necessary for making more and more efficient and profitable all kinds of possible trades and traffic: in drugs, in paedophilia, with new arms or second-hand weapons for solving (?) conflicts by means of warfare, with parts and organs of animals and even humans for making miraculous cures or transplantations, in fine arts smuggling, by means of very clever financial speculations, by demanding a ransom after kidnapping people, etc. They constitute gradually powerful mafias which eventually may succeed to become new elites, as it was usually practised in the past by pirates, colonizers, armies, pawnbrokers, moneylenders and other clever people. It is a process that takes necessarily some time and requires good luck, mainly when it becomes necessary to hide gradually and effectively the illegal and unethical activities, till it becomes possible for them to emerge as respectable business gentlemen dealing with special and important institutionalized transactions or financial affairs.

Dissident humans: they are still very few individuals, though increasing gradually, who are trying to be sufficiently aware of what causes the narrow minded and shortsighted behaviour that impedes many other humans to conceive a sensible appreciation about our raison d'etre as humans; an appreciation that might help HoSaSa to push forward once more the manifestation of evolutionary forces on our unique planet. They are strange creatures that apparently agree in asuming that some time ago, Gaia made possible the emergence of life in the time and in the space, which She has maintained and helped to develop diversely, in spite of five devastating extinctions of species caused by natural events, till suddenly "...when the time seemed right, humankind entered and became the vessel for the unprecedented phenomenon of consciousness: the miraculous state of self-awareness, (but) yet evolution , not for the first time, had blundered. As species we were not yet ready to be the vessel for so powerful a phenomenon, and so, instead of employing the gift to *enhance* Earthly life, we set about destroying it." [Rivers in NI, March 1989]

HoSaSa seems to be neither happy nor satisfied with <u>'its'</u> contradictory performance which may cause soon the most serious crisis that <u>'it'</u> has ever faced, because it will be the culmination of various 'civilized experiences' organized cleverly but arrogantly and selfishly by few fellows located in powerful positions, aggravated by the blind civilization's trends, which remain determined by 'values' conceived by decision makers under the influence of an elitism supported 'intelligently' by:

Technological facilities that are increasing the productivity and efficiency of all kinds of production processes.
Scientific work carried out according to utilitarian values conceived and implemented through peculiarities generated by the technological involvement.
Development of economic means for making possible such kind of technological and scientific work as well as the related educational activities.
Pedagogical actions conceived for training millions of people as human resources or human instruments or human commodities in accordance with possibilities offered by the economic relations eatablished.

Being altogether scarcely concerned with the basic needs of billions of humans who remain unable to play a conscious role on the planet, due mainly to an increasing unemployment complemented by a welfare increasingly restricted.

Anyhow, there are today some millions of people who continue **'pedalling'** faster and faster in the sense determined by the dynamics of this civilization, whose features are efficiently diffused through the **effectiveness of the modern mass media**. Consequently they keep trying to idealize better how their individual and collective requirements might be gratified according to assumed values that in practice are not applicable to them, while their basic needs are only partially and scarcely attended.

It just comes to my mind the reflexion of Bertolt Brecht.

"QUESTIONS of a WORKER WHO READS:

Who built Thebes, that one of the Seven Doors?

In the books appear the names of the kings.

Did the kings dragged the stone blocks?

And Babylon, which was destroyed so many times, who returned always to build it again?

In which houses of the golden Lima were living the builders?

Where did the bricklayers go the night when the Great Wall of China had been finished?

The grandiose Rome is filled by triumphal archs. Who erected them?

Upon who the Caesars triumphed?

In Byzantium , the so chanted, were there palaces for its inhabitants?
Even in the legendary Atlantis, the night when the sea was swallowing it,
those who were sinking, shouted calling their slaves.

The young Alexander (the Great) conquered India. Was he alone?
Caesar defeated the Gauls. Didn't he take with him at least a cook?
Philip of Spain cried when his armada was sunk? Didn't somebody else cry?
Frederick the Second vanquished in the War of Seven Years
who did also vanquish besides him ?

Every page a victory. Who did cook the victory banquet?
Every ten years a great man. Who did pay the expenses?

So many stories.
So many questions.

Six millenia of civilized experiences seem to be altogether a relatively short time for allowing the humans to mature their minds in order to make them sufficiently sensible and reasonable when facing the enormous number of interactions that are taking place in the natural surroundings - among many and diverse living species - and also in the social surroundings - among a lot of diverse personalities. At present, the critical problems that are making difficult the presence of our species in the universe, confirm that it has not been easy for many of us to learn how to live and perform individually as members of a community, to engage in the evolvement of the particular community each one belongs to, to be sufficiently aware of our local, regional and global duties, and also to act accordingly every day in order to help every living circumstance to react for helping us to justify our particular relatively short existences. Though it is evident that the changes in "...the affectivity of human behavior and experience, the control of individual affects... and the structure of all forms of human expression...are indicated in everyday speech by such statements as that the people of our own society are more 'civilized' than they were earlier, or that those of other societies are more 'uncivilized' (or even more 'barbaric') than those of our own" [Elias, 1978, p. 221], it is quite a lot more evident an even dramatically incontestable the fallacy of our thinking intending to sustain a sophisticated, but barbaric civilization, built by means of contradictory and incompatible clever work.

It might be that **Homo Sapiens Sapiens is still in its childhood**, but no one has taken care of the way this living species has been trying to develop its inherent capabilities, then it seems that its potentiality has evolved in many cases for evil, without being noticed that certain false achievements would obstruct seriously its development. In addition it seems, rather unfortunately, that HoSaSa has fallen ill, but until today it has not yet realized that it on its own must identify the causes of its illness in order to be able to organize and carry out its self-organizing cure. How could the humans now conceive a way forward and implement new kind of human actions in order to organize as soon as possible and effectively this healing.

3. NEEDS, VALUES AND ATTITUDES, INHERENT FACTORS OF HUMAN BEHAVIOR

Are from now onward indispensable conceptual possibilities for helping us to identify a **way forward**. Knowing that our reasoning is one outcome of our high intelligence, it could be assumed that it will lead us, whatever the circumstances, to find a way forward in order to cope well with those side effects caused by the civilization's trends which have been generating, rather unconsciously, the crisis that humankind faces 'a fortiori' today. However, "... finding a way forward depends on creating a society which is both capable of learning from the effects of its own actions and then acting on the insights so gained" [Raven, 1995].

But such learning seems to be very hard for many of us who might be slow learners. Though, nature on the Earth has survived during more than three billion years, maintaining relatively stable the interactions among the various elements of the living web, even after those five previous cataclysms that happened during the last 500 millions of years, when mass extinctions of many species occurred in relatively short time periods, because natural laws could determine in each case a particular recovery, though necessarily slower, compared with the previous rate of extinction. Most of us refuse to learn that today these laws are seriously confronted to the arrogance of those humans, who are taking for granted that the blue-green Earth will be a unique sanctuary for human life and forever.

Trying to behave optimistically, it is still possible to argue that there are several "motives behind the increasing desire to protect species and landscapes: such as: <u>1. The ethical</u>. Nature is not simply there for humans to transform and modify as they please for their own utilitarian ends. <u>2. The scientific</u>. ...the rich insect faunas of the tropical rain forest should be preserved for future study. <u>3. The need to maintain genetic diversity</u>. By protecting species we maintain the species diversity upon which future plant and animal-breeding work will depend. 6. Recreational. Preserved habitats and landscapes have enormous recreational value, and ...may have economic value as well. 7. Economic. Many of the species of the world are still little known, (then, we may) have great store houses of plants and animals which, when knowledge improves, may become useful economic resources." [Goudie, 1981, p. 382]. In actual circumstances, the utilitarian motives described above will cause some privileged people that had already tasted some civilized flavours to act in very different and sometimes even opposite ways because every individual or group of individuals is induced to act according to motives that are interpreted through particular cultural values determined peculiarly while facing real difficulties when trying to learn what might be a civilized way of life. Let us imagine, as an example of real complications, the various reactions of radical and flexible conservationists, local people involved in running tourist resorts, animal-rights activists, members of Safari clubs, impoverished farmers, eco-tourists, government officials, etc. involved in finding out how to make sustainable the use of renewable resources in an African region inhabited by impoverished people and many other animals and vegetation partially exploited. Could such motives induce these fellows to act according to a reasonable agreement attained while taking into account at least some aspects, the essential ones, of their biased values? Could they negotiate among themselves their diverse interpretations about the meaning of Material wealth, Social justice, Freedom, Security, Prestige,... and the priority that should be given to each value among all? How could them altogether find out a feasible way forward?

Values, needs and attitudes seem to be three complementary mental configurations that may help every human being from time to time to organize better his/her individual performance, while learning to develop a unique and even peculiar personality, but conditioned or stimulated by relationships established either on purpose or spontaneously. Each configuration being composed by previously selected values, needs or attitudes, interacting and overlaping their domains among them according to peculiar ways determined by every peculiar individual under the unavoidable influence or bias produced by personal interpretations of specific geographical, cultural and social circumstances.

The needs seem to be something inherently necessary for the physical and physiological survival of every individual.

The values seem to be something intrinsically convenient for the psychological survival and development of every individual, learning to determine when and how to deal with his/her particular concerns, and also when and how to establish connections with other individuals in order to participate jointly in dealing with common concerns.

The attitudes seem to be something that helps every individual to adapt him/her emotionally better to the interpretation of the reality as it could be perceived by everyone.

It may be assumed, that intrinsically every person will be ready to play a role, after conceiving relatively well what to do on the terrestrial environment where he/she is located, learning sequentially and recursively to have better expectations, to project individual intentions, to act and perform consequently, to evaluate what he/she has done and eventually to make the necessary corrections for improving what he/she is fulfilling. In addition, being every person an element inside one or several groups, where everyone is willing to confirm the group-role and to achieve consequently its goals, then each individual accept to learn how to communicate particular views and appreciations, to agree or to disagree when facing other's views, to reinforce consensus, to cooperate or to fight, to negociate and to settle differences, to evaluate consensus and disagreements and eventually to find out synergistically possible ways for improving what the group is achieving.

The various actions set down above have always a particular meaning determined gradually and not easily by the individuals, who have been willing to satisfy the needs of every particular group while every member searchs how to satisfy his/her personal needs, ideally in accordance with values determined jointly, if necessary through negotiation of personal appreciations and always necessarily according to peculiar combinations of many different personal attitudes.

Then, any particular societal system inside a community, after reaching successfully an agreement while trying to determine the values that would make possible to satisfy its needs, must design a compromise among them in order to organize the system in accordance with the resources available and the degree of priority given to every value. The structure of one system would be determined by the values ordered according to the priority assigned to them, it could be for example: health, environmental conservation, material wealth, justice, security, knowledge, quality of activities, prestige and freedom [Parra Luna]. The members of another community assigning different degrees of priority to the same values would build a different structure for a similar system.

4. WE MIGHT EXAMINE HOW COULD WARFARE AGAINST HUMANS HELP

To build a new kind of civilization. Recently far more nuclear weapons have been scrapped than produced, strategic nuclear forces have gone down and warheads have been cut [WDM in NI, November 1994, p. 19] However, "the monumental disposal and clean-up task could equal or exceed the original production of the nuclear weapons" [Sivard, 1993 in NI, November 1994, p. 19]. "Fewer weapons are being produced in the world. Military spending is going down. But the flow of arms is maintained - at about 22 billion dollars' worth a year. ...international markets are now awash with second-hand weapons cascading from the old superpowers. It's a buyer dream, with prices down and more suppliers on the scene - although the US has 48% of the seller's market" [SIPRI, 1994 in NI, November 1994, p.19]. "Vast arsenals of NATO and Warsaw Pact weapons are legally and illegally finding their way to areas of tension..." [Sivard, 1993] "Annual military spending in developing countries is $125 billion. Just 4% of this could achieve universal primary education, halve adult illiteracy and educate women to the level of men. Just 12% could provide health care and safe drinking water for all [UNDP,1994 in NI, November 1994, p.19]. "Little has been said about the fact that while the former Hutu government of Rwanda was conducting its genocidal campaign against Tutsis and other political opponents, 'civilized' countries like France and South Africa were busy suplying the murderin regime with military equipment (Human Rights Watch, US 1994)" [Baird in NI, November 1994, p. 4]. "...during the 1980-88 war between Iraq and Iran the British Government had actively encouraged the sale of weapons to Iraq, and possibly Iran, in spite of a declared veto on doing so;...British banks were actively supporting this effort and highly-trained teams helping them do so;...[Hellier in NI, November 1994, p. 10]. "There has been a humanitarian catastrophe in Kosovo in the past year...attributable to Yugoslav Military forces. The main victims have been ethnic Albanians... In such case(s) outsiders have three choices: to escalate the catastrophe, to do nothing or to mitigate the catastrophe" "Nato commander...declared that it was entirely predictable that Serb violence would intensify after the bombing. Kosovo is an illustration of the first choice -- try to escalate the violence with exactly that expectation." " 'A very hard choice ...but we think the price is worth it,' said the US Secretary of State...in 1996 when asked for her reaction to the killing of 500,000 Iraqi children" [Chomsky in GW, May 1999, p. 8 of Le Monde Diplomatic supplement] "...the creation of a European army will require a beefed-up defence industry, which will see contracts and subsidies ladled out to all parts of the European Union... There are almost 200 sovereign states across the globe, many of them controlled by ruthless regimes pusuing local wars and internal repression. There is money to be made provided that not to many questions are asked. ...Nato says it is waging a humanitarian campaign which just happens to be conducted by dropping bombs from 15,000 feet so that there is no risk to its service personnel. Just as there is a third way to tackle poverty, where the rich are assured that they will suffer no collateral damage, so this is a third way war." [Elliot in GW, May 23, 1999, p. 19].

Bearing in mind that there are today millions of very poor people living miserably, and also that the number of them is increasing quite fast because they have become useless instruments for the evolvement of this civilization, somebody might suggest that a quick and effective optional solution to the crisis would be to organize efficiently the globalization of most human concerns with a reasonable number of human beings, i.e. quite less than the 6 billion that inhabit the planet today. Then, a simple and immediate solution would be to reduce drastically and effectively such a huge number of beings, in order to make feasible a proper education and training of a reasonable number of human resources as supervisors of all the automatic devices that would be necessary for providing the elites, their accomplices and also them (as living intruments) everything - comprising food and clothes - that were necessary for making possible to every human on the Earth to live decently, according to his/her rank in the social hierarchy that has become necessary to organize explicitly. Consequently, the annihilation of all those humans who won't be able after all to make any substantial contribution to the civilized way of life, which they wouldn't be able even to know, could be organized properly during the next 11-23 years,. Well this is only a digression that might be useful for saving our magnificent civilization, but it will take some time for being assimilated by the decision makers till they may design and implement this operation reasonably well. It is an option that may help justify better their role on this planet, instead of behaving simply emotionally in front of uncertain perspectives of possible ethnical and ideological conflicts. It could be something similar to all those conquests, colonizations and massacres that have happened in previous years, but now organized efficiently and implemented as a real way forward.

5. ANOTHER OPTION MAY ARISE IF THE HUMANS ACCEPT VOLUNTARILY

To become engaged in **making functional the world system**. Let me accept the suggestions made by Parra Luna that "human beings are simply "bipedal axiological profiles" seeking the fulfillment of their maximum potential" that a social system would be "a set of individuals with needs, inter-related for the achievement of certain ends, which may be or may not be common," that a a social organization would be "a set of individuals with needs artificially and formally and more or less permanently inter-relating for the achievement of more or less common objectives in terms of the power relationships between individuals" that a "businessman managing an organized social group in order to make money must at the same time achieve a series of inevitable and desirable values: security, knowledge, health, social justice, environmental conservation, quality of activity, freedom, prestige" that an university may pursue similar objectives but "firstly, to reach the highest possible level of knowledge." Though, the concept of social system and social organization have been used explicitly rather recently, since the the 50s until today, most social entities have been organized during centuries according to certain values chosen freely or some times arbitrarily by powerful decision makers or leaders, who were interested and motivated in reaching some objectives, which were necessarily values with the highest priority.

Until today, as part of the human adventure (1) most decision makers in the name of the elite that each one represented; (2) leaders of communities, ethnical groups and peoples looking for being able to perform as independent entities; (3) an increasing number of professionals - during the last decades - learning as entrepreneurs to take business risks in the hope of making profits, while organizing and managing a firm, a company or a similar social entitiy; and also (4) farmers who, along centuries have organized empirically and successfully the participation of the whole family in one or another of the possible productive entities that made possible their survival; etc.

Today it is not any more reasonable to act individually or collectivelly in order to reach some objectives chosen in advance, pretending that it is valid, according to 'my' or 'our' priority values to disregard whether my or our performance: (1) has affected or would perturb displeasingly the performance of other individuals and groups and/or (2) has troubled or would disturb annoyingly the life or even only the habitat of other living beings and/or (3) has damaged or would bother inconveniently those natural features of our planet, which are allowing the presence of life on it. Though in the past, in particular during the last 6 millenia, some of our 'clever' congeners have claimed that self-privileging humans in powerful positions, in the name of human supremacy, would have always the right and freedom to decide how to take or employ all kinds of living circumstances, according to utilitarian criteria that have been evolving due to increasing knowledge acquired by people exploring and researching natural and man-made things. Some of them have recognized that their decisions had destroyed natural environments and had caused the extermination of many living beings, comprised the human ones, but even today they have always believed that it was justified, because they accomplished their aims.

Nowadays we are learning, under very difficult conditions engendered and accumulated by human ignorance and also irresponsibility, that it is urgently necessary to assure the sustainability of the human life on the planet which has become unsustainable due to the performance of human beings unaware of such a danger. The whole situation makes indispensable the formation of globally minded humans. A feasible perspective for the human species ought to be conceived and built necessarily by the members of humankind, who are forced now by the inconsistent and incongruous evolvement of the civilizing process to accept the interdependence of human societies and natural environments, learning how to develop effectively their intellectual capabilities and affective attitudes, i.e. their intelligent faculties as a necessary condition:

* for grasping well enough how the trends of the prevailing civilization have become risky factors for the presence of many living species -man included- on the Earth;
* for dealing consistently with the paradoxical features of our technological civilization in order to create humanitarian circumstances that may help really determine how to make feasible its evolvement;
* for determining how everybody might contribute systematically to stimulate the self-organizing potentiality of Gaia, related to the self-organizing possibilities of every 'living cell' (species), in order to discover how to improve on purpose the biodiversity and the cultural diversity.

Then the question remains open: how to start making functional (sustainable) the human world at the beginning of the 3rd millenium. Let me examine some technological and economic circumstances.

"Man is pre-eminently the handy animal: his meeting thumb and forefinger and his free arms have given him a power to manipulate his environment that no other creature possesses" [Mumford, 1944]. "We, as humans, have used, quite extensively, our inherent physical 'tools', well supported by our 'intellectual tools': imagination & reasoning, which our ancestors developed gradually when they evaluated their successful achievements & inevitable failures, helping them to reinforce their intentions. The identification of feedback relations among the dynamics of real processes & the way people could comprehend such dynamics, have no doubt helped to develop our capabilities. Perceptions and interpretations have certainly been extended and amplified in many directions & degrees by means of diverse kinds of tools, built & used by men supported by diverse sorts of techniques in order to improve diverse kinds of human performances. It seems, indeed, evident that the knowledge acquired by human beings about terrestrial environments, the actions needed for transforming them, and the skills and abilities required for such actions have been always improved through technological fulfilments." [Elohimjl, 1998 (2)]. "...all cultures create material technologies to make their lives easier" [Ferkiss, 1993]

We, the humans have survived until today, during several millenia, due to our inherent possibilities, which some of our ancestors develop quite succesfully. But, some other of our ancestors and many of us at present have not grasped well enough how to make technology an instrument for pushing ahead the manifestation of evolutionary forces. Instead, a lot of technological 'achievements' have become tools that are employed by many irresponsible humans for the degradation of the human world. "...Lewis Mumford challenge the notion that technological advances equal human progress. In *The Myth of the Machine* he described the domination of society by a small, powerful elite who use modern technologies to centralize social control." "...the endless stream of consumer goods produced by technological society is a 'magnificent bribe' which masks our understanding of the real human impact of technological change" "...both individual freedom and community will be submerged by the technical imperative --the megamachine (whose correct and efficient operation) needs a concentration of economic and political power, instantaneous communication and a system of information storage capable of keeping track of every event -- this new invention {the computer} alone should suffice" "Although Norbert Wiener's work boosted both computer science and artificial intelligence research, he himself was pessimistic (and eerily accurate) about the uses to which new technology would be put. He feared its use both by military and by profit-conscious industrialists who would use the technology to lay off workers and increase production." "...the new technology would lead to an 'unemployment situation in comparison with which ... even the Depression of the 1930s ...will seem a pleasant joke". "Herbert McLuhan believed that modern information technologies would create a 'global village' (in its electronic interdependence), breaking down barriers of language and distance... (he thought) that technologies (would) shape social and political environments and structure human relationships. But he also feared the loss of democratic control over technological choice and warned of humans becoming 'servomechanisms' of a technology which is controlled by the few at the expense of the many." [NI, December 1996, p. 26]. "...the Web is the creation of the entrepreneurial worldview: It favors high-tech effects and attention-grabbing tricks. The key forces behind it are seeking desperately to transform the medium into the new television, the new movies. Their objective is to get millions to look at their site so that they can make a lot of money.... What passes through the medium is bound to be shaped by those values, not by any significant regard for quality, truth or taste" [Roszak in NI, December 1996, p. 12]

Willingly or not, consciously or not, humankind is confronted now to a very serious crisis of values, "This crisis is mainly generated by humans organizing their performance while being fascinated by the splendour of unbelievable technological achievements and by the great success of business conceived for international trade aiming to allow rich countries to sell everywhere their technological production, and to buy cheaper what the poor countries manage to produce. The success (?) of this trade, supported increasingly by IT, has led some traders to invent as a perspective the Globalization of national, regional and local economies, without recognizing that the Globalization of Misery has arisen explosively" [Elohimjl, 1998 (2)] followed by the Globalization of Corruption, the Globalization of Hipocrisy, the Globalization of Violence. "McLuhan foresaw unity of culture and consciousness, but the defining trend among earth's ecosystems is the opposite: fragmentation. So far as nature is concerned, we live in a global archipelago. To say it more plainly, a world of islands. ...(his) Global Village is also becoming reality, as advances in communication and travel carry Web sites, rap music, Marlboro ads, American TV and English-speaking tourists to every corner of the planet, pushing local cultures and languages to extinction. That McLuhanesque connectivity, with its homogenizing effect, turns out to be just as destructive, in its own way, as the island syndrome." [Quammen, 1997].

The so-called 'technological innovations', since the end of the Second World War, are seen by many professionals engaged in actions for pushing ahead the economic development as if they were magic devices that would stimulate ad infinitum the economic trade and the whole of economic relations as soon as they were introduced in the market. Based on such general assumption several decades ago these people invented the policies of Research and Development which pretty soon became inevitably sorcery and trickery under the pressure of financial interests. Now "...it has become impossible to ignore or simply disregard many diverse and large side effects, rather undesirable ones, produced indirectly through the implementation of R&D policies, because they are concerned with a simplified kind of development, which is, de facto, unaware of ethical and ecological commitments. The R&D policies have provoked unavoidably the emergence and growing of ambiguous and uncertain consequences, which have become the source of very complex problems happening necessarily and also unexpectedly... In fact, it has become strictly necessary to recognize that the kind of 'economic growth' that could be obtained in particular cases by means of generalized investments in science and technology in order to conceive, produce and sell all kinds of technological 'innovations' - determined de facto via information- becomes pretty soon, in actual circumstances, a fallacious growth mainly because the people involved are not concerned at all in determining humanitarian aims supported by humane criteria. This growth surrounded by financial profits for decision makers impede those involved to notice that such aims and criteria should be developed systematically bearing in mind that our Earth has only limited possibilities for all the species that inhabited it. This problematic situation aggravates because the related policies are conceived and implemented without realizing that the technological innovations cannot be inserted indiscriminately everywhere, assuming that sooner or later everything related will be arranged 'automatically', presumably due to some hidden attributes of all technological devices and processes." [Elohim in IF, März 1997, p. 17]

Many governments in the North talk very loudly about liberalization and democracy but most people have not been asked whether the 'globalizing' economy might help them at least to survive, instead this new phenomenon is imposed particularly in the South by means of "a driverless machine (which) assembles a thousand different parts from a dozen different countries to make a motorcar, or processes the bookings for European airlines in Bangalore. (This) machine is, of course, industrial capitalism. The invisible driver is in fact the urge to maximise profits and minimize costs. The biggest of these cost is....the toilers whom capitalism calls 'labour'. So it searches out the cheapest labour it can find - these days largely in the South, which provides the majority of the world's industrial labour. The irony is that the cheaper things are... produced; but low wages and unemployment mean that (the toilers) are less likely to be able to afford them. Capitalism, assisted by modern technology has a predilection for what's called a 'crisis of over-production' - ie waste. Leave aside...the stupefying folly of 'overproducing' from non renewable resources in vulnerable eco-systems. In strictly economic terms the gap between rich and poor widens and capital accumulates to the point where it no longer quite knows what to do with itself. Rich people valiantly spend what they can on luxuries, but the rich are too few to solve this crisis...and luxuries are useless to most of the world's people. The remainder of this excess capital swills around in 'finance houses' and banks getting bored, casting about for something more lucrative to do. ...gambling, 'speculation' on whatever comes to hand... commodities, foreign exchange, bonds, stocks, shares, all kinds of 'instruments created for just this purpose. Foreign-exchange transactions (in the temptingly volatile 'emerging markets' of the South and former Soviet bloc) for example, now amount to more than a thousand billion dollars a day, with only a small proportion relating to any 'real' economic activity at all." "...given the resources , what should be done and by whom?... Governments offered $50 billion (to wipe the shameful spectre of absolute poverty from the face of the earth) to the UN Social Summit in Copenhagen but have not yet redeemed it. Absolute poverty cannot be 'eradicated' by the judicious application of superior wisdom and cash from on high. The 'deep roots' of poverty lie in depriving the landless of their land, the homeless of their homes, and the workless of their jobs. [Ransom in NI, November 1997, p. 10].

Globalization pretends to be legal, but evidently it is not ethical. "This is in effect a casino economy featuring capital flight, tax havens, bank secrecy, speculation on currency and commodity markets, rogue traders. According to UN figures the amount involved in illegal trade is $750 billion a year - more than twice global foreign direct investment by multinationals in 1995."

[Swift in NI, November 1997] "...globalization is all about paying no taxes, states are becoming virtually bankrupt. The evidence of history suggests that when 'finance capital' gains the upper hand then the game is nearly up" [Greider, 1997].An alternative would be "to impose a tax on the transactions of foreign-exchange markets, which could generate two and a half thousand billion dollars a year. Such a 'transaction tax' would.... calm feverish speculation and force goverments to work together internationally." [Ransom in NI, November 1997, p. 10]. "Most of the world's governments are now insolvent, facing a cash crunch to pay both current bills and past creditors." [Swift in NI, November 1997]. "Unless debtor governments agreed to certain conditions, then they would be cast into the outer financial darkness and new loans would be withheld. ...the 'Baker' and 'Brady' plans in Mexico in 1982 devised (in order) to protect the interests of the US private banks that held most of the Mexican debt...were touted around the world as the paradigm of financial probity and became known as 'structural adjustment'...the purpose was to 'liberalize' the country in question: devalue its currency, open it up to world markets, reduce government intervention and flog off as many of its assets as possible at bargain-basement prices. The theory was that this would increase prosperity so that debt would no longer be a problem." [Ransom in NI, May 1999, p. 7] "The IMF rescue packages are intended only to rescue the Western creditors. ...financier George Soros complains that the game of international speculation and bailout played by the Western financial establishment -in which hot money rushes into a country, then pulls out leaving behind a wrecked economy to be cleaned up by local governments and G7 taxpayers - is a menace to world economy stability. ...Asia's bankrupt businesses, insolvent banks and jobless millions are the spoils of 'financial 'warfare' {Chossudovsky). The gains to be won from these financial hit-and-runs are immense. ...first of all, the foreig-exchange reserve of the target countries, (which had accumulated) currency reserves by running trade surpluses....selling more abroad than they purchased....at great cost to the working populations, who labor hard to produce goods, destined to be consumed by foreigners. Asian countries spent nearly $100 billion in accumulated reserves trying -- vainly as it turned out -- to prevent devaluation. Next, there are the bargains... once the target country's currency has collapsed and its firms are strapped for cash. Years of effort ...by the Korean élite to keep business firmly under control of state-supported conglomerates...were undone in a matter of months.... Citygroup and other firms were snatchingup ownership of Asian banks and industries... Asia is today a bargain-hunter's paradise.... Then,... the higher tribute that countries, once in debt peonage to Western creditors must pay on both old and new loans. South Korea under the terms of the IMF bailout, will pay interest on foreign loans that is 25-30% higher than rates on comparable international loans, despite they have been guaranted by the Korean government. Since the crisis began, international lenders have doubled and tripled the interest rates they charge on emerging market debt. Next, there are the people themselves, engulfed in debt, impoverished and committed by the governments to an endless course of domestic austerity and debt repayment. ...millions of newly unemployed, whose desperation will pull wages down world wide. ...entire countries will (turned) into export platforms, where human labor is transformed into the foreign exchange needed to repay debts. Finally the governments engaged in reforming the legal and political systems according to conditions imposed by IMF, which entail a major loss of national sovereignty, but fix the rules of the game: to protect and expand, at all cost, the wealth of the international financial elite. [Frank, in NI, May 1999].

We may still believe that there is some time available for allowing us to test and confirm our pretended supremacy on the Earth, as living beings who may exercise a 'high' intelligence in accordance with utilitarian attitudes though, we know today that they ought to be conditioned by ecological, ethical, and ethological compromises, being indispensable for us to learn: how we may contribute to sustain Gaia in good health; how we should share equitably the terrestrial space with all the other living beings; how we must encourage ourselves -at heart and by heart- with mind and hands, to develop diversely our intrinsic synergistic capabilities, but it seems that such an enormous task has been already overwhelmed by the very clever decision makers who have been performing madly during the last decade in order to sustain a financial system for speculative purposes. "Sustainable development requires the conservation of natural resources. One way of doing this is to make them more expensive. In fact, they are getting cheaper. World-market prices for raw materials - 'commodities' like oil, timber, sugar, coffee or cooper -- have fallen to their lowest levels in living memory. ...partly because there is less demand in times of economic recession, ...partly because of technological change. But it is mostly because of debt in the South... Structural-adjustment policies imposed... mean that poor countries have to earn foreign currency to 'service' their foreign debts before...anything else. This means exporting the only things they can sell on world markets -- raw materials. Because all poor countries have to increase their exports at once, there is a glut on world markets and prices fall -- sometimes by as much as a half, so that twice as much has to be exported to earn the same amount of foreign currency. The rich countries are the beneficiaries, they not only get their debts 'serviced', but cheap commodities keep prices down, profits up and inflation is under control. The losers are the people of the South, and the global (natural) environment" [NI, May 1999]

Suddenly it seems that development has disappeared. " 'Development' was the promise of universal economic growth along the routes pioneered by the leading countries of the West. 'Growth' implied steady economic expansion... in the form of industrialization... People would eat more, grow taller, become more literate, live longer and enjoy fuller lives. Women would enjoy... emancipation. Soon it turned out that 'development' was also 'Americanization'. ...This was the promise of McWorld... ...curries, sushis and couscouses would surrender to hamburgers... (it) would take place gradually fromn the top of society downwards; that is as incomes rose. ...through the 'development' decade of the 1960s...it was a universal article of faith. ...by the mid-1990s the happy face of developmental optimism was masked by gloom. (Then) 'development' came to mean a modest increase in 'industrialization', that is, an improved capacity to produce textiles, sports shoes, furniture, agricultural products,... 'Developing countries' had (in 1996) become 'emerging markets' " [Mason, 1997, p. 407]. Instead of 'sustainable development' some people are now increasingly interested in sustaining the 'DEBT' "...the 1980s were the 'lost decade' for most of the Third World, and their economies didn't grow. Third World debt doubled to almost $1,600 billion by 1990. An ever-increasing proportion of it was simply to pay off old debt and keep the system up and running... the debt continues to increase relentlessly. By the end of 1997 it exceeded two trillion dollars. No one seems to know where it will end, though everyone knows that it cannot continue." [Ransom in NI, May 1999, p. 7]. "The result is a debt of $400 for every man, woman and child in the developing (?) world - where average income in the very poorest countries is less than a dollar a day. In six of the eight years from 1990 to 1997, developing countries paid out more in debt service (interest plus repayments) than they received in new loans - a total transfer from the poor South to the rich North of $77 billion. Debt in the poorest places on earth has become a modern form of slavery." However the richest countries have large 'public' debts, too. But they don't have the IMF or World Bank running their governments." [NI, May 1999, p. 18]. The whole situation is getting worse due to other powerful pressures "A treaty (global pact) among 170 countries (in Colombia) to ensure safe trade in genetically modified organisms (GMOs) has been sabotaged by the United States, which believes that its business interests are threatened" [GW February 28, 1999].

"At the beginning of the last two decades of our century, we find ourselves in a state of profound , world-wide crisis. The deterioration of our natural environment has been accompanied by... the 'diseases of civilization'... On the psychological side...disorders appear to spring from a parallel deterioration of our social environment. There are numerous signs of social disintegration... (included) an epidemic of violent deaths... Along with these social pathologies we have been witnessing economic anomalies... Rampant inflation, massive unemployment and a gross maldistribution of income and wealth have become structural features of most national economies... These problems are closely interconnected and interdependent. They cannot be understood within the fragmented methodology characteristic of our academic disciplines and government agencies. ...most of our leading thinkers use outdated conceptual models and irrelevant variables. ...we need to see our situation in the context of human cultural evolution" "After civilizations have reached a peak of vitality, they tend to lose their cultural steam and decline. An essential element in this cultural breakdown, according to Toynbee, is a loss of flexibility. ...(this) loss in a disintegrating society is accompanied by a general loss of harmony among its element, which inevitably leads to the outbreak of social discord and disruption. ...(but) the society's ability to respond to challenges is not completely lost. Although the cultural mainstream has become petrified by clinging to fixed ideas and rigid patterns of behavior, creative minorities will appear on the scene and carry on the process of challenge-and-response." [Capra, 1982].

The turning point ought to be a new paradigm, a new vision of reality; a fundamental change in our thoughts, perceptions and values, a qualitative change of the way humans may develop their capabilities, while learning why and how the civilized world has become increasingly dysfunctional in relation to evolutionary forces. This point should become really a recovery process, handled by humans who would aim at recognizing and implementing humanitarian and humane features, which are intrinsic to our nature but requiring now to be determined by means of ethical, ethological and ecological values, being at the same time terrestrially, biologically and culturally concerned when carrying all kinds of local actions, which need to be regionally responsible and globally minded

"Ludwig von Bertalanffy advocated that we dare to broaden our loyalty from nation to globe and urged that we become patriots of the planet, endeavouring to think and act primarily as members of humanity, (and) begin pledging our allegiance to humanity and to the earth on which we stand, one planet indivisible or no planet at all. "We must begin protecting the individual and cultural identity of others". He advocated as well a new global morality: "an ethos which does not center on individual good and individual value alone, but on the adaptation of mankind, as a global system, to its new environment". "We are dealing with emergent realities; no longer with isolated groups of men, but with a systematically interdependent global community. It is this level of [reality] which we must keep before our eyes if we are able to inspire large-scale action designed to assure our collective and hence our individual survival." He argued: "Possibly the model of the world as a great organization can help to reinforce the sense of reverence for the living which we have almost lost in the last sanguinary decades of human history". He believed that "the overall fate of the world depends on the adoption by humanity of a new set of values, based on a general systems

WELTANSCHAUUNG (worldview)". "We are seeking another basic outlook: the world as organization. This [outlook] would profoundly change the categories of our thinking and influence our practical attitudes. we must envision the biosphere as a whole ...with mutually reinforcing or mutually destructive interdependencies. [We need] a global system of mutually symbiotic societies, mapping new conditions into a flexible institutional structure & dealing with change through constructive reorganization." " [Davidson, 1983

Turning points have been conceived already. "Mahatma Gandhi...believed that the crushing poverty of rural life in India was worsened by the import of cheap, machine-produced fabric from British textile mills. ...millions of spinners and weavers lost their jobs when Britain flooded the subcontinent with Lancashire cotton goods and restricted the export of textile machinery to India. "It is machinery that has impoverished India... By reproducing Manchester in India we shall keep our money at the price of our blood... How can a country with millions of living machines afford to have a machine which will displace the labour of millions". His model of de-centralized development and village-level democracy was scorned ...as reactionary nonsense in the face of necessary modernization. In the 1920s he adopted the charkha or traditional Indian spinning wheel as the symbol of his campaign against the technology. " " "As an alternative to a Western-imposed 'monoculture of the mind' (what is needed is) an 'insurrection of subjugated knowledge' and a rejection of technologies which undermie and worsen the human condition" said Vandana Shiva who praised passionately traditional cultures and indigenous technologies" [NI, December 1996, p. 26]. "In the North, massive subsidies sponsor largely 'globalized corporate interests. Similar sums spent on local initiatives would create more jobs, improve the environment and distribute resources more equitably. Shifting subsidies away from industrial agriculture to smaller-scale, more diversified farming would promote biodiversity, healthier soils and better food. In the South, a decentralized, renewable-energy infrastructure would strengthen the economies of villages and small towns. The best land...diversified production for local consumption would improve food security and help to reduce absolute poverty. Enormous educational benefits flow from local-language and location-specific knowledge adapted to the bio-region and culture, reversing the sense of inferiority imposed by 'Westernization'." [Faraclas in NI, November 1997, p. 24]. "Drop the debt: The Jubilee 2000 Coalition believes that 52 countries should have their external debts cancelled to mark a fresh start for the new millenium. They are poor, often desperately so, and repayment is making them poorer still. In 1996 there were 984 million people living in these countries. Foreign debt was larger than annual income in 31 of them" [World Bank, Jubilee 2000 in NI, May 1999]. "END THIRD WORLD DEBT AND START TO END

THE THIRD WORLD . We've a once in a millenium opportunity to make an historic gesture. We're calling for a one-off cancellation of the unpayable debts of the Third World. So that the money spent on repayment can be used to help eradicate poverty,. It's a chance to break this vicious chain of debt. A chance to change millions of lives for the better, forever." Christian Aid and the Co-operative Bank are working together to call for the cancellation of poor countries' unpayable debts [NI, May 1999]. " "The evidence is in the archives where paper after paper, signature after signature, show that from 1503 to 1660, 185,000 kilos of gold and 16,000,000 kilos of silver were shipped from America. Plunder?... Genocide?... Pillage?... Dr. Arturo Petri thinks that the growth of capitalism and of the current European civilization was due to that early flood of precious metals from the Americas. (Then) those thosands of kilos of gold and silver (were) the first of several friendly loans granted by Native Americans for Europe's development. Such...capital exports were a Marshalltezuma Plan to guarantee the reconstruction of a barbarian Europe, ruined by deplorable wars against the Muslim foe." "...unfortunately you (European brothers) squandered it on wars, invincible armies, Third Reichs and other forms of mutual extermination, only to end up being occupied by the Yankee troops of NATO. Like Panama, but without a canal. Financially, you were incapable - even after a moratorium of 500 years - either paying back capital with interest or of becoming independent from raw materials and cheap energy imported from the Third World. ...it compels us to claim - for your own good - the repayment of capital and interest which we have so generously delayed reclaimimg for all these centuries. We shall only demand the return of all precious metals advanced, plus a modest interest rate of ten-per-cent per year (we refrain from claiming the despicable floating rates of 20 or even 30 per cent) accumulated over 500 years... What huge piles of gold and silver! How much human blood would weigh the same? To say that in half-a-millenium Europe has not been able to produce sufficient wealth to pay back this modest interest is to admit to the total failure of capitalism". Cuaicaipuro Cuautémoc (a Mexican indigenous leader). " [Reprinted in NI, May 1999 from Renacer Indianista No. 7, and from the translation in Resurgence No. 184].

!e@&A%u$§§o"?ß*#i*!

Words and some more words and even other additional words,
under the assumption that they will help some of us
to structure **values** sufficiently suitable for allowing us
to find out how a **turning point** might emerge in our thinking.
What for?
Well, we noticed only yesterday that we, along several millenia, have been going **nowhere**,
that during the last centuries we decided to go always faster and faster,

that now, since few decades ago, we are going 'there' even faster;
in spite of being increasingly evident that our intelligent and efficient going
has been destroying our planet, our habitat, our living companions and many of us as well.
Why could we behave so inconsistently?
because during our going our values have been degenerated;
the arrogance of several of us has degraded our 'high'.intelligence;
clever intentions of us have 'created' many degenerative processes;
the utilitarian knowledge, that we continue accumulating, has generated new diseases;
our performances have broken many evolutionary options on the planet;
in fact, our presence on the Earth contradict our origin.
We are only grass roots on the Earth,
intelligent, but every one of us still need to learn how to be clever enough,
how to reflect systematically on
"for what purpose am I and are we living?"
bearing in mind that our presence in the time and in space
has been possible due to an improbable event that happened once,
but it won't occur again.
We do not know now whether today is already too late
for searching in our minds how our species might still survive,
while pondering the disastrous situation and catastrophic perspective
that our civilizing processes have generated.
In case there were still some time available for us,
we should build on purpose **"our raison d'etre"**
on this unique planet – **Gaia** – the only one available for us in the Universe;
then, while still being alive
our immediate duty is to constitute ourselves in a **task force**
that would aim at setting right the effects of too many actions of us
which were conceived and carried out intelligently,
but without being conditioned by ethical, ethological and ecological values;
our commitment is to organize a coherent **research-action**
rationally conceived and emotionally motivated,
for helping every human during their relatively short lifes
make a significant contribution, each one on his/her own,
toward the creation of a humanitarian and humane civilization,
conditioned by the needs of all the other living beings,
and finding out courageously how we might help Gaia to be in good health.

6. REFERENCES

[Alexander & Levin in JSI, Winter 1998] Michele G. Alexander and Shana Levin. ***Theoretical, Empirical & Practical Approaches to Intergroup Conflict*** in JSI.(SPSSI) Vol. 54, No. 4, 1998

[AMI, 1970] Maria Montessori. *A centenary anthology*. Amsterdam, 1970

[Baird in NI, November 1994] Vanessa Baird. *Lethal Lies*.

[Barnes, 1994] Graham Barnes. *Justice, Love and Wisdom*. Medicinska Naklada, Zagreb. 1994.

[Burger, 1990] Julian Burger. *The Gaia Atlas of First Peoples*. Anchor Books. Doubleday. 1990.

[Burke, 1984] James Burke. *The Day the Universe Changed*. British Broadcasting Corporation.

[Capra, 1983] Fritjof Capra.*The Turning Point. Science, society & the rising culture*. Flamingo. 1983.

[Chomsky in GW - Le Monde Diplomatic supplement, May 1999] Noam Chomsky. *Nato, master of the world.*

[Daly, 1990] Herman E. Daly & John B. Cobb Jr. *For the Common Good, Redirecting the economy towards community, the environment & a sustainable future*. Green Print. 1990.

[Davidson, 1983] Mark Davidson *Uncommon Sense. The Life & Thought of Ludwig von Bertalanffy*. J.P. Tarcher, Inc. 1983.

[Elias, 1978] Norbert Elias. *The Civilizing Process*. Basil Blackwell. Oxford. 1978.

[Elohim in IF, März 1997] Elohim Jiménez López. *There is plenty of Information Available but - What for?* IF Band 11, Nr.1, März 1997

[Elohimjl, 1998 (1)] *Learning to build a sustainable economy; it ought to be terrestrially, biologically and humanely minded*. 42th Meeting of the ISSS. 1998.

[Elohimjl, 1998 (2)] Elohimjl. *Learning to design societal systems for a global society is quite something*. Cybernetics & Systems. R. Trappl. Austrian Society for Cybernetic Studies, 1998.

[Elliot in GW, May 23,1999] Larry Elliot. *West lines up for the third way war*.

[Faiola in GW, April 1999] Anthony Faiola. *Brazil's Sick Economy infects Ecosystem*. Page of Washington Post inside Guardian Weekly April 25, 1999.

[Faraclas in NI, November 1997] Nicholas Faraclas. *More powerful than guns*. NI No. 296

[Ferkiss, 1993] Victor Ferkiss. *Nature, Technology and Society. Cultural Roots of the current Environmental Crisis*. New York University Press, 1993.

[Frank in NI, May 1999] Ellen Frank. *Double Dealing*. NI No. 312.

[Goudie, 1981] Andrew Goudie. *The Human Impact on the Natural Environment*. Blackwell. 1981

[Greider, 1997] William Greider. *One World , Ready or Not: The Manic Logic of Global Capitalism*, Simon and Schuster, New York, 1997.

[Hellier in NI, November 1994] David Hellier. *Sssshh...*

[Lockie & Geddes, 1995] Andrew Lockie & Nicola Geddes. *Homeopathy*. DK Book, 1995.

[Mason, 1997] Mike Mason *Development and disorder. A History of the third World since 1945*. Between the Lines. Toronto. 1997.

[Mumford, 1944] Lewis Mumford. *The Condition of Man*. Martin Secker& Warburg, London, 1944

[NI, November 1994] *Letal Lies. The Arms Trade*. NI No. 261.

[NI, November 1996]. *The Poverty of Aid*. NI, No. 285.

[NI, December 1996] *Seduced by Technology*. NI No. 286.

[NI, November 1997] *Globalization*. NI No. 296.

[NI, May 1999] *Drop the Debt.* NI No. 312

[Parra Luna] Francisco Parra Luna. *Why Society?* Working Document.

[Quammen in Time, October 27, 1997] David *Quammen. Archipelago Earth.* Our Precious Planet, Time (Special Issue).

[Ransom in NI, November 1997] David Ransom. *Globalization -an alternative view.* NI No.296

[Ransom in NI, May 1999] David Ransom. *The dictatorship of debt.* NI No. 312

[Raven, 1995] John Raven. *The New Wealth of Nations. A New Enquiry into the Nature and Origins of the Wealth of Nations and the Societal Arrangements needed for a Sustainable Society.* Royal Fireworks Press, 1995.

[Rivers in NI, March 1989] Patrick Rivers. *Life in the Balance. The Globe at a Glance.*(article adapted from The Stolen Future: How to rescue the Earth for our Children published by Green Print. NI No. 193, March 1989.

[Robbins in JSI, Fall 1994] *Environmental attitudes conceptualized through Developmental Theory* in JSI, vol.50, No. 3, 1994.

[Roszak in NI, December 1996] Theodore Roszak. *Dumbing us down.* NI No. 286

[Schwartz in JSI, Winter 1994] Shalom Schwartz. *Are there Universal Aspects in the Structure and Contents of Human Values?.* in JSI (SPSSI). vol. 50, no. 4, 1994

[Sivard, 1993 in NI, November 1994] Ruth Leger Sivard, *World Military and Social Expenditures, 1993.*

[SPSSI, April 1995] *Global Opportunities for Volunteer Service.*News from the United Nations. SPSSI Newsletter No. 196.

[Swift in NI, March 1997] Richard Swift. *Endangered species,* Who's next? NI No. 288.

[Time, March 19, 1999] *The Century's Greatest Minds.* Time, March 19, 1999

[UNDP, 1994 in NI, November 1994] *Human Development Report* of United Nations, 1994

[WDM in NI, November 1994].*World Development Movement*

AMI Association Montessori Internationale
GW Guardian Weekly
IF Informatik Forum
JSI Journal of Social Issues
NI New Internationalist
SPSSI Society for the Psychological Study of Social Issues
UNDP United Nations Development Programme

Chapter 18

An exploration into the measurement of social efficiency on the Internet

MOISÉS GARCÍA JARA
Facultad de CC.PP y Sociología. Universidad Complutense de Madrid. Campus de Somosaguas

The ongoing evolution of society and the invasion of each and every one of its components by the informational domain have given rise to new horizons to be envisaged, new risks to be assumed and new vulnerabilities to be studied by social scientists.

The variety of areas of research in social sciences, together with the impact of new communication technologies, has fostered the creation of new lines of action in the realm of cyberspace. Nonetheless, to obtain valid results from such research we must begin with a review of the many theories that have preceded the present study. In my opinion, the various theories can be grouped into two areas: the political and the cyberformal (to be considered in greater depth here).

From the strictly political standpoint, the works of a number of authors stand out as beacons in General Systems Theory. We shall begin with Karl W. Deutsch, focusing on two of his works (1974 and 1980). His vision of political systems could well be applied to the virtual system addressed in this paper. It deals with political systems in terms of communicational messages, formulating dimensions such as system memory, conscience or the selection screens used to control the disturbances in the system mechanics.

I would, moreover, refer to the author's comments on the possibility of defining the ethical responsibility of political systems; in this regard, I would include his formulation of the term "political performance", which covers the application of the terms and scales of values ensuing therefrom in the context of virtuality.

Another author to make a mark on social system theory was David Easton. Both in his book on political systems (1959) and in others of his

The Performance of Social Systems
Edited by Francisco Parra-Luna, Kluwer Academic/Plenum Publishers, 2000

works, Easton conceives the political system to be a body subject to a series of stimuli and responses, along pure behaviourist lines. Building on this argument, Easton subdivides the political system into four dimensions, all of which have a bearing on the broad outlines of this paper: 1) the system itself; 2) the environment; 3) the outputs and, finally, 4) feedback.

In the second category, compiled ad hoc, I include under the term "cyberformal" all the authors who have consolidated and fuelled new lines of research and their application to new social phenomena, such as the Internet in this case.

In the strictly formal sphere, we might cite the eminently mathematical work by Cortés, Przeworski and Sprage (1974), which provides a common language for the co-ordination of the various kinds of systems theory research.

But perhaps the author that has had greatest influence on this and other studies is J.P. Van Gigch (1978), who set out the basic precepts for formal conceptualisation of the performance of a social system, including under that term system morality, the problems arising within systems or the dilemma inherent in the formulation of suitable social indicators to quantify social phenomena.

With regard to strictly "cyber"-related works, we might refer to authors such as M. Castells (1997), who coined the famous term "network society" (prevalence of social morphology over social action) or to papers by researchers such as S. Turkle (1995), R. Shields (1996) or S.G. Jones (1997), which survey more internal aspects of Internet dynamics such as network subcultures, the most intrinsic aspects of the mind and an abstract approach to the predominant or optimum qualities of self on the Net.

From the more specific viewpoint of social system performance, we would mention a series of papers by F. Parra Luna. In several of his books, El balance integrado de la gestión estratégica or Elementos para la teoría formal del sistema social, Parra Luna accurately reflects the precepts of the application of the performance of social systems in all their domains (justice, politics, business, etc.). I shall take his remarks as the fundamental theoretical grounds for this study of Internet performance.

In another, less theoretical, more practical vein, I shall be referring to works by Tretter (1998), Sánchez Súcar (1998) and Nicopoulous (1998), to name a few, on the implementation of performance in a social system (public organisations or institutions).

Drawing on all these approaches and conclusions, I shall attempt to contribute to this field by focusing on the computation of global Internet performance, along with a discussion of what I personally find to be its advantages and drawbacks.

The changes in information brought on by an intangible entity such as the Net have made it possible to apply its eminently technical benefits to traditional sources of information (encyclopaedias, for instance). What are the basic differences between the Internet and such traditional sources? Initially, we can take the major effect to be the continuous updating of information, along with the possibility of introducing many megabytes in a cyberspatial flow with enormous ease. Compared to the observation or search for information in an encyclopaedia, the Internet provides universality of knowledge, intangibility, anonymity and speedier communications between social actors.

In view of these innovations, I shall study the outputs of the virtual system to compute its global performance, in an attempt to gain insight into the quality of its multiplications, along with the optimum quantification of the enhancement of the information ensuing as compared to inputs.

Taking even its very first steps with the self-assurance of a giant, the Internet has, in a very short period of time, absorbed all the domains of real society. In a matter of just a few years it has become the ideal medium for interchanging information, whose technical features enable millions of users to interact, circumventing the barriers of distance and ethnical-cultural differences.

The Internet has developed domains in our lives that we had all but forgotten. It has done so by developing a complex structural organisation, using information as the entity, channel and medium for the interchange of knowledge among individuals in a social system. This development has begun to give real shape to a dream that we have all had at one time or another: unlimited interdisciplinary communication. Since establishing its frontiers and outlining its forms is a task incumbent upon all of us, I shall attempt to be rigorous in my arguments.

In the broad context of Niklas Luhman's theory, and focusing on some of its least explored aspects, I would relate two basic terms: the Internet and Self-referencing Systems. The dependence between the two and their complexity lay the grounds for resolving the issues addressed in this discussion: how the two concepts complement one another, what nuances must be included to perfect the theory and what is the ideal course to take to obtain a better overall view.

This series of issues brings us to the second part of this discussion: in what way the information flowing over the communications channels created by the Internet network enhance or detract from knowledge and the ultimate understanding that it provides users. The question is to observe whether the amount of data that can be obtained from the virtual flow of information improves or worsens the user's biorhythm (making it a slave to data) or, on the contrary, it nourishes the internaut's personality and facilitates learning and sociability through the machine.

A glimpse of the pressing problem that threatens the existence of the network of networks - trash information - can be caught in this complex real-virtual grid. The accessibility of data, along with the advantages that the Internet medium affords compared to the traditional means of dissemination of information with which it competes, makes it possible to instrumentalise cyberspace to convert it into a sewer for intellectual discards that users can place on the network via different channels, legitimate or otherwise. Analysing and drawing conclusions in this regard will aid to combat them and re-steer the illicit use of information, be it for business, commercial or sexual purposes.

In the final part of the paper, I address the main thesis and core of the issue under discussion: Internet synergies and computation of performance. I shall attempt to respond to a series of difficult questions, as clearly as possible and following empirical-scientific precepts: what is the Internet's social efficiency as a self-referencing system? To what extent does cyberspace modify, enhance or transform human values? What good is the Internet to humanity? What are the strong and weak points of such a complex and abstract system and, therefore, its global performance? How and where can the vaccine against systemic degeneration be introduced?

All these questions are geared to an analysis of the virtual system's inputs/outputs. Since the assessment of the Internet's social efficiency is the principal aim, I shall try, as far as possible, to provide a schematic view of the medium's potential to increase effectiveness as well as of the efficiency of its purpose.

1. INTERNET AS A SELF-REFERENCING SYSTEM

To undertake the analysis of such a cubist world as cyberspace, a series of obsolete cybernetic concepts must be re-defined. Any self-observed system multiplies its complexity and observations become vice-ridden when the stimuli accruing saturate the receiver's retention potential. In the global village, i.e., the Internet, the subject shares the ability to record and convey information, overlapping the primary functions of the object. In this compendium of inputs and outputs, unity is to be found in the bond, the link, bringing us to consider the Internet a priori as an organisationally closed and informationally open social system.

As a complex viewpoint, the theory of self-referencing systems has transformed concepts applicable to the field of cybernetics, building a new structure of substantial proportions. The change brought about by this theory is no obstacle to criticising, summarising and applying it to eminently practical cases. The Internet features all the characteristics of a self-referencing system, but from my view there is a series of elements that do not quite fit into the logic of cyberspace, which may lead to logical and practical reconstruction.

The complexity of the perspectives, the abstraction of the internaut and the change in space-time variables make the network one of the most advanced systems as far as communication is concerned. Therefore, the fundamental pillar of my critical review will be the use of communication by social systems as a form of autopoyetic reproduction (within the limits of the social network, never outside it). Communications are not actions in themselves, but rather require three indispensable elements: information, assertion and comprehension (Luhmann, 1989).

The subject and the object in both virtuality and reality are linked by a Godelian sentence (Ibáñez, 1990) in a "virtuous circle" which, if ruptured, can cause their death. In social systems, communication plays the important role of a channel for the conveyance of information but not the leading role in the circle of links created in cyberspace.

The Internet and self-reference are both tri-dimensional, complementary and similar concepts. A self-referencing system is one in which its operations are referred to the operations themselves, whereby it encompasses tautology (Luhmann, 1989). The concept of autopoyesis means essentially the same thing, although the terminology changes: Luhmann (1989) defines autopoyetic systems to be systems that reproduce all their elementary units by means of a network of those very elements, delimiting an environment.

Since the onset of the Internet, all concepts, elements and links lead to virtual self-creation, or rather, personalised modelling and defragmentation on the part of each machine, environment and user that forms a part of this new abstract world. The system came to life in the context of a powerful technical framework, which allowed the association or linking of elements and units, whereby the most complex communications network of all times grew out of nothing. Such a perpetual dilemma confounds reality but provides a glimpse of one of the essential concepts of this analysis: the Internet does not exist, it is just one of a number of media for interchanging information. It survives by regulating, administering and selecting its own operation and elements, continuously opening and closing the circle.

I take two concepts from N. Luhmann's complex and criticised theory that are fundamental to understanding the implications of the Internet phenomenon in the context of systems theory: observation and difference.

1. Observation. In order for a system to survive and co-exist with the rest of its corollaries, it must have a series of established limits. Like any mirror, a social system reflects and refracts all the actions that the actors perform inside its territorial bounds. These actors modify the object, and become observer-manipulators. Observation thus becomes one of the essential elements in self-referencing systems, sharing dynamics and interconnections,

observing themselves, according to Luhmann (1997), as well as their surroundings.

The Internet thus establishes its contextual limits with the rest of the virtual environment. These cyberfrontiers may be classified into three major groups: users, websites and mere technical implications. In practice, in self-referencing systems theory, the staging of the limits of a system highlights the importance of users within the globality of the virtual complex as opposed to real society. The Internet expands on this point of the theory and "re-vindicates the concept of subject" (Ibáñez, 1994). Luhmann's anti-humanism is laid bare when he posits that society is not made up of human beings but of communication, thereby re-defining the traditional concept of subject.

Things change in the virtual world. The role of the subject prevails over the rest of the elements of the system, which is observed by the subject and changed thereby. The subject becomes the action-co-participant of system growth by injecting it with negative entropy (Ibáñez, 1991). It transforms the information that flows through its channels, adapting itself and adapting the object to the subject. In this relation the subject "extracts information from the system through observation and injects negative entropy through action" (Ibáñez, 1991). Therefore, a link is created between the subject and the object (the Internet), extant in the system memory. Information originates in the subject and returns to him/her for analysis, through the use of positive means or negative stimuli. In the Internet medium the subject prevails over communication: since the channel needs the emitter to operate, the act of communication depends on this. Essentially, communication would not be forthcoming without a subject to formulate it: ultimately the personal element (subject) is always requisite to informational development.

2.Virtual differences. On the Internet, once can observe the whole made into a part and vice-versa. There is systemic differentiation. Independent subsystems are formed that co-exist with the whole in unique units (computers) that make up a virtual paradox (web connections) which increases system complexity. This adversity includes selection and contingency; there is risk and order is created. The order created, in turn, registers new levels of security and, being thus "bound" to security, may create and develop vulnerable domains, establishing a new range of risks or "global threats" (Beck, 1998). The whole has an absolute influence over the parts, because the concept of the Internet could not be understood without looking into each subsystem connected to the network. There would be no multiple connection among machines if there were no environment to link them, connecting them in unison, in this case, in cyberspace, idealised on the Internet medium. The result of analysing each of the entities that comprises the virtual medium is the computation of its global effectiveness, one of the essential parameters of the present study.

2. INFORMATION AND COLOSSUS

Information is the bloodstream, the essential referential element of the medium that has revolutionised communications in the Twentieth Century. Data on new connections rise like foam and cyberspace evolves exponentially with no signs whatsoever of decadence. This is an overly positivist view, because account must be taken of all the hidden dimensions of the Internet medium. In this regard, I feel it is vitally important to analyse the existential relationship in the "communications game" (Ibáñez, 1991) and review the global outputs together with the personalised contributions of each member of the Internet social system.

Before going into that, we might come back to the definition of information to ascertain what effect it has on Internet users. The word "information" embraces two meanings: "become informed about" (extract information) and "give form to" (inject negative entropy) (Ibáñez, 1991). This definition covers all the domains of meaning applicable to the information that circulates on the Internet. The ease with which vast amounts of information can be introduced in the virtual flow (websites, FTPs) entails the risk that not all the content circulating over the network meets the ethical and moral standards it should, with the resulting deterioration of quality into mere quantity, at times hazardous.

The ongoing production of information may cause the social system to collapse, thereby questioning its reactivating function. The growth of the social system has promoted the compilation of vast amounts of information. Nonetheless, a serious, in-depth study is in order of how this "power" affects the most hidden streams of cybersociety, including thereunder the subgroups or residual groups formed in cyberspace. Any terminological analysis should blend macrosocial vision with a millimetric appraisal of the micro-whole.

A social system as complex as the Internet poses the materialisation or practical observation of most of the postulates of systems theory and the paradigms of sociocybernetics. The social system stretches over and is bound by this interacting space, deciding what it communicates, about what it communicates and what it must refrain from communicating.

Indeed, the theory advocated by N. Luhmann may be criticised on the grounds of the procreation of negative information in the system, which eliminates any possible levelling method, since it is unable to select, process and much less eradicate virtual infection. Luhmann sustains that information must be accepted or rejected, but that it is not simply "there" (1989). I must say that I believe the contrary. On the Internet, data are often entered in the information stream via uncontrolled channels, whereby the system becomes saturated with trash, which is detrimental to the structure of the net itself. Therefore, information is indeed there but what is sought is a way to avoid it, eliminate it or, at least, control it as far as possible. N. Luhmann contends that the system processes information, assuming responsibility for that action only, which forms a part of the process, but not for the information itself (1989). From my standpoint, just the opposite is true: the responsibility lies in the quality of the information entered and the system is responsible for enforcing such quality. This vision is overly deterministic; on occasion, horizons would need to be expanded and the other extreme analysed as well, the extreme consisting of those that comprise the demand for action, in this case, net users.

Reaction to a stimulus or control of the bilateral input/output relationship is the backbone of the role of information within the social entity, so users can handle it more effectively and in order for the social system, in this case the Internet, to have a good framework for application in real life. The net, generally speaking, has multiplied the potential for disseminating culture among people and therefore has become a virtual incentive for culture and knowledge in general. The Internet provides today's society with a range of possibilities that had never before been available, from eminent technical and communicational resources, to the possibility of disseminating vast amounts of information around the world in a minimum amount of time.

As in all complex systems, there is normally a subgroup or subculture that attempts to constitute the negative aspect of the medium, taking advantage of the potential to disseminate information of a lucrative or even depraved nature for an absurdly low price.

The above may be illustrated with a series of practical examples of uncontrolled developments in the Internet world. Tendencies that have a detrimental effect on the social entity as a whole may arise in the systemic framework. Two models may be taken as an example: spamming and mail-bombing. I have chosen these two kinds of attacks to show how easily the net can be saturated with trash material. The spammer is a subject who engages in abusive advertising, illicitly obtaining a large number of electronic addresses. By falsifying electronic mail, spammers post - generally 900 number - ads for sex or easy-money schemes to all possible e-mail accounts or available newsgroups.

The other example is mail-bombing. This technique is open to any browser who knows the right address from which to use it. Hence, any net user can provoke the collapse of all kinds of systems with extraordinary ease, by merely opening the proper website to commit the misdeed. Mail-bombing is a technique consisting of sending an unlimited number of messages to an account on any server, simply by specifying the body of the message and the number of copies to be sent to the account (from 1,000 to an undetermined upper limit). This technique occasions financial loss, i.e., the cost of recovery from the attack, and may even cause the system to collapse if it is not correctly configured. This is what would happen if the administrator has the e-mail directory on the system's hard disk, which would cause the entire complex to "crash".

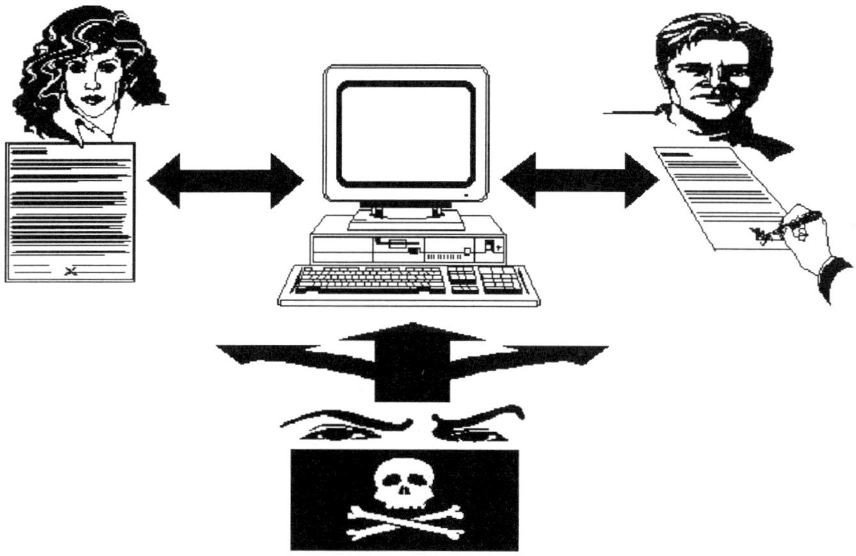

Figure 1: The impact of the "dark side" of the Internet

These two techniques, along with any number of others, constitute the dark side of the Internet. The net has a much larger scope than these subcultures can attain. In the context of education and communication, the potential is unlimited. Such attacks can be perpetrated with a linear-unique protocol and in real time, whereby "communication, bereft of any informational quality, would continue, even where communication becomes controversial or unpleasant" (Ibáñez, 1991), which is what occurs in most IRC cybernaut chat rooms.

But this is generally only a small patch of land in cyberspatial holistics; there is, therefore, potential to be developed that can change the perception of social reality and even modify people's habits and customs, enhancing and developing individuals' unexpected and hitherto undiscovered aptitudes.

3. THE INTERNET'S SOCIAL EFFICIENCY

Social efficiency. These simple words summarise the globalising effect and cultural meaning of the Internet information network. All social systems are characterised by the multiplication of the quality of outputs in relation to the inputs entered a priori. Measuring all the system inputs and outputs is a difficult and complicated endeavour on the Internet, but in order to delimit its social effectiveness and, therefore, efficiency, we must attempt to be strict in this regard.

According to Parra Luna (1992), to measure the efficiency of a social system, it is important to distinguish between system effectiveness per se and the effectiveness of its governing bodies.

The first element of indispensable action is systemic effectiveness. This concept may be pictured schematically through the analysis "... of what the system produces globally and the structural resources at its avail to do so". In the case in question, the delimitation of all the content globally produced by the Internet, is an impossible task, not to say a total waste of time. What can, however, be controlled is data entry through the technical means used by users to enter their knowledge on the net. A normal browser's Web page is subject to the same conditioning factors of quality and effectiveness as the front page of the Washington Post. Each user assumes the role of a home-bound editor. This new social role provides for ongoing and automatic modification of net contents. This is one of the essential elements of the cybernetic explosion; and hence the importance of control and analysis of this phenomenon.

The second element has to do with the bodies governing the system. In the field of virtuality, the context for application would be "the assessment of the transformation conducted by the system's transforming body (OT)" (Parra Luna, 1992). In my modest opinion, all the content placed on the net is conditioned in one way or another by the technical resources available. For this reason, the drivers of virtual systemic change or metamorphosis will be the individuals who subject all other internauts to a permanent policy of security. I refer to the Internet's "established group", discussed below.

Like any social system, the Internet has avail to a substantial flow of information. In order to attempt to measure net performance and, in turn, the social effectiveness of the system, we must design a formal model that clearly addresses its inputs (X), outputs (Y) and transformation (T).

These requirements lead to the implementation of the following formula:

$$T=Y/X.$$

In the conceptualisation of outputs (Y), we have followed the so-called referential value pattern (Parra Luna, 1983) which comprises the following nine values: order, freedom, justice, knowledge, prestige, health, material wealth, quality, conservation of nature. Another series of values, not covered in the present study for lack of time and the breadth of the analysis required, includes, inter alia, the quality of activities. This value addresses the loss of human contact among the actors involved, which may breed self-concentration and an ongoing symbiosis with the machine. The value conservation of nature is of particular relevance to this study, since the Internet medium entails a savings of materials used in the traditional manufacture of books and encyclopaedias, such as paper, thereby reducing the felling of trees for wood pulp, etc. This cost-ecological impact ratio is enhanced by the net at barely any expense in terms of other natural resources, thereby confirming the conservationist benefits of the new technology.

In the discussion below, I am going to attempt, albeit subjectively, to sketch the Internet inputs and outputs as a sort of virtual conceptual diagram:

A/Outputs: What should happen in the complex Internet system? This question is difficult to answer, but in my opinion the first variable to bear in mind is the value Freedom (L), defined, virtually, in terms of the following parameters:

- The ability to browse freely through Internet contents.
- The possibility of having free advertising space (WWW).
- The role of users as personalised editors through homepages.
- The freedom to obtain information through innovative tools (FTP, News, IRC, etc.).
- Subjective anonymity that the net affords. This group might be expanded to include the psychological phenomenon of the virtual splitting of personality, alongside the effect of the anonymity provided by telematic file transfer applications.

Certain values are considered to be essential to the survival of a system, one of which is the knowledge (N-c) value, which would be the browser or navigator's understanding of the basic processes and tools to move around the Internet (navigators, e-mail, file transfer, computer structure, etc.), in addition to the knowledge users acquire from the stream of information.

Similarly, an analysis should be conducted of the quality of the information put out by the net after users enter it in the stream, considering the mutation-multiplication process applied to data through cyberspace. What is pursued in this category is the quality value, on the grounds of (Cant) quantity - (C-i) quality obtained, channels used (WWW, FTP, chat rooms, IRC, e-mail, etc.) and regarding the Internet user as a "perpetual consumer of information".

Another of the values to be borne in mind in our description of the Internet medium is the (distributive) justice value (J) insofar as the net is "fair" to all its users or contains the potential to virtually discriminate, with a priori selection of the best suited internaut candidates (equality) (I).

The value prestige would also have to be taken into account in this assessment of possible outputs, including the following dimensions under this term:

- The prestige (P) that the Internet affords the users themselves (in their private circle or beyond it).
- The prestige afforded by the knowledge of the existence of subgroups (Sub) created around the net.
- The prestige the cybercultures have for the normal user (P-c).
- The additional prestige gained from being up-to-date (latest hardware and software developments) (H&S), consequences of the virtual world, etc.).
- The differentiation (D) among users on the grounds of their knowledge of new and updated web sites on the net.

Finally, the following elements are also considered:

- The order value (O), one of the most important values for the spatial development of the net. New users connecting to the Internet conduct an extensive analysis of what such a connection involves. One of the most radical values in this assessment is the anarchy prevailing on the net. Under this category we need to ascertain whether the Internet really is the most anarchical medium known, or whether, on the contrary, there is some kind of "technical order" lurking in some remote corner.
- Illicit media for divulging information. What opinion do internauts have of this phenomenon and how much information is actually entered on the net in this way. (M-i).
- Harmful trash information (B).
- Channels through which trash information is entered (C-b).

B/ Inputs: What is entered on the Internet system? A priori, there is no reply to this difficult question, since delimiting all the system inputs would be an impossibly arduous task. In any case, the following is an attempt to fragment what I feel are the basic inputs into smaller categories:

a/ Time (measured in hours connected).

2/ Sundry costs: (telephone line, server fee, etc.).

3/ Depreciation cost of hardware-software (average life for desktop equipment, 5 years).

4/ Possible discomfort deriving from spending long hours in front of a screen browsing on the Internet.

Although the inputs could be summarised in the terms mentioned above, I am going to proceed to a more detailed breakdown of the (x) factors that affect and delimit system inputs, together with initial input, which I shall use below in the descriptive graphs:

A) Initial user knowledge which becomes information through the following channels: WWW, e-mail, chat rooms, etc. (C-u).

B) Anomalies (inconvenience) in network connections, poor connections, etc. (A).

C) Health: user reactions to the computer medium, deriving from physical inactivity: headache, nervousness, general discomfort, addiction, etc. (H).

D) Selection applied by the internaut to the stream of information: subject, content selection, etc. (S).

As remarked at the beginning of this paper, the Y/X ratio is conditioned by an element of change: the Transformer Body (T). In this case, we are going to define the intermediary elites (cyberspace technocrats) as virtual catalysts. Here we refer to the "established groups" (Ibáñez, 1994) which virtually direct and condition the existence of the system, steering its development and expansion. These groups monitor even the most personal scopes of the information that circulates over the net, including user particulars. We mean the notorious hackers, crackers or members of virtual gangs: phreakers, carders, trashers, viriis, etc.

4. METHODOLOGY

This survey is eminently exploratory, conducted with the intention of investigating some of the dimensions of the alluring and promising world of the Internet. The indicators designed in this investigation will serve as the outline for a much broader study, which will attempt to encompass all the social aspects of the net.

Taking basic methodological premises as a point of departure, the method used for collecting information was a cross-user descriptive survey, targeting internauts over the age of 18 with some cyberspatial experience and willing to make a contribution to the virtual community (knowledge of the degree of global efficiency of the Internet).

The novelty of the technique, together with the youth of the medium, led to the posing of serious difficulties that had to be solved when entering a questionnaire on the Internet:

- The scant representativeness of the sample. The principles of probabilistic sampling are not met.
- The large number of biases, which makes the researcher's work more difficult.
- There are an estimated 50 million people connected to the Internet, and therein lies one of the major problems. Since it is a new and barely studied phenomenon, there is no reliable way of knowing the exact number of Internet users (all data circulating on the net itself are estimates).

- This instability in the collection of information is due to the fluctuation of net browsers, making it impossible to foresee the size of the sample, whereby the sample is self-selective.

Aside from the limitations arising when running research on the net, certain aspects favour conducting such surveys in this medium:

- The sample is enhanced by the large number of replies that can be collected, and the concomitant savings in reproduction and questionnaire coding costs. The survey is more effective, as it collects data over a short period of time at a very low cost.

- The collection of information on the net is promising. Opinions can be obtained from anywhere in the world, in near-interactive circumstances. The web site is accessible to all cybernauts aware of its publication, either via information browsers (Lycos, Yahoo!, Alta Vista, etc.) or third parties, who inform them of the existence of the data collection page.

Two web pages were devised to collect information, one for inputs and the other for outputs. After the information was collected the questionnaires were subjected to strict review and weeding, eliminating duplicate questionnaires (sent more than once) and possible incomplete replies. The collection, weeding and coding process lasted from 31 March to 31 June 1998.

Users hitting the web page found a closed questionnaire consisting of 12 questions, divided into several units. The most important feature of these questionnaires was the implementation of scales:

- A fixed scale was used in all items, ranging from "extremely high" to "extremely low". In the questions relating to the quantification of subgroups, the following scale was used:

Extremely high. Awareness of 8 groups.
Very highAwareness of 7 groups.
High Awareness of 5-6 groups.
Neither high nor low. Awareness of 4 groups.
LowAwareness of 3 groups.
Very low Awareness of 2 groups.
Extremely low Awareness of 1 group or none.

To measure order, the scale was inverted, with the lowest score (extremely low - total lack of order) considered to be the best result, i.e., the freer and more anarchical the Internet is - and, therefore, the more difficult to control - the better.

The effectiveness of the questionnaire published on the Internet depended largely on the publicity it was given on the web. Banners were placed on friendly homepages to increase the dissemination of the data collection page and the web page's URL was entered on information browsers such as Yahoo!, Alta Vista, Lycos, Ozu, Olé, Hot-Bot and so forth.

The still low penetration of the Internet - less than 3% of the population has access - affords a relatively small sample of browsers. There is no absolute guarantee that the respondents constituted a representative sample of the internaut population. Generally speaking, only those users who hit a collaborating site, saw the banner and decided to participate in the survey, had the opportunity to fill in the questionnaire, all of which represents a serious obstacle. The waste of time and the reluctance to provide personal particulars are the major handicaps encountered in this sort of surveys. Pending confirmation, the direction of possible bias is assumed, among others, to be towards collaborating website visitors and the most frequent Internet users.

The survey universe consisted of all cybernauts connected at the right time and day to fill in the questionnaire.

5. ANALYSIS OF RESULTS

Taking account of all the parameters discussed in the preceding sections, in this final phase, we shall proceed to quantify the global effectiveness of the Internet.

To do so, the series of indicators shown below was subjectively designed, in an attempt to clearly and concisely define the social efficiency of the net (Figure 1).

INPUTS:

1. Initial user knowledge (C-u).
2. Anomalies (A).
3. Health (H).
4. Selection (S).
5. HW&SW depreciation costs (H&S).
6. Sundry costs (C).

Table 1. Outputs

1.Freedom (L)	10. Updates (H&S)
2. Level of Knowledge (N-c)	11. Differentiation (D)
3. Quality of information (C-i)	12. Order (O)
4. Quantity (Cant)	13. Illicit means (M-i)
5. Justice (J)	14. Trash (B)
6. Equality (I)	15. Trash channels (C-b)
7. Internet prestige (P)	
8. Subgroups (Sub)	
9. Cyberculture prestige (P-c)	

The table below shows the computation of the social effectiveness of the Internet, broken down by the inputs (X) and outputs (Y) listed above.

Table 2.

OUTPUTS (Y) Range 0-100		INPUTS (X) Range 0-100	
L	62	C-u	44
N-c	46.0	A	53
C-i	71	H	29
Cant	55	S	44
J	45	H&S	49
I	60	C	46
P	72		
Sub	49		
P-c	58		
H&S	58.5		
D	41.5		
O	32		
M-i	52		
B	45		
C-b	51		
	Y= 53.20		X= 44.16

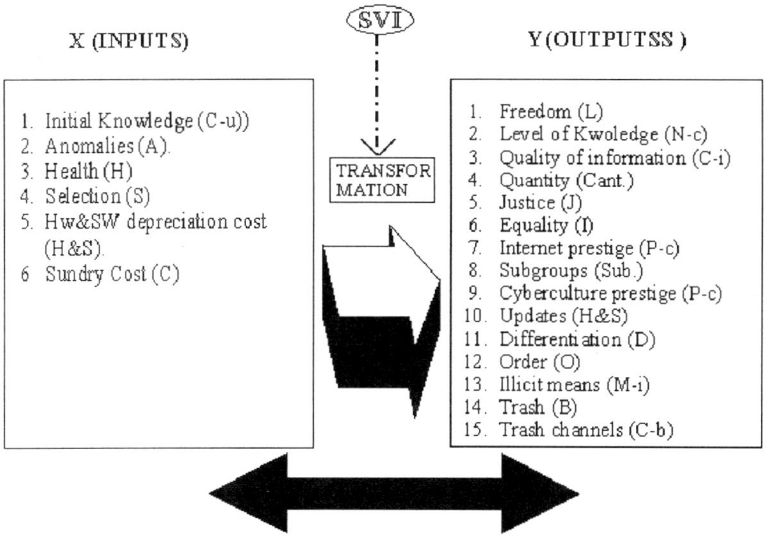

Figure 2: Inputs and Outputs.

COMPUTATION OF <T>

T=Y/X=53.20/44.16=1.20

The value obtained for the Internet's global performance (1.20) shows that the Internet is a tremendously productive social system, since it increased input information by 20%.

The two following sections contain descriptive graphs relating to the calculation of the Internet's global efficiency:

- Section 1: Graphs 1,2,3,4 show the general data trends.

-Section 2. Graphs showing user profiles, on the basis of the following scale: High user (Y-X), Average user (Y-X) and Low user (Y-X).

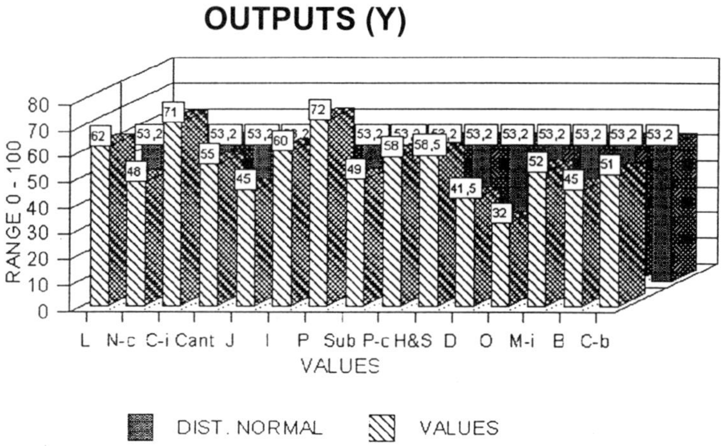
OUTPUTS (Y)
RANGE 0 - 100
L N-c C-i Cant J I P Sub P-c H&S D O M-i B C-b
VALUES
DIST. NORMAL
VALUES

1)

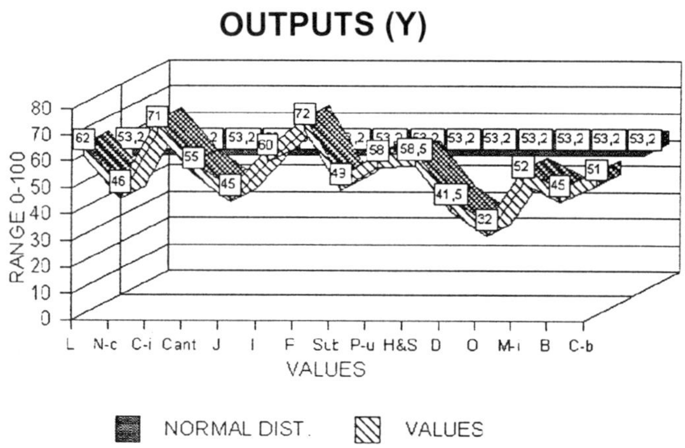
OUTPUTS (Y)
RANGE 0-100
L N-c C-i Cant J I F Sub P-u H&S D O M-i B C-b
VALUES
NORMAL DIST.
VALUES

2)

SECTION 1

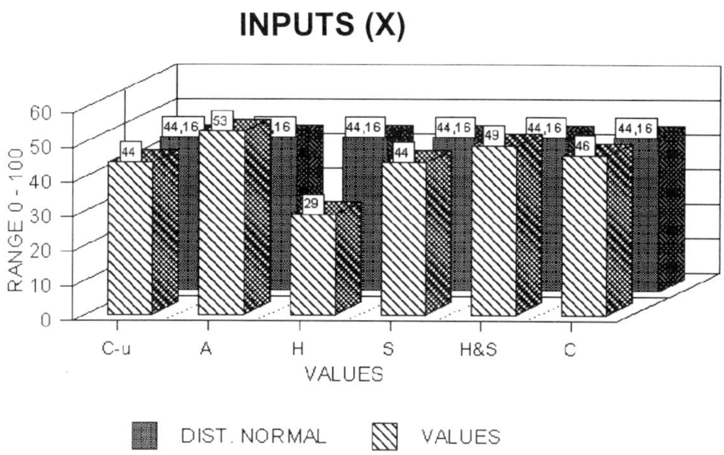
INPUTS (X)
RANGE 0 - 100
60
50
40
30
20
10
0
44,16
53
44
29
49
46
C-u
A
H
S
H&S
C
VALUES
DIST. NORMAL
VALUES

3)

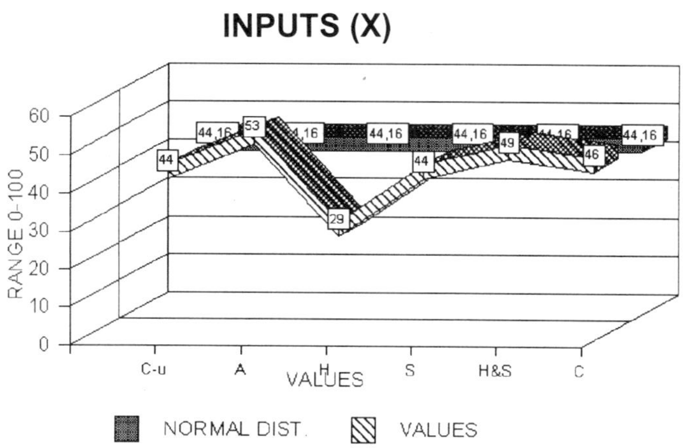
INPUTS (X)
RANGE 0-100
60
50
40
30
20
10
0
44,16
53
44
29
49
46
C-u
A
H
S
H&S
C
VALUES
NORMAL DIST.
VALUES

4)

SECTION 1

SECTION 2:
1\ HIGH USER (X-Y):

HIGH USER (X)

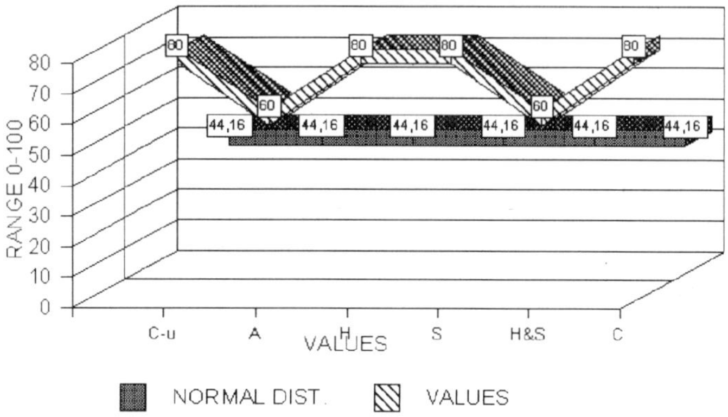

HIGH USER (Y)

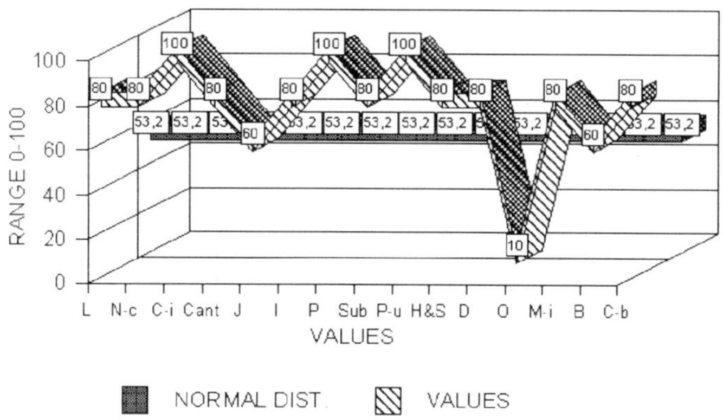

2\ AVERAGE USER (X-Y):

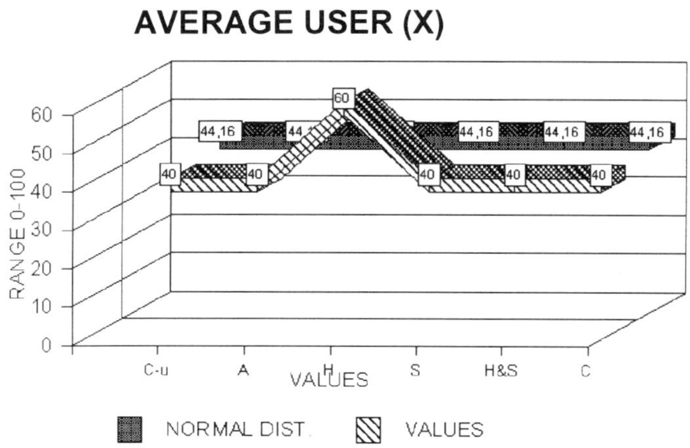

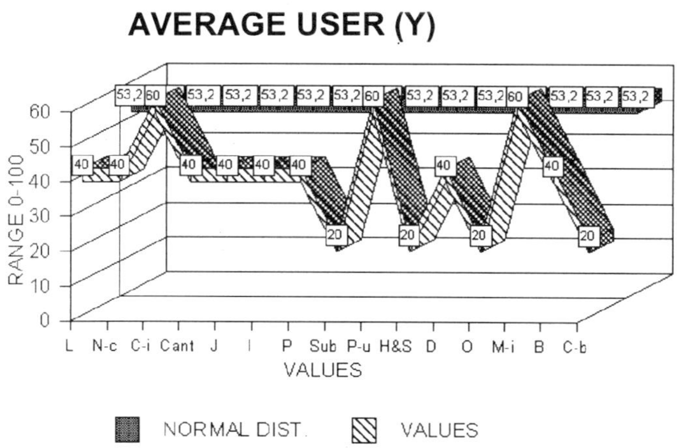

3\ LOW USER (X-Y):

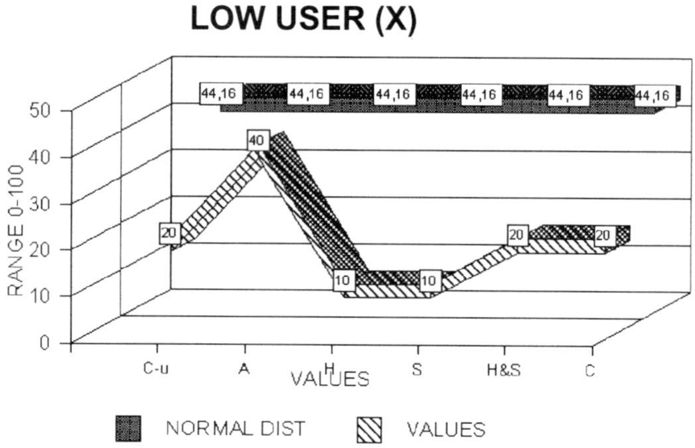
LOW USER (X)
44,16
44,16
44,16
44,16
44,16
44,16
40
20
10
10
20
20
50
40
30
20
10
0
RANGE 0-100
C-u
A
H
S
H&S
C
VALUES
NORMAL DIST
VALUES

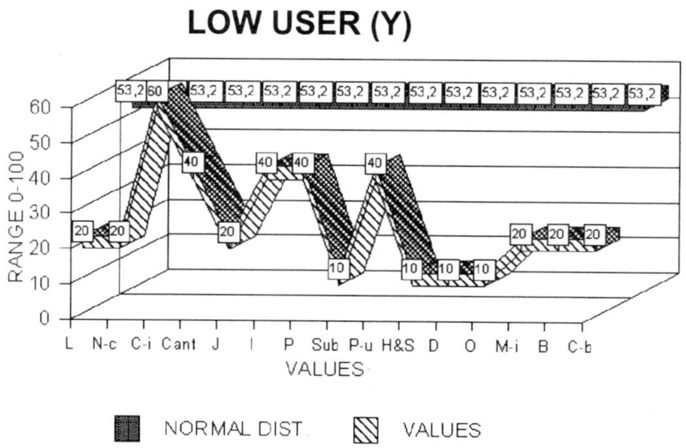
LOW USER (Y)
53,2
60
53,2
53,2
53,2
53,2
53,2
53,2
53,2
53,2
53,2
53,2
53,2
53,2
53,2
40
40
40
40
20
20
20
10
10
10
10
20
20
20
60
50
40
30
20
10
0
RANGE 0-100
L
N-c
C-i
Cant
J
I
P
Sub
P-u
H&S
D
O
M-i
B
C-b
VALUES
NORMAL DIST.
VALUES

6. CONCLUSIONS:

It is abundantly clear that the Internet will be one of the information systems to revolutionise the Twenty First Century. Globality affects all aspects of society and as a society forged in a pristine order, cyberspace will renovate human knowledge and understanding, acting like a global village for universal culture.

I have shown in a series of graphs how the data from this survey may establish overall lines of action for future policy in the virtual domain, but the following summary of the advantages and drawbacks to the net affords a more specific delimitation of concepts:

ADVANTAGES

A) Generally speaking, the quality of information is very high.

B) Vast amounts of information flow over the Internet. Such flow is impossible to calculate or control. The potential is infinite.

C) Two values stand out above all the rest in the definition of the "internaut philosophy": equality and freedom. In this case, the advantages of net anonymity should be highlighted, although this may be overvalued by many users.

D) The Internet enjoys high prestige, both internally and externally. This includes the category of prestige afforded by a knowledge of the latest Internet cybercultures.

E) The Internet is directly related to new HW & SW developments: the more new developments, the greater is the potential for net improvements.

F) Material resources are turned to enormous advantage. Generally speaking, performance is high.

G) The cultural quality of the information flowing over the Internet is high.

H) The Internet is acknowledged to be one of the most anarchical social systems known. Users consider this to be a positive feature for increasing variety.

I) Speed should be highlighted as one of the major features of the net. Large amounts of information can be moved in record time.

DRAWBACKS

1/ Information can be entered on the Internet by illicit and normally abusive and harmful methods, unknown to most users. This will affect the quality of information in the short and long term.

2/ Large amount of trash information flows over the Internet, which may eventually saturate the net to the point of its losing virtually all its present prestige.

3/ The harm done by such trash information to the Internet and its users is extensive.

4/ Users show an exceptional need to consume information, which may reach nearly pathological levels in extreme cases.

5/ People become easily irritated when their computer does not operate properly or when the net is shut down or data are not transmitted as rapidly as they wish.

The following excerpt from G. Balandier, on the description of "ritual", seems to me to be equally applicable to the cyberspatial world, and therefore to the Internet, by way of illustration of the effectiveness of a yet-unknown medium:

> ...highly complex communication and action system. At the same time it has a symbolic structure, a value structure, a teleological structure and a role structure. Since it governs culturally defined communicational behaviour, it is subject to a general code, reinforces its pertinence and effectiveness by multiple repetitions and thematic variations, which reduce the ambiguities of the "noise" which could blur its meaning .[1]

The Internet stores all the social knowledge of the users of advanced technology of recent times. It changes readily and speedily multiplies knowledge. Technology advances full steam ahead. In just a few years' time we shall know things about this world that are now hidden to us.

Who can say whether, within a few decades, the Internet will be so widespread that we won't need to leave our homes for anything at all? This may seem like an exaggeration, but in this progress sociocybernetics, along with social science, the study of virtual social systems and the understanding of cultures, will be the two major pillars of scientific development. The reality to come will affect most people's level of awareness and in this context, sociology will have to study new cybersocial, rather than just social, realities.

7. REFERENCES

Luhmann,N.(1989). Individuo, Individualidad,

Individualismo. En: V.Muñoz, N.Luhmann, G.Teubner, D.Zolo (Eds), La teoría de los sistemas Autoreferenciales. Zona Abierta 70/71.Madrid.

Luhmann, N.(1997) Sociedad y Sistema: la ambición de la teoría. Paidos. Barcelona.

Luhmann,N.(1981) Hacia una teoría científica de la sociedad. Revista Anthropos 173/174. Barcelona.

Luhmann, N (1996) Introducción a la teoría de sistemas. Anthropos. Barcelona.

Ibañez, Jesús.(1994) El regreso del sujeto: la investigación social de segundo orden.Siglo XXI. Madrid.

Parra Luna,F (1992) Elementos para una teoría formal del sistema social. Editorial Complutense. Madrid.

[1] Translated from the Spanish, not the original French text. T.N.

Parra Luna,F (1993) El Balance integrado de la gestión estatégica. Deusto. Bilbao.

Nuevos avances en la investigación social. La investigación social de segundo orden. Suplementos Anthropos 22. Barcelona.

Balandier, G (1993) El desorden: la teoría del caos y las ciencias sociales. Gedisa Editorial. Barcelona.

Sanchéz Súcar,A (1998)The enviromental balance and the perfomance of social system.ISA. WG01. 14th Wolrd Congress of Sociology.

Tretter,F(1998) Towards increasing the perfomance of a Hospital:complex dynamics of management tasks and mappinf technique. WG01. 14th Wolrd Congress of Sociology.

Castells,M (1996) The rise of the network society. Blackwell Publishers.

G.Jones, S (1997) Virtual Culture: identity&Communication in cybersociety.Sage Publications. London.

Turkle, S (1997) Live on the sreen: identity in the age of the Internet.Touchstone Edition. Nex York.

Shields,R (1997) Cultures of Internet: virtual spaces, real histories, living bodies. Sage Publications. London.

Beck, U (1998) La sociedad del Riesgo. Paidós.Barcelona

Cortés et al. (1974: System Analysis for Social Scientists. John Wiley and Son. Nueva York.

Deutsch, K. W. (1974) Politics and Goverment. How people decide thir fate. Houggton Mifflin Company. Boston

- Los nervios del Gobierno: Modelos de comunicación y control político.(1980) Buenos Aires. Paidós.

Easton, D (1970) Esquema para el Análisis Político. Amorrortu. Buenos Aires.

Van Gigch (1978) Aplied General Systems Theory. Harper y Row. Nueva York.

Nicolopoulos, P (1998) An operational model for working out the perfomance of social systems. WG01. 14th World Congress of Sociology.

Index

D

E

F

G

H

I

Q

R

S

T

V